MACROMOLECULAR COMPLEXES

MACROMOLECULAR COMPLEXES
Dynamic Interactions and Electronic Processes

Eishun Tsuchida, Editor

Department of Polymer Chemistry
Waseda University
Tokyo 169
Japan

VCH

Eishun Tsuchida
Department of Polymer Chemistry
Waseda University
Tokyo 169
Japan

Library of Congress Cataloging-in-Publication Data

Macromolecular complexes: dynamic interactions and electronic
  processes/Eishun Tsuchida, editor

   p. cm.
  Includes bibliographical references and index.
  ISBN 0-89573-784-1
  1. Macromolecules. 2. Complex compounds. I. Tsuchida, E.
(Eishun), 1930–
QD381.8.M28   1991
547.7–dc20   90-48754
                                       CIP

British Library Cataloguing in Publication Data
Macromolecular complexes.
  1. Polymers. Dynamics

  I. Tsuchida, Eishun
  547.7045413

  ISBN 0-89573-784-1

  3-527-27988-1

Printed in the United States of America
ISBN 0-89573-784-1 VCH Publishers
ISBN 3-527-27988-1 VCH Verlagsgesellschaft

Printing History:
10 9 8 7 6 5 4 3 2 1

Published jointly by:

VCH Publishers, Inc.	VCH Verlagsgesellschaft mbH	VCH Publishers (UK) Ltd.
220 East 23rd Street	P.O. Box 10 11 61	8 Wellington Court
Suite 909	D-6940 Weinheim	Cambridge CB1 1HW
New York, New York 10010	Federal Republic of Germany	United Kingdom

Contents

Part I   Introduction

Chapter 1. **Macromolecular Complexes and Their Dynamic Interactions and Electronic Processes**   3
E. Tsuchida

Part II   Dynamics of Weak Coordination Systems

Chapter 2. **Coordination Dynamics of Macromolecular Porphinatoiron Complexes**   15
E. Tsuchida

Chapter 3. **Dynamics of Electron-Lattice Interaction in Macromolecular Complexes**   39
K. Horie

Chapter 4. **Ultrafast Dynamics of Macromolecular Complexes**   61
H. Masuhara and A. Itaya

Part III   Soft Interaction and Multiple Interaction

Chapter 5. **Multistep Complexation of Macromolecules**   93
Y. Kurimura

Chapter 6. **Multiple Interaction in Molecule Transport through Macromolecular Complexes**   119
H. Nishide and E. Tsuchida

Chapter 7. **Soft Interaction of Macromolecule-Lanthanide Complexes**   143
Y. Okamoto and J. Kido

Chapter 8. **Local Structure of Macromolecule-Metal Complexes in Solution**   175
H. Ohtaki and H. Masuda

Part IV Multielectron Transfer Processes

Chapter 9. **Intracomplex and Intramolecular Electron Transfer in Macromolecules** 213
M. A. Cusanovich

Chapter 10. **Bioelectrocatalysis at Enzyme-Modified Electrodes** 229
M. Senda and T. Ikeda

Chapter 11. **Redox Behavior of Tetraheme Protein—Cytochrome c_3** 251
K. Fan, H. Akutsu, and K. Niki

Chapter 12. **Electrochemical Pulse Analysis of Multielectron Transfer in Electroactive Macromolecular Complexes** 275
N. Oyama and T. Ohsaka

Chapter 13. **Catalytic Activity and Multielectron Transfer of Metal Clusters** 315
A. Nakamura and N. Ueyama

Part V Sequential Potential Field Constructed in Macromolecular Complexes

Chapter 14. **Sequential Potential Field and Electron Transfer in Metal Complex Clusters** 331
N. Toshima

Chapter 15. **Geared Cycle in Photoexcited Multielectron Transfer in Macromolecule–Metal Complexes** 353
M. Kaneko

Chapter 16. **Sequential Potential Fields in Electrically Conducting Polymers** 379
T. Yamamoto

Index 397

I Introduction

Macromolecular Complexes and Their Dynamic Interactions and Electronic Processes

Eishun Tsuchida

Department of Polymer Chemistry, Waseda University, Tokyo 169, Japan

1-1. Introduction

Recently there has been growing attention to the molecular functions of macromolecular complexes as a new frontier in material systems. Macromolecular complexes are defined as complexes of macromolecules. A typical example is macromolecular metal complexes composed of macromolecules and metal ions (Fig. 1-1). Their synthesis represents an attempt to give inorganic functions to an organic polymer. Intermacromolecular complexes and molecular complexes between macromolecules and organic molecules are also included in the category of macromolecular complexes. Because the macromolecular complexes are combinations of macromolecules, small organic molecules, and metal ions at the molecular level and because they are structurally labile, there are unlimited possibilities for providing a wide variety of previously unknown molecular functions.

In the macromolecular complexes, physicochemical properties and chemical reactivities of the complex moieties are often strongly affected by interactions with the polymer matrices, which surround the complex moieties as illustrated in Fig. 1-1. These interactions are weak but significant and act multiply and dynamically: They not only construct the macromolecular complexes themselves, but also control their higher-order structure based on a dynamic conjugation between the complex moieties and the polymer matrices. Thus a weak and soft profile in the coordination reactions, a cooperative interaction between the complex moieties and a multiplied or enhanced interaction are observed as characteristic features of the macromolecular complexes.

The polymer matrices also provide specific microenvironments around the complex moieties that contribute to regulation of the electronic state in the complex moieties. An integrated structure or a sequential structure of the complex moieties brings about

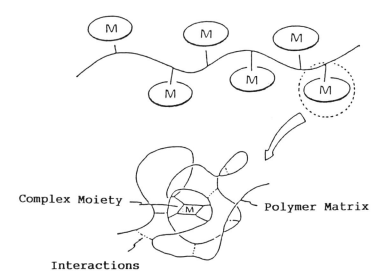

Complex Moiety

Polymer Matrix

Interactions

Figure 1-1 ▪ Schematic illustration of a macromolecular complex.

new concepts of the multielectron transfer step and sequential potential field, which also characterize the macromolecular complexes.

It is the purpose of this book to describe the dynamic interactions and the electronic processes characteristically observed for macromolecular complexes, to demonstrate the differences between the macromolecular complexes and the corresponding low-molecular complexes, and to present their potential applications as frontier material. This chapter describes new concepts or ideas that are realized by utilizing macromolecular complexes, the outline of this book, and the issues to be discussed.

1-2. Dynamic Interactions in Macromolecular Complexes

Interactions in macromolecular complexes consist mainly of various secondary binding forces, such as coulombic or electrostatic force, hydrogen bond, charge-transfer interaction, hydrophobic interaction, and so on (Table 1-1). The features of macromolecular complexes are rapid formation, reversibility, and multiplicity in binding in comparison with covalent bonds. Because they are plural, these secondary binding forces cooperatively play an important role in the macromolecular complexes. The overall interaction appearing in the macromolecular complexes is characterized as "soft interaction" or "multiple interaction," that is, the interaction contains multi-bonding hands, each of which is weak, kinetically active, and reversible, and that are integrated along a backbone matrix. The soft multiple interactions construct the macromolecular complexes themselves and govern the molecular-level functions of

Table 1-1 ▪ Characteristics of Secondary Binding Forces

Classification	Constituents	Physical State	Distances	Strength	Directions
Coulombic force	Ion–ion	Solid, solution	Long range	Strong	Attraction and repulsion (isotropic)
Dipole–dipole interaction	Polar molecules	Solid, solution	Short range	Weak	Attraction (anisotropic)
Hydrogen bond	Proton acceptor– proton donor	Solid, solution	Short range	Weak	Attraction (anisotropic)
Charge-transfer interaction	Electron acceptor– electron donor	Solid, solution	Short range	Medium weak	Attraction (anisotropic)
Hydrophobic interaction	Hydrocarbon molecules	Aqueous solution	Short range	Weak	Attraction (isotropic)

the macromolecular complexes such as molecular recognition, active and facilitated transport, and chemical selectivities.

These interactions are dynamically observed in the macromolecular complexes. Different from low-molecular complexes with stable coordination structure and static configuration, the macromolecular complexes bear labile, unsaturated, and strained coordination structure caused by dynamic change in their configuration (Table 1-2). These dynamic processes in the macromolecular complexes were recognized previously, but were too complicated to be discussed quantitatively and clarified until the present time.

Table 1-2 ▪ Characteristics of Dynamic Interactions in Complex Systems

Characteristics	Macromolecular Complex	Low-Molecular Complex
	Nature of Coordination Sphere	
Strength of ligand field	Relatively weak	Strong
Configuration	Distorted	Symmetrical
Ligand structure	Higher order	Simple
Saturation of coordination	Often unsaturated	Saturated
	Metal–Ligand Interaction	
Strength of interaction	Relatively weak	Strong
Mode of interaction	Multiplied and cooperative	Simple
	Nature of Dynamic Process	
Rate	Very fast	Fast–slow
Response to micro-environment	Sensitive	Insensitive– sensitive

Dynamics of the specific interaction in the macromolecular complexes now can be investigated by use of new spectroscopic methodologies, because the complex moieties are spectroscopically very active and become good indicators for monitoring dynamic behavior.

For example, in the iron porphyrin complexes combined with macromolecules, the distorted complexes are transformed reversibly into the corresponding stable com-

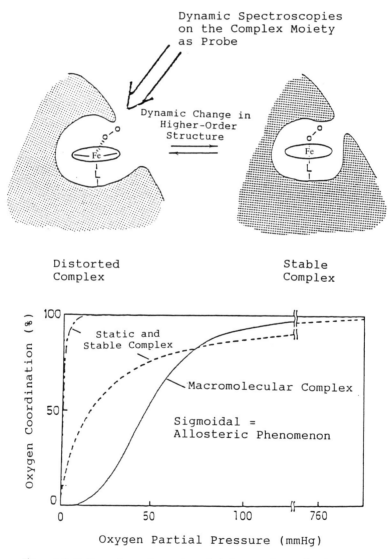

Figure 1-2 ■ Dynamic spectroscopic analysis for a weak coordination system.

plexes, accompanied by conformational change of the macromolecules that provide the microenvironment around the iron complex moieties (Fig. 1-2). Conversely, the dynamic conformational change of the macromolecules induces the rearrangement of the coordination structure. These processes are well known as the allosteric phenomena usually observed in biological systems: the sigmoidal oxygen-binding curve or the very efficient in vivo oxygen-transporting system. These dynamic processes will be discussed quantitatively by use of dynamic or time-resolved spectral analyses in Part II. Oxygen coordination to the complexes dynamically conjugated with the macromolecular structures is compared with natural hemoglobin.

In a solid macromolecular complex system the specific interactions due to unstable structure and dynamic change of polymer-matrix conformation are especially important. The effects of molecular structure and the motion of matrix polymers on photochromic reactions and on photochemical hole burning are described also in Part II. The principles and most current advances in time-resolved transient UV absorption measurements are reviewed and electronic structure and dynamics of excited states of the macromolecular complexes are discussed.

In Part III, soft interaction and multiple interaction are discussed: Kinetic and equilibrium parameters for formation, substitution, and electron transfer of the macromolecular complexes in solution are systematically listed. The profile of multi-step complexation of macromolecules is characterized. Multiple interaction of gaseous molecules, oxygen and nitrogen, with metal complexes fixed in the solid macromolecule, is described in terms of coordination kinetics and molecular diffusion. Facilitated transport of small molecules in the macromolecular complexes is demonstrated. Complexation of macromolecules with lanthanide ions that have soft coordination structure is described. Lanthanide ions are effective as a fluorescent probe to elucidate dynamic motion and environmental effects of macromolecules in aqueous solution. Solute–solute and solute–solvent interactions in macromolecule-metal complexes are also shown by the molecular dynamics and local structure of the complex moieties.

1-3. Electronic Processes of Macromolecular Complexes

Metal complexes composed of metal ions and organic ligands show characteristic electronic processes based on d-orbital configurations of the metal ions. The combinations of five d-orbitals provide a variety of electronic configurations that enhance the mobility of electrons through complexation. Therefore the complex moiety acts as the most important electron-transfer medium. d-orbitals contain many electrons and form highly symmetrical configurations in comparison with p-orbitals, and they yield conducting band or high-spin configurations due to their degenerated orbitals. In addition, the nature of these d-orbitals depends on the configuration and/or electronic states of complexed ligands. In the macromolecular complexes these characteristics of the metal complexes are enhanced by the microenvironment of the polymer matrices and the macromolecular structure.

For organic molecular complexes, electrons also show higher mobility based on polarization and charge transfer. These molecular complex systems also work effectively as an electron-transfer medium. The molecular complex reveals high electron

Table 1-3 ▪ Characteristics of Metal Complexes and Molecular Complexes

Characteristics	Metal Complexes	Molecular Complexes
	Structural Properties	
Coordination number	Multicoordination 1:1–6	Mainly 1:1
Configuration	Specified	Relatively labile
Stability of complex	Strong–weak	Relatively weak
	Electronic Configuration	
Bonding orbitals for complexation	$d(p)$-orbitals	p-orbitals
Electronic structure	Closed or open shell	Closed shell
	Electron Transfer Process	
Oxidation number	Widely variable	$+1$ to -1
Redox reaction	Quite facile	Facile
	Magnetic Properties	
Spin multiplicity	Large	Small
Strength of interaction	Strong	Weak

conductivity when at least one component molecule in the complex has a π-conjugated system. These kinds of complexes are extensively investigated as organic conductors. The electronic process in these molecular complex systems depends on the combination of component molecules similar to those in the metal complex system. In other words, stereostructure and higher-order structure govern the electronic processes in both metal and molecular complexes. Table 1-3 summarizes the electronic and coordination characteristics of these complex systems.

Electronic functions of macromolecular complexes are caused by electronic processes in the complex moiety. Of course, macromolecules maintain the complex moieties, in a specific environment, which enhances stability and feasibility as functional materials. But the much more important effects of the macromolecules on electronic processes of the complex moieties are: Molecular environments of macromolecules control the electronic configuration and mobility. Alignment of the complex moieties along the polymer matrix yields an integrated electronic process. Interfacial fixation of the complex moieties forms a sequential potential field (Fig. 1-3). Bulk and quantum effects, for example, formation of electron-transfer channels, charge-separating processes induced by photoexcitation, tunnel effects, strong suppression of perturbation around the complex moiety, and control of the electronic configuration, are also expected for the macromolecular complexes.

In Part IV, multielectron transfer processes in the macromolecular complexes, including metalloenzyme systems, are reviewed from the viewpoint of long-distance electron transfer or cooperative electron transfer. The related field of macromolecular mixed-valence complexes is also described. Interfacial electron transfer of metalloenzymes on electrodes is an effective method for studying multielectron transfer

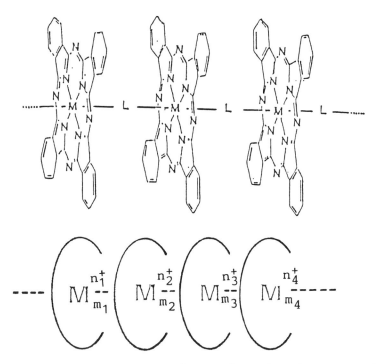

Figure 1-3 ■ Sequential alignment of a complex moiety.

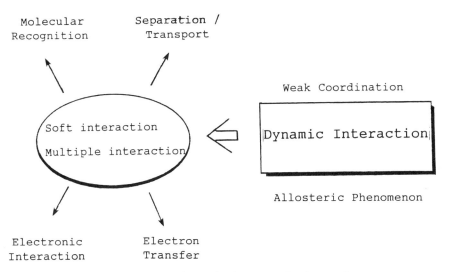

Figure 1-4 ■ Dynamic interaction in macromolecular complexes.

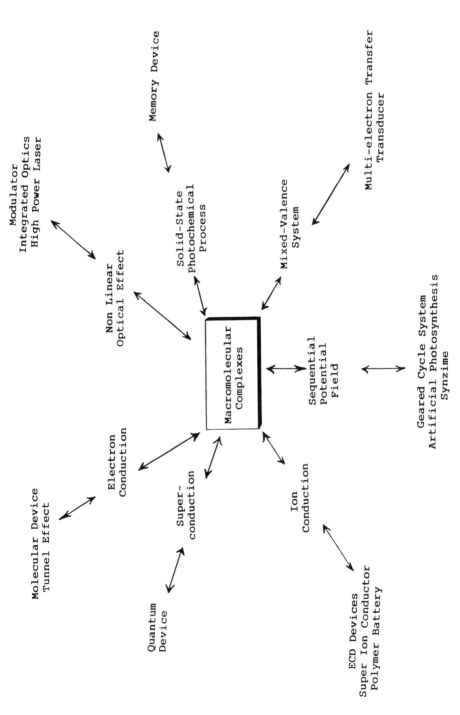

Figure 1-5 ■ Potential applications in the field of macromolecular complexes.

processes and their various pathways. Combination of metalloenzymes with electrodes leads to the new research area of bioelectrocatalysis.

The electron-transfer process of cytochrome c is described by a spectroelectrochemical study. The role of protein structure on the electron transfer is clarified. The multielectron process in electroactive macromolecular complexes coated on an electrode is elucidated by electrochemical pulse methods and kinetic profiles of electron transfer in the macromolecular domain are characterized. Catalytic actions of iron–sulfur clusters combined with macromolecules are described in terms of multielectron transfer processes. They provide a model system for the electron chain of hydrogenase and nitrogenase.

It is also important to discuss the effects based on anisotropic arrangement or accumulation of complexes in connection with their electronic densities and valence states. Electronic interaction occurring among complexes surrounded by macromolecular chains leads to the creation of higher functions.

The concept of a sequential potential field is reviewed in Part V in connection with electron transfer in biological and artificial macromolecular complexes. A new aspect of redox reactions is shown for macromolecular complex catalysts as well as the metal cluster catalysts bound to macromolecules. The geared cycle constructed with metal complexes on macromolecules is demonstrated by photoexcited electron-transfer reactions. A reaction system that models photosynthesis is included. The electroconducting profile of conjugated macromolecular complexes is reviewed from the viewpoint of conduction through sequential fields. Current applications of these polymers as electron-transfer catalysts and sensors are also described.

1-4. Potential Applications and Future Scope

A brief overview of some of the aspects of macromolecular complexes has been presented. Two main characteristic features of macromolecular complexes, dynamic interactions and electronic processes, have been emphasized. Most of these higher-order functions in macromolecular complexes are caused by conjugation of the dynamic interactions and the electronic processes (Fig. 1-4).

Potential applications derived from research studies on macromolecular complexes are illustrated in Figs. 1-4 and 1-5. They include advanced functions that are derived from molecular-level features of the macromolecular complexes, and that should be realized in the near future. Most of them are now under development and promise a great impact on related scientific and technological fields. For example, some of the macromolecular complexes can be used as high-performance molecular devices, such as superconductors, organic ferromagnets, and nonlinear optics (Fig. 1-5). It is necessary to promote studies of the macromolecular complexes as new materials.

II Dynamics of Weak Coordination Systems

Chapter 2

Coordination Dynamics of Macromolecular Porphinatoiron Complexes

Eishun Tsuchida

Department of Polymer Chemistry, Waseda University, Tokyo 169, Japan

2-1. Introduction

Hemoglobin (Hb) serves to transport molecular oxygen in a living body. Hb is a conjugated protein with molecular weight 64,450 and it consists of four polypeptide chains (globin proteins), each of which has a protoporphinatoiron IX (protoheme). That is, Hb is typical example of a macromolecular complex. Hb is the protein that is easily isolated in a pure form from blood. Its primary, secondary, and tertiary structures were investigated accurately earlier and a clear view of the structures of Hb is now shown.[1,2] The globin chain has a compact globular conformation and the porphinatoiron moiety is embedded in the "pocket" of the globular protein. The porphinatoiron coordinately bonds to the imidazole of the histidine residue of the pocket-forming segment of globin and becomes the oxygen-coordinating site

$$+ O_2 \rightleftharpoons \tag{2-1}$$

The hydrophobic character of the space inside the pocket, caused by the surrounding hydrophobic amino acid residues, seems to relate to the oxygen-coordinating function of Hb.

On exposure of the Hb solution to oxygen, the porphinatoiron complex (deoxy type) forms a 1:1 oxygen-coordinated complex (oxy type). Typical curves of oxygen coordination versus the partial pressure of oxygen (P_{O_2}) under physiological conditions are shown in Fig. 2-1 for Hb and myoglobin (Mb). The oxygen-coordinating affinity of Hb is remarkably reduced at low P_{O_2}. However, once oxygen begins to coordinate, the oxygen-coordinating affinity successively increases and the coordinating curve becomes sigmoidal (i.e., a cooperative coordination). On the other hand, the oxygen reservoir Mb, which consists of a single globin protein and a porphinatoiron, has a greater oxygen affinity than Hb at any P_{O_2} and its curve is of a hyperbolic type. Hb absorbs oxygen with 95% of the coordinating saturation in the lungs (P_{O_2} = ca. 110 mmHg) and desorbs oxygen (the coordination is reduced to 72%) in the terminal tissue (P_{O_2} = ca. 40 mmHg) and gives it to Mb in the tissue. Thus Hb effectively transports molecular oxygen from the lungs to the terminal tissue although there is only a small difference in P_{O_2}.

If the oxygen-coordinating site, the porphinatoiron complex, is isolated from Hb and Mb and exposed to oxygen in a solution, the porphinatoiron complex is immediately and irreversibly oxidized to its ferric [iron(III)] complex and it does not act as an oxygen transporter.

The reasons for the reversible oxygen coordination by Hb and Mb are considered to be as follows.[3-5]

(i) The porphinatoiron complex has a five-coordinate structure whose sixth coordination site is vacant, which allows it to additionally coordinate molecular oxygen:

$$Fe(II)PL + O_2 \rightleftharpoons LPFe\text{-}O_2 \qquad (2\text{-}2)$$

(ii) The porphinatoiron complex is dispersed and diluted to suppress the irreversible oxidation via the μ-dioxo dimer:

$$LPFe\text{-}O_2 + Fe(II)PL \longrightarrow LPFe\text{-}O_2\text{-}FePL \longrightarrow PFe(III)\text{-}O\text{-}Fe(III)P \qquad (2\text{-}3)$$

(iii) The porphinatoiron complex is surrounded by the hydrophobic environment, causing the proton-driven oxidation to be retarded:

$$LPFe\text{-}O_2 + H^+(H_2O) \longrightarrow Fe(III)PL + HO_2^* \qquad (2\text{-}4)$$

In Eqs. (2-2)–(2-4), FeP, Fe(II)P, and Fe(III)P represent porphinatoiron, -iron(II), and -iron(III), respectively, and L is an axial ligand such as an imidazole derivative.

Globin protein forms the five-coordinate porphinatoiron complex and "tucks it away" separately, that is, globin protein protects the porphinatoiron complex from irreversible oxidation [Eq. (2-3)] by embedding it separately in the macromolecule, and the hydrophobic domain of the globular protein excludes water molecules and suppresses the proton-driven oxidation [Eq. (2-4)].

Since a decade ago, much research has been directed to mimic oxygen transporters like Hb by synthesizing various modified porphinatoiron derivatives.[4-7] These syn-

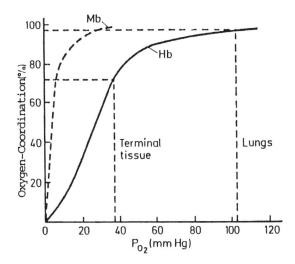

Figure 2-1 ■ Oxygen-coordination curves of Hb and Mb.

thetic porphinatoirons have been successful in oxygen-coordinating reversibly in organic solvents, but in aqueous media they have been irreversibly oxidized. Until recently, only Hb and Mb were known oxygen transporters in aqueous media.

The purpose of this chapter is to discuss how to understand the effects of the macromolecular structure of Hb in its highly efficient oxygen-transporting function by using synthetic porphinatoiron complexes. Because the oxygen coordination of Hb's porphinatoiron is conjugated with environmental and dynamic contributions of the globin protein in aqueous media, the discussion here will be centered not only on the study of the coordination site (porphinatoiron complexes) but also on effects of the environment (globin proteins) surrounding the complex moiety. In this chapter, first we briefly describe tuning of the oxygen-coordinating affinity and dynamics in modified porphinatoirons and then our attempt to replace the roles of the globin protein with a macromolecular assembly.

2-2. Modified Porphinatoiron Complexes and Weak Coordination of Oxygen

Much recent work has been aimed at overcoming the requisites for the reversible oxygen coordination to a porphinatoiron complex (i)–(iii) and has been partially successful in aprotic solvents.[4,5] In aprotic solvents the proton-driven oxidation [(iii), Eq. (2-4)] is excluded, so that the problems of reversible oxygen coordination are how to form the five-coordinate complex as the deoxy state (i) and how to inhibit the irreversible oxidation via dimerization [(ii), Eq. 2-3)]. The successful approach was an elegant steric modification of porphyrins: Porphyrins have been substituted in a fashion that satisfies (i) and (ii). Some interesting porphinatoirons have been produced by clever synthetic techniques.[5-7] A typical example is Collman's "picket-fence" porphinatoiron: tetrakis($\alpha,\alpha,\alpha,\alpha$-$o$-pivalamidophenyl)porphinatoiron (**1**).[6,8] **1** has steric bulkiness constructed with the pivalamide groups on one side of the porphyrin plane and the other side unencumbered. The imidazole ligand is allowed to coordinate to the unhindered side of the porphinatoiron and the other side remains a pocket for oxygen coordinating. Moreover the fence would discourage the dimerization of a μ-dioxo complex. The **1** imidazole complex could bind molecular oxygen reversibly in dry benzene at room temperature for over a week. This result mean that a skeleton structure and a special environment around the oxygen-coordinating site are important to form the reversible oxygen-coordinated complex.

1

Especially in **1** the pivalamide fences are believed to provide a distal moiety with a weak hydrogen bonding with a coordinated dioxygen[9]:

$$LPFe—O—O\cdots H—N= \qquad\qquad (2\text{-}5)$$

In fact, the modified porphinatoirons with reversible oxygen-coordinating capability often bear amide residues in their substituent groups on the porphyrin plane and those without amide residues gave reversible oxygen-coordinated complexes only in amide-type solvents. Thus there remains a question as to whether the distal amide residue is essential for reversible oxygen coordination.

Oxygen coordination to tetrakis(2',4',6'-triphenylphenyl)porphinatoiron (**2**) was reported.[10] This porphinatoiron has two "pockets" formed by phenyl substituents on both faces of the porphyrin. The pocket structure minimizes the solvation of the coordinated oxygen, but the lifetime of the reversible oxygen coordination is much reduced in comparison with that of **1**.

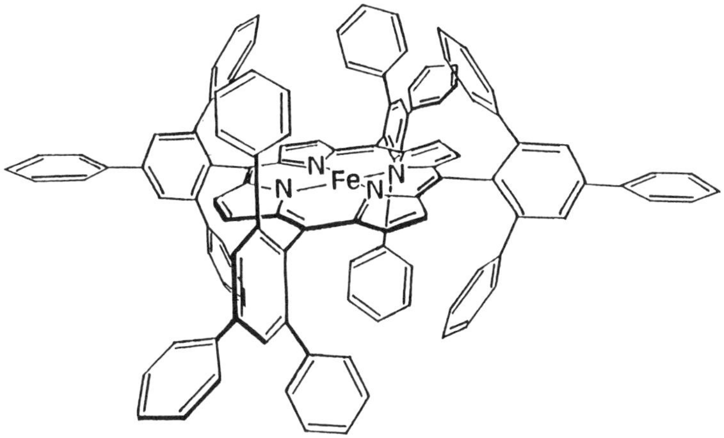

2

Recently we synthesized a new porphinatoiron derivative substituted with ester groups on both faces: tetrakis(2,6-bispivaloyloxyphenyl)porphinatoiron (3).[11] Oxygen coordinates reversibly to the 3-imidazole complex in toluene. The irreversible oxidation product of 3 was not a μ-oxo dimer but was an iron(III) species, which indicates perfect inhibition of the irreversible oxidation step via dimerization [Eq. (2-3)]. But the oxygen-coordinating equilibrium constant was ca. 10 times smaller than that of the corresponding 1 complex. This indicates that oxygen coordinates reversibly, but weakly, to a porphinatoiron complex without amide residues. It is considered for 1

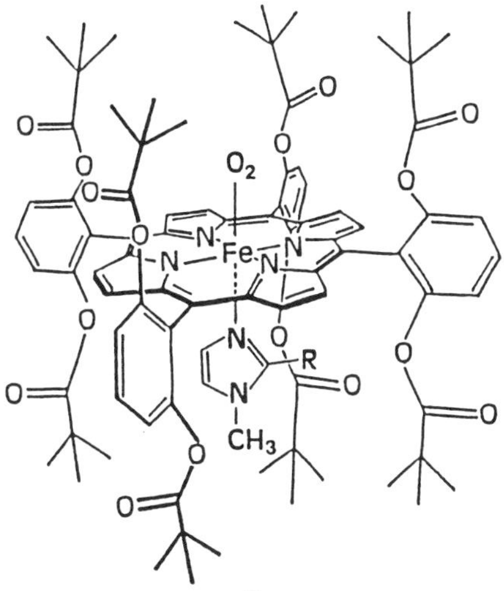

3

and its derivative complexes that the porphinatoiron-coordinated oxygen weakly interacts with the amide groups constructed upon the porphyrin plane, which restricts rotation mobility of the coordinated oxygen and enhances the oxygen-coordinating affinity.

Modification at the porphyrin plane recently has become more skillful and complicated. For example, a capped and strapped porphinatoiron (4) that underwent reversible oxygen coordination in toluene was built up.[12] Although 4 has an extremely long lifetime compared with other porphinatoiron complexes, the oxygen-coordinating affinity was too reduced and only a small amount of the oxygen-coordinated complex existed at room temperature. The latter result was ascribed to locking or unfavorable steric interactions of the parent porphinatoiron in the "domed" configuration.

4

Diporphinatometal derivatives 5 were first synthesized by Chang.[13] The imidazole complex of diporphinatocopper-iron 5a [M_1 = Cu(II), M_2 = Fe(II) in 5] reversibly forms its oxygen-coordinated complex and its lifetime is fairly long and comparable to that of the 1 complex in dry benzene. Molecular oxygen coordinates to the porphinatoiron through opening of the face-to-face structure of diporphyrin. It was considered that the inert porphinatocopper, tightly linked to the porphinatoiron, protects the porphinatoiron-coordinated oxygen.

Diporphinatodiiron 5b [M_1 = M_2 = Fe(II) in 5] showed the same coordination, whereas two molecules coordinate to it due to the diiron structure.[14] The coordination equilibrium curve for the 5b-imidazole complex appeared sigmoidal, whereas the curve for the 5a-imidazole complex was hyperbolic. The cooperative parameter n was estimated: The relationship between the partial oxygen pressure and the degree of coordination with $n = 1$ corresponds to a hyperbolic coordination curve; for $n > 1$ it is understood that the coordinating sites cooperatively interact with each other and the coordination occurs at a stroke (cooperative phenomenon). n corresponds to the strength of the cooperative interaction between the sites (n for Hb = 2.8, n for Mb = 1.0). The cooperative parameter n estimated for the coordination to **5b** was 3.4, which means a strong cooperativity in the coordination, whereas that for **5a** was unity.

In Hb the planar structures of porphinatoirons slightly differ from each other depending on the fifth and sixth coordinating ligands. In the low-spin porphinatoiron complexes, the central iron ion lies in the porphyrin plane and gives the plane a strictly planar structure. This is the cases of oxyHb and deoxyHb with higher coordinating reactivity [relaxed (R) state of deoxyHb]. In contrast, the central iron ion of the high-spin porphinatoiron complexes is pulled up above the plane and the plane is slightly distorted, which occurs in the case of deoxyHb with lower coordinating reactivity [tense (T) state of deoxyHb]. It is considered to be related to the cooperative coordination by Hb that Hb contains deoxy-porphinatoirons that are in equilibrium between the T and R states that differ in their coordinating reactivity.

To explain the pseudo-cooperative coordination phenomenon for **5b**, the following porphinatoiron–porphinatoiron interaction was proposed by a coordination-kinetic and X-ray crystallographic study.[14] For **5b**, planar porphinatoirons link covalently in a parallel and stable face-to-face structure [probably corresponding to the relaxed (R) state of Hb]:

$$(2\text{-}6)$$

For the **5b**-imidazole complex, two imidazole ligands coordinate to the porphinato-iron only from the outward-facing side: This leads to the five-coordinate high-spin porphinatoiron complex where the iron ions lie slightly out of the plane and the porphyrin planes are distorted [probably corresponding to the tense (T) state of Hb]. When oxygen coordinates to one of the porphinatoiron–imidazole complexes of **5b**,

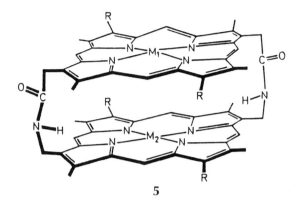

5

the T structure changes to that of the six-coordinate low-spin complex and the porphyrin ligand combined with oxygen becomes strictly planar. This structural change of the first reacted porphinatoiron induces the distorted porphyrin facing it to return to the planar R state, which facilitates the oxygen coordination to this second porphinatoiron.

The cooperative coordination of the synthetic **5b** with oxygen was produced by triggering of a structural change of the coupled coordination sites (porphinatoirons) and was a good model of the cooperative coordination of Hb where the conformational change of the globin protein induces the reactivity change of porphinatoiron.

The above mentioned studies on the modified porphinatoirons are of great significance because they have demonstrated steric and environmental effects on oxygen coordination. However, in these studies the oxygen-coordinating ability itself for these complexes in aqueous media unfortunately was not described.

Water-soluble modified porphinatoirons were recently reported.[15-17] For example, cyclodextrin was used as the sterically protective group covering one face of the porphyrin plane.[15] In example **6**, the secondary hydroxy groups (larger bottom) of α-cyclodextrin were combined with protoporphinatoiron IX with three urethane

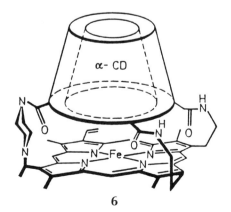

6

bonds. The ^{13}C-NMR spectral data for the corresponding iron-free porphyrin derivative supported its cyclodextrin-capping structure; the carbon signals of the cyclodextrin were shifted to a higher field by the diamagnetic ring current of the suspended porphyrin. A hydrophobic effect of the capping cyclodextrin was also confirmed by a fluorescence spectrum of the corresponding porphinatozinc derivative in aqueous media. **6** formed a five-coordinate deoxy complex with a sterically bulky water-soluble imidazole ligand, due to the one face-hindering structure provided by the capping cyclodextrin.

The imidazole complex of this cyclodextrin-capped porphinatoiron formed a semistable oxygen-coordinated complex in a cooled aqueous medium, whereas non-modified porphinatoiron-imidazole complexes were rapidly and irreversibly oxidized under the same conditions. It was considered that the toroidal shape structure with a hydrophobic inside moiety of the capping cyclodextrin keeps the sixth coordination site vacant for oxygen coordination, prevents the irreversible oxidation via dimerization, and provides a hydrophobic cavity for the coordinated oxygen to retard the proton-driven oxidation.

However, the lifetime decreased with the temperature of the aqueous medium and the oxygen-coordinated complex was scarcely observed at room temperature. The steric and hydrophobic effects of the modifying group on the porphyrin plane were limited in aqueous media.

As previously described, the modification of porphinatoiron has become more skillful and complicated in recent years. But often, the visible absorption spectra of these highly modified porphinatoirons were not similar to those of the parent porphinatoirons and the central iron ion was removed from the porphyrin planes even under mild conditions. This means that a strain is introduced into the porphyrin ligand plane by the strapping or capping structure over the plane and by substitution with excessively bulky groups on the plane. There is a limitation for the approach: To give a porphinatoiron complex oxygen-coordinating ability under severe conditions, such as in aqueous media, the porphyrin ligand itself must be modified.

2-3. Macromolecular Assembly of Complex with Lipids

In Hb and Mb the globin protein protects the porphinatoiron complexes that are tucked separately into a hydrophobic domain of the protein. In order to construct a hydrophobic environment in an aqueous solution, it is possible to use the macromolecular assembly "liposome" of a phospholipid instead of the globin protein.

Recently we synthesized an amphiphilic porphinatoiron derivative having four alkanephosphocholine groups, 5,10,15,20-tetrakis[$\alpha,\alpha,\alpha,\alpha$-O-[2',2',-dimethyl-20'-(2''-trimethyl(ammonioethyl)phosphonatoxyeicosanamido]phenyl]porphinatoiron (lipid-heme, **7**)],[18,19] that significantly increased the compatibility of the porphinatoiron complex with a phospholipid and also increased oxygen-coordinating ability. With this porphinatoiron complex we succeeded in utilizing the hydrophobic region of the liposomal assembly of the phospholipid as the efficient matrix for phorphinatoiron instead of the globin protein. That is, the stereo structure and the amphiphilic property of **7** enhance its compatibility with a phospholipid bilayer and form a very stable macromolecular assembly of the complex with the lipid. A hydrophobic

7

environment to retard the proton-driven oxidation [Eq. (2-4)] is constructed around porphinatoiron in aqueous media by embedding porphinatoiron in a phospholipid bilayer, and the porphyrin plane of the porphinatoiron complex is oriented parallel to the bilayer, which prevents the oxidation via a μ-dioxo dimer [Eq. (2-3)] (see Fig. 2-2).

Because the hydrophobic–hydrophilic balance as well as the stereostructure of the complex are adjusted to a lipid bilayer, their liposomal assembly was easily prepared by modifying the normal method of liposome preparation.[19] Only 20 mol of dimyristoylphosphatidylcholine (DMPC) was enough to solubilize 1 mol of the complex in water. The gel permeation chromatography (GPC) and the ultracentrifugation of the DMPC/lipid-heme solution showed that the complex was completely entrapped within the liposomal assembly.[19]

An electron microscopic photograph showed that the assembly was a single-walled unilameller liposome with a diameter of ca. 400 A or a multilamellar liposome with a diameter of > 1000 A. The closed vesicle structure and the stability of the assembly were also confirmed by the leakage of the encapsulated fluorescent compound.[20] The encapsulated fluorescent compound within the liposome leaked slowly across the bilayer; the rate of leakage was DMPC liposome/lipid-heme = DMPC liposome. This also means that the phospholipid forms a stable liposome assembly with the porphinatoiron complex.

Liposomes undergo distinct structural changes at a certain temperature, that is, the phase transition temperature (T_c), when heated or cooled.[21] These changes do not affect the gross structural features of liposomes, that is, they remain as roughly spherical closed smectic mesophases of the phospholipid bilayer, and correspond to the orientation of the alkyl chains of the lipid. Below T_c, lipids in the bilayers are in highly ordered gel states, with their alkyl chains in all-trans conformations. Above T_c

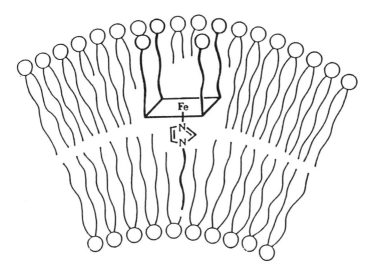

Figure 2-2 ■ Macromolecular assembly (liposome) of the amphiphilic porphinatoiron complex with phospholipid.

lipids become crystal states as a consequence of gauche rotations and kink formation. By differential scanning calorimetry on the liposome/lipid-heme, the endothermic peak for the gel–liquid crystal phase transition was observed at the same temperature as the corresponding liposome itself: for example, $T_c = 24°C$ both for the DMPC liposome and the DMPC/lipid-heme liposome.[20] This suggests that the orientation of the phospholipid in the liposome/lipid-heme assembly is equivalent to the liposome itself and that the compatibility of the porphinatoiron complex with the phospholipid is large enough to form a stable liposome.

The incorporation of porphinatoiron in the liposome was also studied by fluorescence spectral measurement using porphinatozinc as the fluorescent probe.[19] For homogeneous alcohol solutions of the porphinatozinc, the fluorescence intensity increased with decreasing solvent polarity (methyl, ethyl, and butyl alcohol), as expected. The intensity was much larger (more than double) for the liposomal porphinatozinc assembly than for alcohols, indicating that the porphinatozinc is incorporated into the hydrophobic region of the lipid bilayer and is molecularly dispersed in the liposome matrix.

The porphyrin plane of the porphinatoiron complex **7** is assumed to be oriented nearly parallel to the phospholipid bilayer matrix (Fig. 2-3), which was confirmed by an electrooptical measurement.[22] At first, electric birefringence of the liposomal (diameter > 1000 A) solution was measured: The liposome has a larger refractive index in a perpendicular direction to the electric field, which means that the longer axis of the liposome (oval-shaped under electric field) is aligned parallel to the electric field. Dichroism of the porphinatoiron embedded in the bilayer was monitored under an electric field. The transient absorbance change was much larger when the incident light was polarized parallel with the electric field. From these results, it was con-

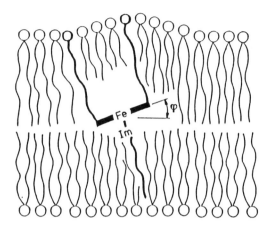

Figure 2-3 ■ Orientation of the porphinatoiron complex in the phospholipid bilayer.

cluded that the angle (φ) between the porphyrin plane and the phospholipid bilayer is small (Fig. 2-3, Table 2-1.). It is assumed that the steric and amphiphilic structure of the complex, which has four substitutent groups built up on the porphyrin plane, keeps the porphyrin plane in parallel to the lipid bilayer.

The stability and miscibility of the complex with phospholipids were also estimated from surface pressure–surface area isotherms of the lipid monolayer on a water surface.[20] The curve for the monolayer of the complex assembly coincided with that for the lipid monolayer itself, which reveals good packing of the lipid molecules with the complex in the monolayer system also. These results clearly indicate the high compatibility of the complex with phospholipids that is caused by the introduction of the alkanephosphocholine groups into the porphyrin plane of the complex.

Next we attempted to improve the stability of the liposomal assembly as the matrix of the porphinatoirons, which will bring about a highly oriented physically and mechanically stable structure for the porphinatoiron assembly. To accomplish this, we made the liposomal assembly stable by polymerization of the lipid bilayer.[23] The double bond of the phospholipid derivative was rapidly polymerized under UV

Table 2-1 ■ Orienting Angle (φ) of the Porphinatoiron Plane to the Bilayer Matrix

Porphinatoiron/Bilayer[a]	φ (deg)
1/EYL	19
7/EYL	3.9
7/EYL + surfactant	28
9/8	2.6
9/8 (copolymer)	1.4
9/8 (copolymer) + surfactant	2.2

[a]EYL: egg yolk lecithin; surfactant: Triton X-100.

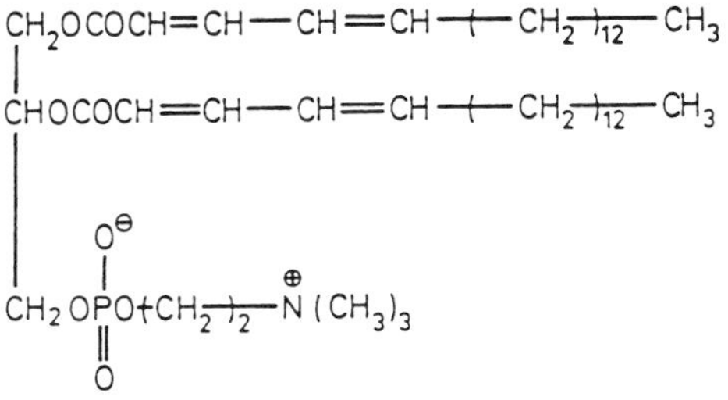

$$CH_2OCOCH=CH-CH=CH-(CH_2)_{12}-CH_3$$

$$CHOCOCH=CH-CH=CH-(CH_2)_{12}-CH_3$$

$$CH_2OPO(CH_2)_2-\overset{\oplus}{N}(CH_3)_3$$

8

irradiation, because of its assembled and oriented structure, to give a covalently bound and very stable lipid bilayer.

The complex was embedded in polymerized liposome of 1,2-bis(octadecadienoyl)-*sn*-glycero-3-phosphocholine (**8**). The liposome [ratio, **7/8** = 1/(20–50)] was allowed to polymerize under UV irradiation to give poly-1,2-bis(octadecadienoly)-*sn*-glycero-3-phosphocholine liposome/lipid-heme.[24] The reduction of the Fe(III) derivative to the deoxy porphinatoiron [Fe(II)] occurred spontaneously during polymerization, suggesting a termination reaction by capturing the propagation radical with the complex.[25] The degree of polymerization of the polymerized lipid monomer was reduced by the presence of the complex to around one-third to one-fifth of that without the complex.

To fix the porphinatoiron complex more precisely in the polymer matrix with respect to its orientation, porphinatoirons substituted with four alkyl groups that have both a polymerizable double bond and a hydrophilic residue at their top positions (**9**,

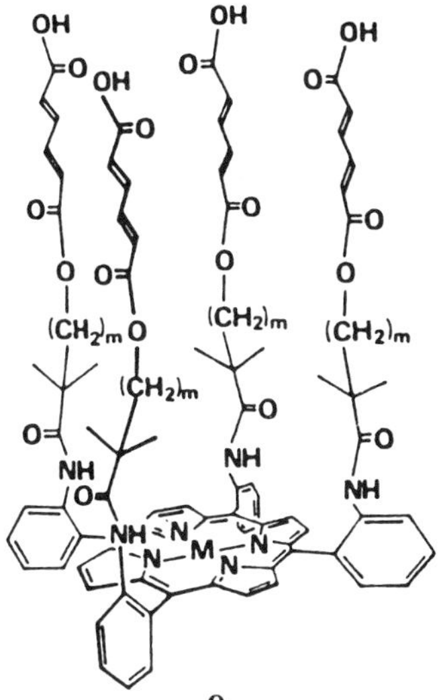

9

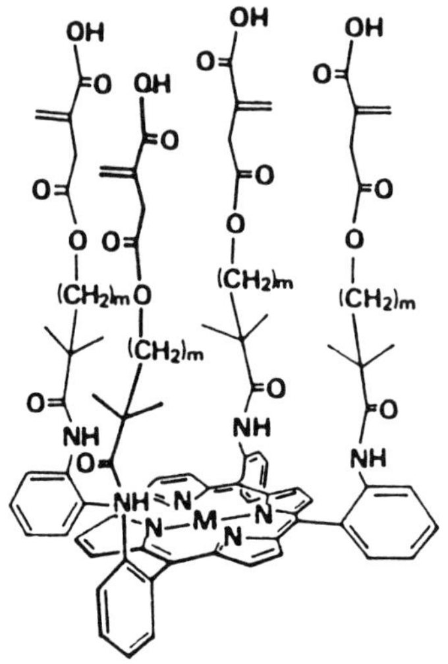

10

10) were synthesized.[26] Because of the hydrophobic–hydrophilic balance as well as the stereostructure of the porphinatoiron complexes, they also have high compatibility with the phospholipid matrix and form a stable lipid assembly. The porphyrin plane of the complexes was also oriented nearly parallel to the bilayer matrix, which was confirmed by the electrooptical measurement (Table 2-1).

Complex **9** was ligated with an alkylimidazole derivative that possessed a double bond (**11**). The **9/11** complex also forms a stable lipid assembly with **8** and is efficiently copolymerized under UV irradiation in the oriented liquid crystal state to give a polymerized bilayer that covalently contains the iron-porphyrin complex residues (Fig. 2-4).[27] **10** was also efficiently fixed in the matrix through the copolymerization. The **9/8** copolymerization rate was much faster than that in the homogeneous solution of **9** and **8**, which indicates in situ polymerization or fixation in the oriented or ordered structure of the polymerizable double bonds.

The covalent fixation of the porphyrin complexes in the bilayer matrices was confirmed by ^{13}C-NMR, GPC, solvent extraction, and so forth.[28] The angle for the porphyrin plane of the covalently fixed complex was also zero, which was not influenced by chemical stimulation such as surfactant addition (Table 2-1).

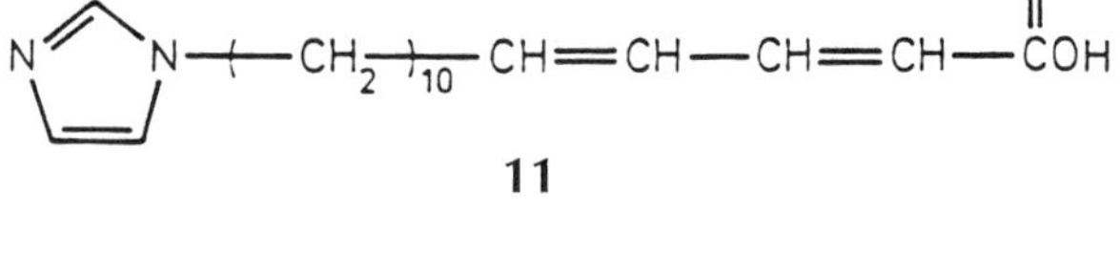

11

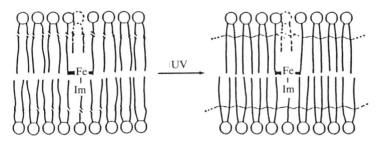

Figure 2-4 ▪ In situ polymerization of the porphinatoiron complex in the oriented assembly.

Laser photoreaction of the covalently and orientedly fixed porphyrin in the matrix was analyzed at 4.5 K to confirm molecular dispersion of the complex in the matrix, to study molecular motion of the fixed complex, and to estimate a weak interaction of the excited complex with the surrounding polymer matrix.[29] Photochemical hole-burning spectroscopy of the **9/8** copolymer at 4.5 K showed a very sharp eight line spectrum, which clearly means that the porphyrin is molecularly dispersed in the polymer matrix with high compatibility and that thermal motions of the complex moiety are severely restricted through multicovalent bond fixation within the oriented matrix structure.

2-4. Assembly Structure and Oxygen-Coordination Dynamics

On exposure to oxygen the red and transparent solutions of the macromolecular assembly of the porphinatoiron complexes with the lipids were rapidly and reversibly changed to the brilliant red solution attributed to their oxygen-coordinated complexes. The oxy–deoxy coordination cycle could be repeated more than a thousand times under physiological conditions.

The volume of the oxygen coordinated to the porphinatoiron complex was measured volumetrically to be 20 ml per 1 mol of porphinatoiron complex at 37°C under oxygen atmosphere (1 atm, 100 ml solution, porphinatoiron complex = 10 mM). The complex-free lipid solution (100 ml) physically uptook 2.1 ml of oxygen: The oxygen coordinating capacity of the porphinatoiron complex solution is ca. 10 times that physically uptaken by water.

Because the oxygen-coordinated complex for the macromolecular porphinatoiron **7** is formed at high concentration even in an aqueous medium, the coordinating bond property of oxygen-iron was easily estimated. IR difference spectra for the solution under $^{16}O_2$ and $^{18}O_2$ gave the $O - O$ stretching frequency of the porphinatoiron-coordinated dioxygen at 1161 cm^{-1}, which differed from that of gaseous dioxygen (1556 cm^{-1}) but was similar to that of superoxide O_2^- (1145 cm^{-1}).[30] Mössbauer parameters for the oxygen-coordinated ^{57}Fe complex of **7** ($\delta = 0.28$ mm s^{-1}, $E_Q = 2.15$ mm s^{-1})[31] meant that the iron ion of the complex is in an iron(III) low-spin state based on the charge-separated structure $Fe(III) - O_2^-$ that has been reported for oxyHb.

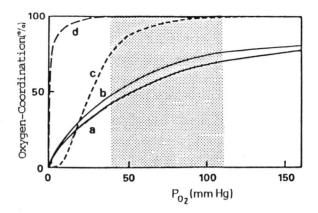

Figure 2-5 ■ Oxygen-coordination equilibrium curve for the macromolecular porphinatoiron complex (**7**) under physiological conditions.

The oxygen-coordination equilibrium curves (Fig. 2-5) show that the porphinato-iron complexes embedded in the macromolecular assembly bind oxygen in response to the oxygen partial pressure.[19,24] The oxygen-coordinating affinity (P_{50}: the oxygen partial pressure at 50% oxygen coordination to porphinatoiron; the reciprocal of oxygen-coordinating equilibrium constant) was determined by the equilibrium curves (Table 2-2).

Table 2-2 shows that the P_{50} value can be controlled with the structure of the porphinatoiron complex (**12–14**).[32–34] Also notice in Table 2-2 that the equilibrium curves for the porphinatoiron complexes situated on the right hand for the P_{50} values are larger in comparison with Hb in red blood cells. This result indicates that the

Table 2-2 ■ **Equilibrium and Rate Constants for the Oxygen Coordination to the Macromolecular Porphinatoiron Complexes under Physiological Conditions**

Porphinatoiron/Lipid Assembly	P_{50} (mmHg)	$10^{-4}k_{on}$ ($m^{-1} s^{-1}$)
12/DMPC liposome	55	3700
7/DMPC liposome	53	9800
13/DMPC liposome	55	1.3
14/DMPC liposome	866	245
9/**8** copolymer	48	1.8
1/DMPC liposome	28	0.79
1/DPPC liposome	24	0.24
Hb in red cell	27	1.1
Stripped Hb	0.35	3300
1 in toluene solution	38	11000

12

macromolecular porphinatoiron assembly has weaker oxygen-coordinating affinity than that of Hb in red blood cell and delivers oxygen to Hb in their mixture system. This moderately weak oxygen-coordinating affinity is one of the advantages of the macromolecular porphinatoiron complex as an oxygen-transporting fluid.

The crystal structure around the porphinatoiron complex of Hb has been well established.[1] The skeleton structure of the protein forms a pocket above the porphinatoiron plane, thereby maintaining the oxygen-coordination site vacant for oxygen penetration. The pocket is constructed with hydrophobic amino acid residues that also contribute to suppressing oxidation of the iron(II). A pathway for oxygen coordination in Hb has been proposed and the effect of the protein on the dynamics of the oxygen coordination has been discussed by use of data measured extensively with stopped-flow, flash-photolysis, and T-jump techniques.[35,36] The protein forms a dense wall between the external medium and the porphinatoiron complex, and a calculated three-dimensional map of the protein conformation suggests that fluctuation of the protein conformation reduces the energy barrier for oxygen penetration and provides a pathway for oxygen coordination.[37,38]

13

The kinetic profile of oxygen coordination by the macromolecular porphinatoiron assembly was studied by a stopped flow and flash-photolysis method as follows (Table 2-2): Oxygen coordination and dissociation occurred reversibly and were completed within seconds. The oxygen-coordination rate constant k_{on} is summarized in Table 2-2. The k_{on} values of the macromolecular **1** complexes were similar to that of Hb in the red blood cell and were about 10^3 times smaller than those of homogeneous systems, that is, the **1** complex in toluene and stripped Hb. This means that the oxygen coordination for the liposome-embedded porphinatoiron complex and Hb encapsulated in membrane (red blood cell) showed a similar feature. The oxygen-coordination is assumed to be largely retarded by the diffusion process of oxygen in and through the phospholipid membrane.

The oxygen-coordination rate was larger for the DMPC than for the dipalmitoylphosphatidylcholine(DPPC)-liposome (**1**). This may be explained as follows: The DPPC-liposome is below its T_c (41°C) and the DMPC-liposome is above its T_c (23°C) under the experimental condition (at 25°C), so that the membrane fluidity and probably also the oxygen permeability of the DMPC-liposome is larger than that of DPPC.

The oxygen-coordination rate constant was also affected by the steric structure and orientation degree of the complex (Table 2-2). The k_{on} values for the porphinatoiron complexes **7** and **12** orientedly embedded in the lipid bilayer matrices, were 10^3 times larger than those for the **1** complex physically dispersed in the matrix, and was similar to that of stripped Hb. **7** and **12** have four long alkanephosphocholine or alkanes-

14

teroid groups on the porphyrin plane and are embedded in the lipid bilayer with high compatibility. Thus the micro-stereostructure and physicochemical environment of the porphyrin plane provide an oxygen-coordinating pathway from the outer water phase to the porphyrin iron, just like the heme pocket of Hb, which is constructed with the higher-ordered structure of globin protein.

But k_{on} was much reduced for the **9/8** copolymer; the coordination reaction was suppressed by the covalent fixation, or too much immobilization, of the complex sphere.

The total potential energy profile for the oxygen-coordination reaction was constructed for the liposome-embedded porphinatoiron above and below T_c, based on their activation parameters, enthalpy, and entropy changes (Fig. 2-6).[39] For the oxygen-coordination reaction for the liposome-embedded complex, the oxygen molecule encounters a potential barrier when it approaches from the solvent phase to the internal complex. The potential barrier is composed of four steps, that is, the bound water region at the surface of the liposomal assembly, the polar head-group region and fatty acid residue region (oriented alkyl chain) of the liposome phospholipid, and the pocket over the porphinatoiron plane. The highest barrier is the oxygen penetration step through the oriented alkyl chain phase of the phospholipid bilayer. On the other hand, for the liposome-embedded **7** complex the porphyrin pocket above the porphyrin plane canceled out the potential barrier observed for the liposome-embedded **1** complex.

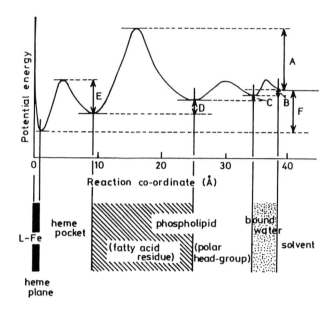

Figure 2-6 ■ Potential energy profile for the oxygen coordination of porphinatoiron complex embedded in the macromolecular assembly. Potential energy (kcal mol^{-1}): For DMPC($< Tc$)/**1**, $a = 13$, $B = -0.57$, $C = -0.50$, $D = -0.76$, $E = 6.7$, $F = -2.5$; For DMPC($> Tc$)/**1**, $A = 12$, $B = -0.57$, $C = -0.50$, $D = -0.76$, $E = 6.7$, $F = -7.4$; For DMPC($> Tc$)/**7**, $A < E$, $E = 6.1$, $F = -7.0$.

The oxygen-coordination profiles of the macromolecular assemblies of **1** and **7** were compared with that of Hb. The former is a model with which to discuss oxygen coordination to Hb in red blood cells and the latter is a model for Hb.

Figure 2-7 shows the temperature dependence of the oxygen affinity, that is, the reciprocal of the P_{50} value and the reciprocal of the temperature.[40] The plots for the oxygen coordination to the porphinatoiron complex of the EYL-liposomal system gives a linear relationship. On the other hand, the temperature dependence for the DMPC-liposomal system has a breaking point at about 23°C. ΔH for the oxygen coordination of the EYL and the DMPC systems above 23°C are ca. -15 kcal/mol, comparable to the corresponding porphinatoiron complex in toluene and Hb; on the contrary, ΔH values for the DMPC system below 23°C are much larger. The phospholipid environment has a large effect on the oxygen-coordinating affinity through its enthalpy contribution. The breaking point of the temperature dependence of oxygen coordination by DMPC system agrees with T_c of the DMPC-liposome.

This phenomenon may be explained as follows. Above T_c the phospholipid molecules are in the liquid crystalline state, which provides an environment just like organic solvents, such as toluene, around the complex. The porphinatoiron complex is in a relaxed (R) state and the oxygen-coordinating affinity and the enthalpy values are comparable with those of the corresponding complex in toluene. On the contrary,

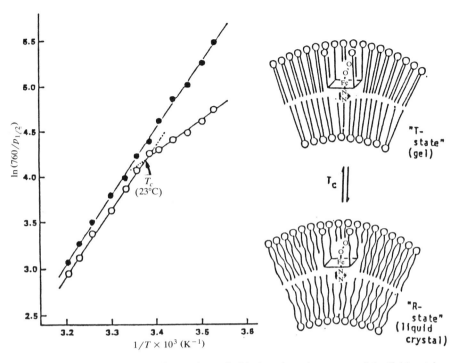

Figure 2-7 ■ Oxygen-coordination conjugated with the oriented structure of the lipid matrix.

below T_c the phospholipid molecules are in the crystal state, which probably induces an orientation of the laurylimidazole ligand and a structural distortion of the porphinatoiron complex because the bulky porphinatoiron molecule is embedded in the phospholipid bilayer; the complex is in a tense (T) state, and this reduces the oxygen-coordinating affinity. The R and T states of the porphinatoiron complex for the liposome-embedded complex are not the same as the R and T for Hb; the latter are caused by globin protein. But the oxygen-coordinating character of the porphinatoiron complex embedded in the macromolecular assembly resembles that of Hb.

The oxygen-coordinating affinity of Hb decreases (P_{50} increases) with the decrease in pH of the medium. This phenomenon is called the *Bohr effect* of Hb; Hb releases oxygen more efficiently when the pH of the medium decreases in the presence of carbon dioxide. On the other hand, EYL-liposome/7 coordinates oxygen more strongly at lower pH (Fig. 2-8)[41,42]. This behavior is in contrast to the Bohr effect of Hb and is explained as follows. The EYL-liposomal/assembly contains a small excess of laurylimidazole, and the noncoordinated imidazole is situated in the bilayer directing the hydrophilic imidazole group outward. The outward directing imidazole is protonated at lower pH and destroys the packing structure of the bilayer, which lets the complex form a relaxed (R) structure. This pH dependence of P_{50} was canceled

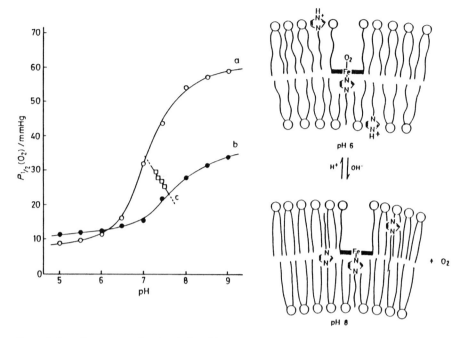

Figure 2-8 ▪ Oxygen-coordination induced by pH of the macromolecular porphinatoiron complex solution.

for the EYL-liposome/**7** with higher EYL concentration and for the porphinatoiron fixed in the polymerized **8** matrix.

The macromolecular assembly of the porphinatoiron complexes coordinates oxygen selectively, rapidly and reversibly even in aqueous media just as Hb does. In this oxygen-coordination reaction to the porphinatoiron complex the kinetic and equilibrium profiles of the coordination were much affected by the physicochemical environment, steric structure, and degree of orientation of the complex.

2-5. Conclusion and Future Studies

Highly efficient oxygen transport by hemoglobin is effected by conjugated interactions of the porphinatoiron complex and the protein macromolecule. Some of these interactions can be partially realized by using much simpler synthetic macromolecular porphinatoiron complexes. The dynamic and weak interactions were quantitatively analyzed by spectroscopy to elucidate the effects of orientation structure, physicochemical environment, strained microstructure, and their dynamic combination, which causes characteristic oxygen-coordination phenomena of the macromolecular porphinatoiron complexes.

The oxygen-coordinating reaction of the macromolecular assembly of the porphinatoiron complex (liposomal and polymerized **8/7**) is now being examined in vivo.[43] For example, half-blood substitution tests with the macromolecular complex solution were conducted in dogs: the porphinatoiron complex indeed transports oxygen in vivo. But after several hours the animals died, presumably because of a disturbance in the blood coagulation system. Further experiments are being carried out by our group to overcome this problem. The practical advantages of disease-free, readily transportable, and easily stored "artificial red blood cells" are quite obvious.

Apart from the artificial red blood cell, we believe the macromolecular porphinatoiron complexes have various potential applications. For example, they will be developed for isolating pure oxygen from air, that is, oxygen-enriching systems containing an oxygen-coordinated complex.[44] Other applications include their use as catalysts for superoxide removal in medical systems, selective oxygenation of organic compounds in drug metabolism, photodynamic cancer therapy and so on. Although this chapter covers only the oxygen-coordinated complexes of porphinatoirons, there are many porphinatometals and other metal complexes with selective gaseous molecule-coordination ability.

The combination of a porphinatometal with a lipid assembly or highly ordered macromolecular matrix offers possibilities unattainable in other ways. Electron transfer or tunneling via the porphinatometal complexes[45] and a photochemical hole-burning system with super highly concentrated memory capacity using the porphyrin moiety[29] have been reported in preliminary studies. They have a potential to be applied as new devices at the molecular level. Investigations of the macromolecular porphinatometal complexes are expected to offer new and unique fields of chemistry.

References

1. M. F. Perutz, H. S. Muirhead, J. M. Cox, and L. C. Goaman, *Nature*, **219**, 131 (1968).
2. A. L. Lehninger, *Biochemistry*, Worth, New York (1975).
3. J. H. Wang, *J. Am. Chem. Soc.*, **80**, 3168 (1958).
4. R. D. Jones, D. A. Summerville, and F. Basolo, *Chem. Rev.*, **79**, 139 (1979).
5. E. Tsuchida and H. Nishide, *Top. Curr. Chem.*, **132**, 64 (1986).
6. J. P. Collman, *Acc. Chem. Res.*, **10**, 265 (1977).
7. T. G. Traylor, and P. S. Traylor, *Ann. Rev. Biophys. Bioeng.*, **11**, 105 (1982).
8. J. P. Collman, R. R. Gagne, T. R. Halbert, and J. C. Marchon, *J. Am. Chem. Soc.* **95**, 7865 (1973).
9. C. K. Chang, B. Ward, R. Young, and M. P. Kandylis, *J. Macromol. Sci. Chem.*, **A25**, 1307 (1988).
10. K. S. Suslick, M. M. Fox, and T. J. Reinert, *J. Am. Chem. Soc.*, **106**, 4522 (1984).
11. T. Komatsu, E. Hasegawa, H. Nishide, and E. Tsuchida, *J. Chem. Soc. Chem. Commun.*, **1990**, 66 (1990).
12. J. E. Baldwin, J. H. Cameron, M. J. Crossley, I. J. Dagely, and T. Klose, *J. Chem. Soc. Dalton Trans.*, **1984**, 1739 (1984).
13. C. K. Chang, B. Wang, and C. B. Wang, *J. Am. Chem. Soc.*, **103**, 5236 (1981).
14. E. Tsuchida, S. Wang, M. Yuasa, and H. Nishide, *J. Chem. Soc. Chem. Commun.*, **1986**, 23 (1986).

15. K. Eshima, Y. Matushita, E. Hasegawa, H. Nishide, and E. Tsuchida, *Chem. Lett.*, **1989**, 381 (1989).

16. E. Tsuchida, H. Nishide, and H. Yokoyama, *J. Chem. Soc. Dalton Trans.*, **1984**, 2383 (1984).

17. H. Nishide, M. Yuasa, E. Hasegawa, and E. Tsuchida, *Macromolecules*, **20**, 1913 (1987).

18. Y. Matsushita, E. Hasegawa, K. Eshima, and E. Tsuchida, *Chem. Lett.*, **1983**, 1387 (1983).

19. E. Tsuchida, H. Nishide, M. Yuasa, E. Hasegawa, and Y. Matsushita, *J. Chem. Soc. Dalton Trans.*, **1984**, 1147 (1984).

20. M. Yuasa, K. Aiba, Y. Ogata, H. Nishide, and E. Tsuchida, *Biochem. Biophys. Acta*, **860**, 558 (1986).

21. G. B. Ansell, R. M. Damson, J. N. Hawthorne, *Form and Function of Phospholipids*, Elesevier, Amsterdam (1973).

22. M. Yuasa, H. Nishide, E. Tsuchida, and A. Yomagishi, *J. Phys. Chem.*, **92**, 2987 (1988).

23. H. Ringsdorf, L. Gros, and H. Schupp, *Angew. Chem.*, **20**, 305 (1981).

24. E. Tsuchida, H. Nishide, M. Yuasa, E. Hasegawa, Y. Matsushita, and K. Eshima, *J. Chem. Soc. Dalton Trans.*, **1985**, 275 (1985).

25. E. Tsuchida, H. Nishide, and M. Yuasa, *J. Macromol. Sci. Chem.*, **A24**, 333 (1987).

26. H. Nishide, M. Yuasa, Y. Hashimoto, and E. Tsuchida, *Macromolecules*, **20**, 461 (1987).

27. E. Tsuchida, H. Nishide, M. Yuasa, T. Babe, and M. Fukuzumi, *Macromolecules*, **22**, 66 (1989).

28. E. Tsuchida, H. Nishide, M. Yuasa, E. Hasegawa, K. Eshima, and Y. Matsushita, *Macromolecules*, **22**, 2103 (1989).

29. E. Tsuchida, H. Ohno, M. Nishikawa, H. Hirooka, and K. Arishima, *J. Phys. Chem.*, **92**, 4255 (1988).

30. M. Yuasa, K. Yamamoto, H. Nishide, and E. Tsuchida, *Bull. Chem. Soc. Jpn.*, **61**, 313 (1988).

31. E. Tsuchida, H. Maeda, M. Yuasa, H. Nishide, H. Inoue, and T. Shirai, *J. Chem. Soc. Dalton Trans.*, **1987**, 2455 (1987).

32. E. Tsuchida, E. Hasegawa, T. Komatsu, T. Nakata, and H. Nishide, *Chem. Lett.*, **1990**, 389 (1990).

33. E. Hasegawa, M. Fukuzumi, H. Nishide, and E. Tsuchida, *Chem. Lett.*, **1990**, 123 (1990).

34. E. Tsuchida, T. Komatsu, K. Babe, T. Nakata, and H. Nishide, *Bull. Chem. Soc. Jpn.*, **63**, 2323 (1990).

35. R. H. Gibson, *J. Physiol.*, **134**, 112 (1956).

36. E. Antonini and M. Brunori, *Hemoglobin and Myoglobin in Their Reactions with Ligands*, North-Holland, Amsterdam (1971).

37. R. H. Anstin, K. W. Beeson, L. Eisenstein, and H. Frauenfelder, *Biochemistry*, **24**, 5355 (1975).

38. J. A. McGammon and M. Korplus, *Acc. Chem. Res.*, **16**, 187 (1983).

39. M. Yuasa, H. Nishide, and E. Tsuchida, *J. Chem. Soc. Dalton Trans.*, **1987**, 2493 (1987).

40. E. Tsuchida, H. Nishide, M. Yuasa, and M. Sekine, *Bull. Chem. Soc. Jpn.*, **57**, 766 (1984).

41. E. Tsuchida, H. Nishide, and M. Yuasa, *J. Chem. Soc. Chem. Commun.*, **1986**, 1107 (1986).

42. M. Yuasa, Y. Tani, H. Nishide, and E. Tsuchida, *J. Chem. Soc. Dalton Trans.*, **1987**, 1919 (1987).

43. E. Tsuchida and H. Nishide, in *Liposomes as Drug Carrier*, G. Gregoriadis, ed., Wiley, New York, (1988), page 569.

44. E. Tsuchida, H. Nishide, and M. Ohyanagi, *J. Phys. Chem.*, **92**, 6461 (1988).

45. E. Tsuchida, H. Nishide, and M. Kaneko, *J. Phys. Chem.*, **90**, 2283 (1986).

Dynamics of Electron-Lattice Interaction in Macromolecular Complexes

Kazuyuki Horie

Research Center for Advanced Science and Technology, University of Tokyo, 4-6-1 Komaba, Meguro-ku, Tokyo 153, Japan

In solid polymer complex systems, the specific interactions due to unstable structure and dynamic change of polymer conformation are especially important. The effects of molecular structure and molecular motion of matrix polymers on photochromic reactions and on photochemical hole burning (PHB) are discussed in the present chapter. These methods, that is, photoreactive probe method and PHB, provide new techniques for investigating molecular relaxation processes, free-volume distribution, electron–phonon interaction, and low-energy excitation modes of matrix polymers. The PHB is also important as a key technology for creating an ultrahigh density optical data storage system in future molecular devices.

3-1. Photochromic Reactions of Azo Compounds and Free-Volume Distribution

The various types of reaction fields affecting the chemical reactions and electron-transfer processes can be illustrated in Fig. 3-1 as a function of immobility and molecular order. Both solutions and amorphous solids are regarded as random systems. In amorphous solids the reactivity of the chromophore may vary due to microscopic heterogeneity and restricted molecular mobility of the system, whereas the reaction field is dynamically averaged in solution. In crystals that occupy an opposite position from solution, the reactivity is mainly controlled by topochemical requisites, which are relatively fixed distances and orientation between potentially reactive groups. Between solution and crystals, there exist partially ordered and

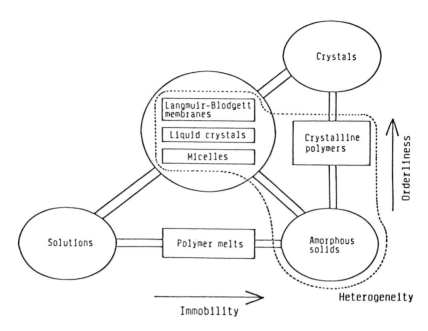

Figure 3-1 ■ Various types of reaction fields affecting chemical reactions and electron transfer processes.

partially mobile systems, such as micelles, liquid crystals, and Langmuir–Blodgett membranes, that have a large potential for realizing various functions of molecular complexes.

The region surrounded by a dotted line in Figure 3-1 corresponds to reaction fields where the reactions that are expected to be unimolecular in solution do not proceed as first-order processes. Among them, the amorphous solid has attracted increasing interest in recent years[1] because microheterogeneity due to a heterogeneous distribution of the free volume of the matrix or conformation of guest molecules would lead to heterogeneous progress of the reaction even in a macroscopically and morphologically homogeneous system.

The trans to cis photoisomerization of azobenzene (AZB),[2] 1,1'-azonaphthalene (AZN),[3] 9,9'-azophenanthrene (AZP), and stilbene (STL) in polycarbonate film was measured over a very wide temperature range (4–400 K) as typical examples of photochromic reactions in polymer solids. Fig. 3-2 shows the case of 1,1'-azonaphthalene, where the initial rate of photoisomerization does not decrease much with decreasing temperature, but the final fraction of the cis form, $\chi_{c_\infty} = [c]_\infty/[t]_0$, decreases markedly with the decrease in temperature. The temperature dependence of equilibrium or limiting cis fractions, χ_{c_∞}, is illustrated in Fig. 3-3 for all cases.

In solution there is no substantial temperature dependence for equilibrium points. Whereas the influence of the thermal backward reaction can be neglected below room temperature, this means that the ratio $\Phi_{t \to c}/\Phi_{c \to t}$ is independent of temperature. The activation energies for both processes must be equal or must be zero. But in

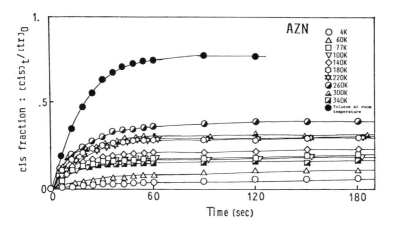

Figure 3-2 ■ Change in the fraction of *cis*-1,1'-azonaphthalene, $[\text{cis}]_t/[\text{tr}]_0$, during trans to cis photoisomerization of 1,1'-azonaphthalene in polycarbonate film at 4–340 K.

polycarbonate film below room temperature, photoisomerization stops at a conversion smaller than the equilibrium value in solution and the limiting conversion decreases with decreasing temperature. The marked decrease in χ_{c_∞} with increasing temperature above room temperature is due to the predominant occurrence of thermal cis to trans isomerization.

For azobenzene, χ_{c_∞} decreases gradually with decreasing temperature T and then decreases markedly for $T < T_\gamma$, where $T_\gamma \simeq 170$ K is the γ-transition temperature of polycarbonate. It is noteworthy that even at liquid helium temperature (4 K) *trans*-azobenzene photoisomerizes up to 17% conversion, and the rate at 4 K at the mobile reaction sites is almost the same as that at room temperature and in solution. The

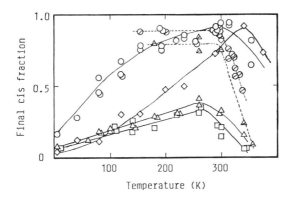

Figure 3-3 ■ Final cis fraction χ_{c_∞} for photoisomerization of ABZ ($\bigcirc, \oslash$), AZN ($\triangle, \triangle$), AZP ($\square$), and STL ($\lozenge$) in PC (solid lines) and in solution (dotted lines).

Table 3-1 ▪ Sweep Volume for Molecular Motion (nm^3)

Mechanism	Phenyl	1-Naphthyl	9-Phenanthryl
Inversion	0.12	0.16	0.23
Rotation	0.25	0.44	0.91

final conversions $\chi_{c\,\infty}$ for azonaphthalene and azophenanthrene in polycarbonate film are much smaller than $\chi_{c\,\infty}$ for azobenzene due to the larger sweep volume necessary for the photoisomerization. However, several percent of the photoisomerization occur at 4 K even for these chromophores with bulky naphthyl or phenanthryl groups.

This occurrence of photoisomerization even at 4 K can be explained only by considering the existence of free-volume distribution in the polymer matrix, where 17% of azobenzene, 7% of azonaphthalene, and 5% of azophenanthrene would be surrounded by a local free volume larger than the critical volume necessary for isomerization to occur. The inversion mechanism for photoisomerization of the azocompounds, where the existence of lone pairs on nitrogen atoms enables the change of hybrid orbital from sp^2 to sp and to sp^2 again, will also explain the small critical volume needed for photoisomerization to occur. This is supported by the temperature dependence of $\chi_{c\,\infty}$ for stilbene (STL) in polycarbonate (Fig. 3-3), where rotation of phenyl groups is necessary for the photoisomerization. The values of sweep volume evaluated for both inversion and rotation mechanisms of photoisomerization are summarized in Table 3-1.

Typical first-order plots for trans to cis photoisomerization of azocompounds in polycarbonate at various temperatures are shown in Fig. 3-4. The deviation from the straight line in the first-order plots becomes more marked with decreasing temperature, but at 4 K the slope for *trans*-azobenzene in the mobile sites is ascertained to be almost the same as that at room temperature.

A kinetic model that considers the free-volume distribution and the temperature dependence of its fluctuation is proposed.[2] The reaction sites in polymer solids below T_g are divided into two parts by critical fractional free volume f_c. The first part is a mobile part where the rate is equal to the rate in solution k_0. The second is the part where the rate is controlled by the mobility of microenvironment with an activation energy proportional to $f_c - f$, where f is local free volume. The extent of photoisomerization in the polymer matrix is given by

$$[t]/[t]_0 = \int_0^\infty G(f)\{A/(A+B)\exp[-k(f)t] + B/(A+B)\}\,df \qquad (3\text{-}1)$$

where $G(f)$ is Γ-distribution function of f, A and B are apparent rate coefficients for trans $\rightarrow$ cis and cis $\rightarrow$ trans photoisomerizations, and $k(f)$ is a rate constant for the individual reaction site that depends on f given by

$$k(f) = k_0 \exp[-\Delta E/k_B T] \qquad (3\text{-}2)$$

$$\Delta E = C(f_c - f) \quad \text{for } f < f_c$$

$$= 0 \qquad\qquad \text{for } f > f_c \qquad (3\text{-}3)$$

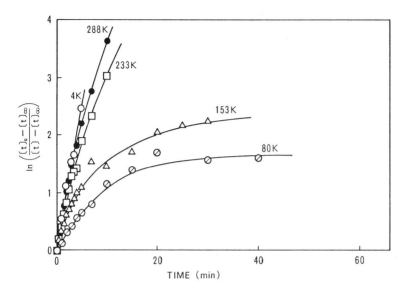

Figure 3-4 ▪ First-order plots of trans to cis photoisomerization of azobenzene in polycarbonate film at temperatures indicated beside the curves.

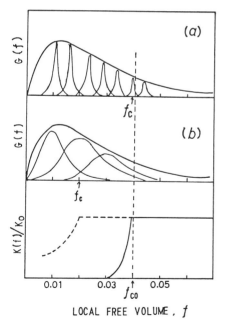

Figure 3-5 ▪ Schematic illustration of fractional free-volume distribution, $G(f)$, at (a) 4 K (b) at higher temperature, and (bottom) reactivity $k(f)/k_0$ according to the free-volume fluctuation model. The solid line in reactivity corresponds to (a) 4K; (b) the dashed line corresponds to a higher temperature.

Table 3-2 ▪ Parameters for Photoisomerization According to Local Fluctuation Models

	AZB	STL	AZN
Isomerization mechanism	Phenyl inversion	Phenyl rotation	1-Naphthyl inversion
Sweep volume (nm^3)	0.12	0.25	0.16
Critical free volume (nm^3)	0.20	0.32	0.27
β (T^{-1})	8×10^{-5}	8×10^{-5}	8×10^{-5}
C (J mol^{-1})	2.0×10^5	7.2×10^5	5.0×10^5

The value of f_c decreases with increasing T,

$$f_c = f_{co} - \beta T \tag{3-4}$$

due to the fluctuation of local free volume.

The schematic illustration of the distribution function $G(f)$ and reactivity $k(f)/k_0$ is given in Fig. 3-5. The simulation according to Eqs. (3-1)–(3-4) reproduces experimental results well, and the parameters for best fit conditions are summarized in Table 3-2. Thus, the free-volume fluctuation model is supposed to describe the characteristics of photoisomerization of azocompounds in polymer films. From the inverse point of view, photoisomerization reactions probe the free-volume distribution in amorphous polymer solids.

3-2. Photochemical Hole Burning of Porphyrins in Polymer Matrices

Photochemical hole burning (PHB) is a phenomenon in which very narrow and stable photochemical holes are burned at very low temperatures into the absorption bands of guest molecules that are molecularly dispersed in an amorphous solid by narrow-band excitation with a laser beam (Fig. 3-6). Proton tautomerization of free-base porphyrins and phthalocyanines and hydrogen bond rearrangement of quinizarin are typical photochemical reactions that provide PHB spectra. Recently PHB has attracted considerable interest not only as a tool for high-resolution solid-state spectroscopy,[4–6] but also as a means for frequency-domain high-density optical storage.[7–9]

The efficiency of hole formation and the temperature dependence of hole profiles burned at 4 K are important aspects of PHB phenomena. In order to increase the thermal stability of photochemical holes, aromatic polymers composed of stiff aromatic main chains without side chains were introduced as matrices.[9, 10] Phenoxy resin (aromatic polyhydroxyether) proved to be an effective matrix for the temperature stability of the hole of free-based tetraphenylporphin (TPP).[11] Recently the formation and observation of PHB holes were realized even at liquid nitrogen temperatures (80–100 K) for TPP in phenoxy resin[12, 13] and for water-soluble TPP in poly(vinyl alcohol).[14]

The effects of molecular structure and molecular motion of matrix polymers on the efficiency of hole formation and the temperature stability of holes are presented in

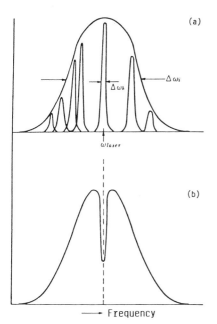

Figure 3-6 ■ The principle of photochemical hole burning (PHB). (a) Absorption spectrum with a broad inhomogeneous width $\Delta\omega_i$ in a solid as a superposition of absorption of each molecule with a homogeneous width $\Delta\omega_h$. (b) A hole formed by laser irradiation at the frequency of ω_{laser}.

Table 3-3 ■ **Factors Affecting the Efficiencies in PHB**

Phenomena	Affecting Factors	Molecular Structure and Properties
1. Efficiency of hole formation		Chemical structure of guest molecule
(a) Initial quantum yield Φ	Rates of elementary processes of excited states / Rates of photochemical process	Structure of microenvironments
(b) Deviation from first-order plots	Shape of the hole / Heterogeneity of matrix	Electron–phonon interaction
(c) Temperature dependence	Hole broadening (dephasing) / Debye–Waller factor / Spectral diffusion	(Dipole density of microenvironment)
2. Hole width $\Delta\omega_h$	Dephasing T_2	Low-energy excitation mode (Nonacoustic contribution, reversible process)
3. Hole density	Hole width $\Delta\omega_h$ / Inhomogeneous line width $\Delta\omega_i$ / Interaction between adjacent holes	Molecular motion and structural relaxation of microenvironment (Irreversible process)
4. Side hole formation	Debye–Waller factors / Phonon frequency	
5. Hole stability at elevated temperatures	Hole broadening (dephasing) / Debye–Waller factor / Spectral diffusion	Crystallinity and dipole density of microenvironment
6. Hole recovery at 4 K	Spectral diffusion	

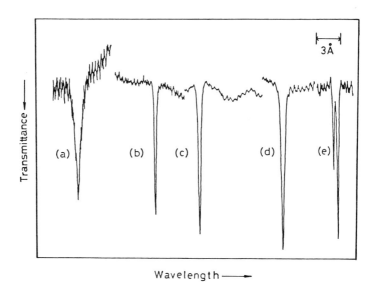

Figure 3-7 ■ Typical PHB spectra of tetraphenylporphin (TPP) in various polymer matrices at 4.2 K. (a) PMMA; (b) PET; (c) PI; (d) PhR; (e) LCP.

the present section by using TPP as a guest molecule. The factors affecting hole formation, hole width, hole density, side hole formation, and thermal stability of the holes are summarized in Table 3-3.

Typical hole spectra of TPP at 4.2 K in poly(methyl methacrylate) (PMMA), poly(ethylene terephthalate) (PET), aromatic polyimide (PI), phenoxy resin (PhR), and a side-chain liquid crystalline polymer (LCP) burned at 4 K by using a helium–neon laser (632.8 nm) or a ring dye laser (about 647 nm) are shown in Fig. 3-7. The efficiency of hole formation depends much on the method of sample preparation. TPP in phenoxy resin prepared by solvent cast, evacuation above T_g of the phenoxy resin, and hot pressing gave the deepest hole during the laser irradiation.[11] The TPP in the hot-pressed samples of polystyrene (PS) and PMMA showed almost the same efficiency of hole formation as that in the phenoxy resin.

The quantum efficiency for hole formation, Φ, was calculated from the initial slope of the change in hole depth $\Delta A/A_0$ during laser irradiation. The results at 4 K are summarized in Table 3-4. It is clear that Φ for TPP at 4 K is independent of the chemical structure of matrix polymers as long as samples are prepared with optimal conditions. This is consistent with the intramolecular nature of the hole-burning phototautomeric reaction of free-based tetraphenylporphin. The low values of Φ for TPP in the poly(ethylene terephthalate) film, which was drawn five times and annealed, and for TPP chemically bonded to the side chain of the LCP would be due to the aggregation of TPP molecules.

The formation of satellite holes in PHB is illustrated in Fig. 3-8 for TPP in the phenoxy resin. By laser irradiation at 632.8 nm, several satellite holes in both higher- and lower-energy regions could be observed.[9] The energy differences given per

Table 3-4 ▪ Quantum Efficiency of Hole Formation Φ of TPP in Various Polymers

Polymer	Condition of Sample Preparation	Φ
PhR	Hot press of solvent cast sample	1.2×10^{-3}
PMMA	Hot press of solvent cast sample	1.2×10^{-3}
PS	Hot press of solvent cast sample	1.2×10^{-3}
PET	Undrawn	5.4×10^{-4}
	5 times drawn (95°C)	6.2×10^{-4}
	5 times drawn and annealed at 220°C for 1 min	1.2×10^{-4}
PI	Imidized at 150°C for 2 hr and 200°C for 5 hr	3.6×10^{-4}
	Imidized for 50% drawn PAA at 150°C for 2 hr and at 200°C for 5 hr	7.0×10^{-4}
LCP	TPP (1%) is chemically bonded to the side chain	1.9×10^{-4}

centimeter in Fig. 3-8 correspond to vibrational frequencies in the excited singlet state of TPP, and can be compared to those of chlorin and porphin obtained from fluorescence excitation spectra.[15]

The hole broadening due to electron–phonon coupling, decrease in the Debye–Waller factor, and hole filling due to laser-induced structural relaxation diminish the efficiency of hole formation at higher temperatures than 4 K. Typical photochemical holes of TPP in the phenoxy resin formed by 1–3 min irradiation of

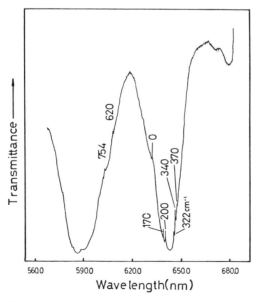

Figure 3-8 ▪ Satellite holes in PHB of TPP in PhR burned with a He–Ne laser at 632.8 nm at 4.2 K.

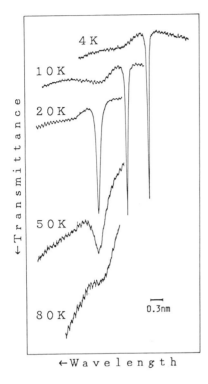

4 K

1 0 K

2 0 K

←Transmittance

5 0 K

8 0 K

0.3nm

←Wavelength

Figure 3-9 ■ Hole profiles formed at various temperatures for TPP in the phenoxy resin. The holes were formed by 0.75 mnW cm^{-2} dye laser irradiation for 1 min (4–50 K) or 3 min (80 K).

0.75 mW cm^{-2} dye laser at various temperatures are illustrated in Fig. 3-9. To our knowledge, the hole formation above liquid nitrogen temperature has been observed only in this system,[13] in TPP in epoxy resin, and in the sulfonated TPP in poly(vinyl alcohol).[14]

The changes in hole depth $\Delta A/A_0$ during dye laser irradiation at various temperatures are shown in Fig. 3-10,[13] where ΔA is the difference in absorbance produced by hole formation and A_0 is the absorbance before irradiation. The values of quantum efficiency for hole formation Φ and hole multiplicity $R = \Delta\omega_i/\Delta\omega_h$ are also given in Fig. 3-10, where $\Delta\omega_i$ is an inhomogeneous line width of absorption spectrum and $\Delta\omega_h$ is a homogeneous line width that is a half hole width. The Arrhenius plots of Φ (Fig. 3-11) show breaks above 30 K, and above the breaks, Φ decreases steeply with increasing temperature. The apparent temperature dependence of -1.5 kJ mol^{-1} for $T > 30$ K is the same order of magnitude as that for the local molecular motion of the matrix polymer, suggesting that the simultaneous occurrence of hole filling during hole burning by laser irradiation, due to local structural relaxation around the guest molecules, is the dominant factor for diminishing Φ at $T > 30$ K.

The temperature stability of hole profiles has been studied by cycle annealing experiments.[9,11] Typical examples are shown in Fig. 3-12. The hole, formed by 30 min irradiation of a 1.8 mW cm^{-2} He-Ne laser at 4.2 K, was measured at a certain temperature shown beside the spectra after annealing at that temperature for 30 min (upper column). Then the spectrum was measured again after cooling down to 4 K

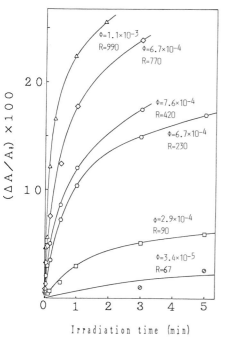

Figure 3-10 ■ Change in the hole depth $\Delta A/A_0$ during 0.75 mW cm^{-2} dye laser irradiation at various temperatures (△: 4 K; ◇: 10 K; ○: 20 K; ○: 30 K; □: 50 K; ∅ 80 K).

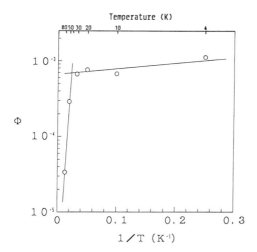

Figure 3-11 ■ Temperature dependence of apparent quantum yield for hole formation Φ plotted versus $1/T$ for the TPP/phenoxy resin system.

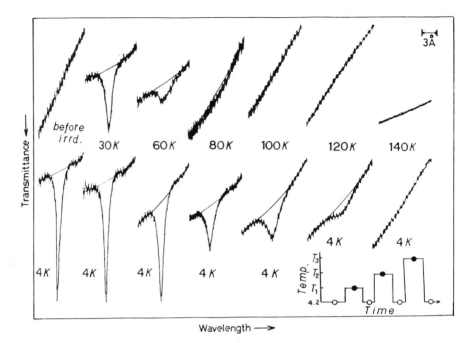

Figure 3-12 ■ Typical photochemical holes of TPP in PhR during cycle annealing. The upper column corresponds to the measurements at the annealed temperature shown beside the spectra after 30 min annealing; the lower column corresponds to spectra measured at 4.2 K after cooling from the annealed temperature.

(lower column). The changes in relative hole depth $\Delta A/(\Delta A_0)$, based on its initial values measured just after irradiation at 4 K are summarized in Fig. 3-13 for TPP in the phenoxy resin, polyimides, and PMMA. The hole measured at annealing temperature becomes shallow with the increase in the annealing temperature T and disappears for $T > 60$ K in PMMA and polyimide and for $T > 100$ K in the phenoxy resin. But the hole partially recovers when the sample is cooled and measured again at 4.2 K. This reversible hole disappearance could be caused by hole broadening that is induced by dephasing due to electron–phonon interaction and by the change in the Debye–Waller factor. Figure 3-13 shows the partial hole recovery at 4 K after cycle annealing up to 110 K for TPP in PMMA, 120 K for TPP in the phenoxy resin, and 130 K for TPP in polyimide. The irreversible disappearance of holes observed in cycle annealing even after cooling is totally a ground state phenomenon and is thought to depend on the change in the microenvironment of the chromophore due to the molecular structure of matrix polymers. Thus stiff molecular structure of the aromatic polyimide provides better restored hole depth measured at 4 K for TPP in polyimide than for TPP in PMMA.

The effects of drawing and successive thermal treatment of poly(ethylene terephthalate) (PET) film on the temperature stability of the hole of TPP are shown in Fig.

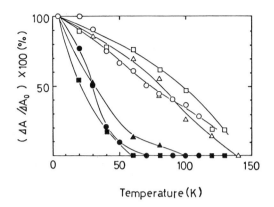

Figure 3-13 ■ Changes in relative hole depth $\Delta A/(\Delta A_0)$ during the annealing cycle measured at the annealed temperatures (▲, ■, ●) and at 4.2 K (△, □, ○) for TPP in PhR (▲, △), PI (■, □), and PMMA (●, ○).

3-14.[16] The recovery of hole depth measured at 4 K after cycle annealing was markedly improved by five-time drawing of the PET film including molecularly dispersed TPP. This suggests that the orientation of the film suppresses the structural relaxation of the microenvironment of the TPP molecule in the amorphous region of PET. However, thermal treatment of the drawn film at 220°C for 1 min relaxed the local structure, leading to the same extent of hole recovery as that for the original undrawn PET sample. The dichroism measurement showed that the rotational relaxation of the TPP molecule in the drawn PET film hardly occurs during the thermal treatment at 220°C for only 1 min. Thus, the structural relaxation process affecting the spectral diffusion or hole filling in PHB is supposed to be of very local character in the vicinity of the chromophore molecule.

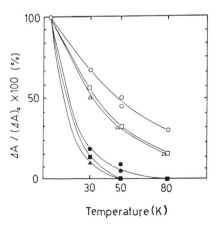

Figure 3-14 ■ Change in relative hole depth $\Delta A/(\Delta A_0)$ compared to the initial hole depth burned at 4.2 K, during cycle annealing measured at the annealed temperatures (▲, ■, ●) and at 4.2 K (△ □, ○) for TPP in undrawn (▲, △) five times drawn (●, ○) PET films and annealed (at 220°C) PET film after five times drawing (■, □).

3-3. Electron-Lattice Interaction and Low-Energy Excitation Modes

The most attractive aspect of PHB from the spectroscopic point of view is that PHB gives information on a homogeneous absorption spectrum of a molecule in the solid state as a hole formed in the usual broad inhomogeneous absorption band. A photochemical hole burned by laser irradiation consists of three parts: a zero-phonon hole, a phonon side hole, and a pseudo-phonon side hole. A sharp zero-phonon hole and a side hole at the high-energy side are due to the excitation and hole burning of guest molecules that have a zero-phonon line at the irradiation wavelength. A pseudo-phonon side hole is formed at the lower-energy side, and the energy difference between the zero-phonon hole and the pseudo-phonon side hole E_s is the same as the difference between zero-phonon hole and the side hole. The pseudo-phonon side hole is caused by excitation of the photoreactive molecules that have a phonon side band at the irradiation wavelength. The mechanism of pseudo-phonon side hole formation is illustrated in Fig. 3-15. Because coupling between a guest molecule and an amorphous polymer matrix is usually weak, a side hole at the higher-energy side is small. However, a pseudo-side hole can be formed to a large size and easily detected as is shown in Fig. 3-9 (4–20 K), owing to the large number of guest molecules that have absorption of phonon side bands at irradiation wavelength.

The homogeneous line width of $S_1 \leftarrow S_0$ transition, $\Delta\omega_h$, is given by

$$\Delta\omega_h = 1/(2T_1) + 1/T_2 \qquad (3\text{-}5)$$

where T_1 is the fluorescence lifetime and T_2 is the dephasing time. The dephasing term T_2^{-1} becomes important in amorphous matrices, so that $\Delta\omega_h$ reflects the interaction between the electronically excited state and the surrounding matrix. The

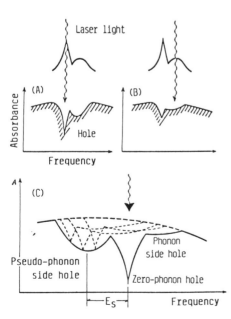

Figure 3-15 ■ Mechanism of the formation of pseudo-phonon side hole. (A) Excitation at zero-phonon line; (B) excitation at phonon-side band; (C) overall hole profile.

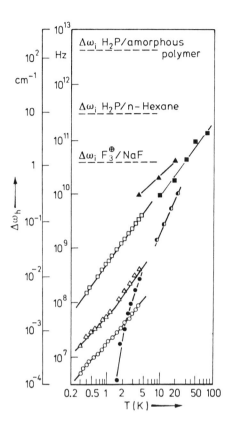

Figure 3-16 ■ Temperature dependence of homogeneous line width $\Delta\omega_h$ for PHB of free-base porphin (H_2P), phthalocyanine (H_2Pc), and TPP. Dashed lines correspond to inhomogeneous line width $\Delta\omega_i$ of absorption spectra. □: H_2P/MTHF; △: H_2P/diglycerol; ○: H_2P/PE; ▲: H_2Pc/PMMA; ◖: H_2Pc/PE; ●: H_2P/n-decane; ■: TPP/PhR.

temperature dependence of homogeneous line width $\Delta\omega_h$ for PHB of free-base porphin,[5,17] TPP,[13] and phthalocyanine[18] is summarized in Fig. 3-16. The values of $\Delta\omega_h$ were determined from the hole width Γ, extrapolated to initial time $t = 0$ by using $\Gamma = 2\Delta\omega_h$. In organic crystals, such as n-decane, a very narrow $\Delta\omega_h$ of the guest molecule was obtained, but it increases with increasing temperature T following a T^7 power law. The $\Delta\omega_h$ seems to be proportional to $T^{1.0-2.0}$ in amorphous matrices, though its absolute values are larger than those in crystals. It is noteworthy that free-base porphin in polyethylene shows a very narrow hole width despite the $T^{1.0-1.3}$ dependence corresponding to the amorphous matrix.[17] This fact may be due to a nonpolar structure without side groups and to the incorporation of the PHB molecule in the boundary region between paracrystallites of the polyethylene chain. The $T^{1.4}$ and $T^{1.7}$ dependences up to liquid nitrogen temperatures are reported for TPP in the phenoxy resin[13] and sulfonated TPP in poly(vinyl alcohol).[14]

A linear relationship between saturated hole width at 4 K and the dipole density of matrix polymers was also obtained[19] (Fig. 3-17), suggesting the importance of dipole interaction on the dephasing process of the excited state electron.

The extent of electron–phonon interaction and the low-energy excitation modes of amorphous polymers can also be evaluated from PHB experiments. Low-temperature

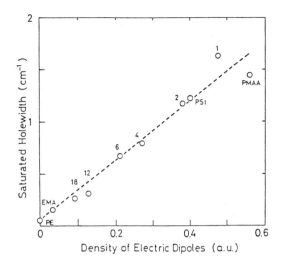

Figure 3-17 ▪ Hole width in saturated regime versus bulk density of electric dipoles in each polymer matrix.[19] PMMA = poly(methacrylic acid), PE = poly(ethylene), and EMA = random copolymer of PE and PMMA (12 wt% of MAA). Numbers correspond to the carbon number of the alkyl group of the side chain in each poly(alkyl methacrylate).

properties of amorphous materials have attracted interest for many years. Many properties of amorphous materials, for example heat capacity at low temperatures, are different from those of crystals owing to the large density of states for low-energy excitation modes in comparison with crystals. The large density of states for low-energy excitation mode leads to excess heat capacity at cryogenic temperatures and the thermal properties differ from those expected by the Debye theory. Low-energy excitation modes in amorphous polymers were investigated by heat capacity measurements[20] and neutron inelastic scattering measurements.[21,22] It was noticed recently, by comparing the PHB hole profiles of quinizarin and TPP in PMMA and some other matrices, that the energy difference between the zero-phonon hole and the pseudo-phonon side hole, E_s, does not depend on the nature of the guest molecules, but is specific to the host matrices.[23]

The values of E_s for various systems are summarized in Table 3-5, together with the energy of low-energy excitation modes estimated from heat capacity measurements E_c and that measured by neutron inelastic scattering measurements E_1. The E_s values are constant for certain polymers irrespective of the nature of guest molecules. The E_s values for PMMA and PS agree with the low-energy excitation mode E_c determined by the heat capacity measurements. The origin of this low-energy excitation mode for PMMA is considered to be the rotation of the ester-methyl group of PMMA.[20] The E_s values for polystyrene and epoxy resin (EpR) almost agree with the peak of the density of states at low energy determined by neutron inelastic scattering measurements.[21,22] Thus one concludes that the energy difference between the zero-phonon hole and the pseudo-phonon side hole E_s is a characteristic

Table 3-5 ▪ **Energies of the Low-Energy Excitation Mode Determined by PHB, E_s, Heat Capacity, E_c, and Inelastic Neutron Scattering E_I Measurements**

Chromophore/Amorphous Polymer Matrix	E_s (cm^{-1})	E_c (cm^{-1})	E_I (cm^{-1})
TPPS/PVA	23.5		
DAQ/PVA	23.0		
TPP/PMMA	13.1	12^a	
DAQ/PMMA	13.6		
TPP/PhR	15.1		
TPP/EpR (ethylenediamine)	14.6		15^b
TPP/EpR (hexamethylenediamine)	17.1		
TPP/PI	13.3		
TPP/LCP	12.6		
TPP/PET (undrawn)	10.9		
TPP/PET (5 times drawn)	11.8		
TPP/PS	10.1	9.1^a	12^c

^{a}Reference 20.
^{b}Reference 22.
^{c}Reference 21.

parameter of the matrix polymer reflecting the phonon frequency corresponding to the low-energy excitation mode of the amorphous polymer.

The Debye–Waller factor DW(T), given by

$$DW(T) = S_0(T)/\left(S_0(T) + S_p(T)\right) \qquad (3\text{-}6)$$

where $S_0(T)$ and $S_p(T)$ are the integrated intensity for a zero-phonon line and a phonon side band respectively, indicates the coupling strength between a guest and a matrix, and its temperature dependence is affected by low-energy excitation modes of a matrix coupled with guest molecules. As mentioned in the previous section, relatively high temperature PHB, that is, the hole formation and observation above liquid nitrogen temperature, was reported for TPP in phenoxy resin[12,13] and sulfonated TPP (TPPS) in poly(vinyl alcohol)[14] systems. One of the requisites for high-temperature hole formation is supposed to be a large Debye–Waller factor and its small temperature dependence. So the information on the Debye–Waller factor is also important for high-temperature PHB.

Figure 3-18 shows a hole profile burned for TPP in the phenoxy resin at 30 K and its profile after cooling down to 4 K. The hole depth $\Delta A/A_0$ grew large after cooling down to 4 K, but the hole width did not change. Because the structural relaxation in the system does not occur during cooling down to 4 K, the hole depth growth after cooling is caused by the change of the Debye–Waller factor. If the oscillator strength for the sum of integrated intensities of a zero-phonon line and a phonon side band is assumed to be constant, $S_0(T) + S_p(T)$ should be constant irrespective of temperature T. Then the temperature dependence of DW(T) can be determined by the measurement of the temperature dependence of $S_0(T)$. When the hole area burned at higher temperatures and its change after cooling down to 4 K were measured, the ratio

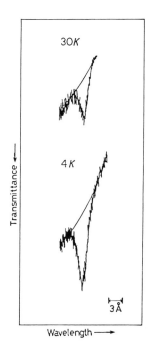

30K

4K

Transmittance ←

Wavelength ⟶

├──┤
3 Å

Figure 3-18 ■ A hole profile burned in the TPP/PhR system at 30 K with 3.6 mW cm^{-2} laser power and its profile after cooling down to 4.2 K.

$S_0(T)/S_0(4)$ roughly corresponds to DW(T) due to the weak coupling between a photoreactive molecule and amorphous polymer matrix, as suggested by the photon echo technique [DW(4) ≃ 0.9 for TPP/PMMA].[24] Figure 3-19 shows the temperature dependence of the Debye–Waller factor thus determined from the PHB measurements. Results in Fig. 3-19 agree well with the results measured by the photon echo technique for TPP in PMMA and for carbonated TPP in poly(vinyl alcohol).[24]

The temperature dependence of the Debye–Waller factor is expressed[25] by

$$DW(T) = \exp\left[-C\int \coth(\hbar\omega/2k_BT)P(\omega)g(\omega)\,d\omega\right] \qquad (3\text{-}7)$$

where C is a constant, k_B is Boltzmann's constant, $(1/2)\coth(\hbar\omega/2k_BT)$ is the

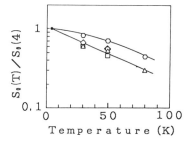

$S_0(T)/S_0(4)$

1

0,1

0   50   100

Temperature (K)

Figure 3-19 ■ Temperature dependence of Debye–Waller factor calculated by $S_0(T)/S_0(4)$ for TPPS/PVA (○), TPP/PhR (△), TPP/LCP (◇), TPP/PET (□), and TPP/PS (○).

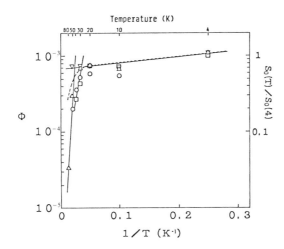

Figure 3-20 ■ Temperature dependence of hole formation efficiency Φ for TPP/PhR ($\triangle$), TPP/EpR ($\triangledown$), TPP/PMMA ($\square$), and TPP/PS ($\bigcirc$). The temperature dependence of the Debye–Waller factor for TPP/matrix systems estimated from $S_0(T)/S_0(4)$ is represented by the dashed line.

average number of phonons in the lattice mode $\hbar\omega$ at temperature T, $P(\omega)$ is Stokes' shift, and $g(\omega)$ is the density of states that have an energy $\hbar\omega$. When the dominant lattice mode $\hbar\omega$ contributing to $g(\omega)$ has large energy, DW(T) shows small temperature dependence resulting from this relationship. The system that has a large E_s value is expected to show small temperature dependence of DW(T). The poly(vinyl alcohol) matrix systems have a larger E_s value than the other matrices and show small temperature dependence. The fact is consistent with the theory. The TPP in the epoxy resin and TPP in PET have different E_s values according to the different sample preparation conditions. This suggests that the degree of the orderliness of the system affects the low-energy excitation modes.

The temperature dependence of hole-formation efficiencies Φ of TPP in various polymer systems are plotted against $1/T$ in Fig. 3-20 together with the temperature dependence of the Debye–Waller factor DW(T) for TPP in these matrix systems estimated from $S_0(T)/S_0(4)$.[26] As can be seen from Fig. 3-20, the temperature dependence of Φ agrees well with the temperature dependence of DW(T) below 30 K and deviates from that of DW(T) above 30 K depending on the nature of the system. The TPP/PMMA and the TPP/polystyrene systems show a rapid decrease of Φ above 30 K. This suggests that structural relaxation in the system begins to occur, leading to hole filling. In the TPP/phenoxy and the TPP/epoxy resin systems, the value of Φ did not decrease rapidly until the DW(T) decreased rapidly. This fact reveals the excellent thermal stability of these systems.

Hole formation at 80 K was observed not only for sulfonated TPP in poly(vinyl alcohol) and TPP in the phenoxy resin, but also for TPP in the epoxy resin, and was not observed in other systems. The E_s value for poly(vinyl alcohol) is extremely large and the DW(T) for this system decreases slowly with the increase in temperature. The E_s values for the phenoxy and epoxy resins are comparatively large and the DW(T) decreases relatively rapidly as in the other matrix systems, but a hole could be formed at 80 K in these two systems. The difference between TPP in phenoxy resin or TPP in epoxy resin systems and the other polymer matrix systems with $E_s < 13.6$ cm^{-1} would be caused by the difference in the rates of structural relaxation in these systems. The TPP in phenoxy and epoxy resin systems are supposed to have slow rates of structural relaxation. This has been ascertained by the slow decrease in Φ with increasing temperature, as mentioned in the previous paragraph. This would be related to the presence of hydrogen bonding in the phenoxy resin and the presence of hydrogen bonding and cross links in the epoxy resin. For TPP in epoxy resin systems, the E_s value grew large and the temperature dependence of the Debye–Waller factor became smaller when the hardener was changed from ethylenediamine to hexamethylenediamine. This suggests that the structure of the cross-linked polymer affects the low-energy excitation mode and improves the thermal property of the system. Thus, the important factor for high-temperature PHB is supposed to be not only large E_s value and small temperature dependence of the DW(T) but also low rate of structural relaxation in the system at elevated temperatures. The high-temperature PHB would be realized by the system containing hydrogen bonding and the cross-linked structure.

The value of $E_s = 30$ cm^{-1} was reported recently for bacteriochlorophyll a in the antenna complex of the photosynthetic system.[27] The PHB of porphins fixed in a polymerized lipid bilayer membrane was also studied.[28] The PHB measurements are now expanded to the biological fields for elucidation of the weak interaction between guest molecule and host matrices in macromolecular complexes.

References

1. K. Horie and I. Mita, *Adv. Polym. Sci.*, **88**, 77 (1989).
2. I. Mita, K. Horie, and K. Hirao, *Macromolecules*, **22**, 558 (1989).
3. T. Naito, K. Horie, and I. Mita, *Eur. Polym. J.*, **26** (1990) in press.
4. J. Friedrich and D. Haarer, *Angew. Chem. Int. Ed. Eng.*, **23**, 113 (1984).
5. R. M. Macfarlane and R. M. Shelby, *J. Luminescence*, **36**, 179 (1987).
6. W. E. Moerner, ed., *Persistent Spectral Hole-Burning: Science and Applications*, Springer, Berlin (1988).
7. A. R. Gutierrez, J. Friedrich, D. Haarer, and H. Wolfrum, *IBM J. Res. Develop.*, **26**, 198 (1982).
8. W. E. Moerner, *J. Mol. Electronics*, **1**, 55 (1985).
9. K. Horie, K. Kuroki, I. Mita, and A. Furusawa, in *Polymers in Information Storage Technology*, K. L. Mittal, ed., Plenum, New York. (1990), p. 145.
10. K. Horie, K. Hirao, K. Kuroki, T. Naito, and I. Mita, *J. Fac. Eng. Univ. Tokyo*, **39**, 51 (1987).
11. K. Horie, T. Mori, T. Naito, and I. Mita, *Appl. Phys. Lett.*, **55**, 935 (1988).
12. A. Furusawa, K. Horie, K. Kuroki, and I. Mita, Preprints *Jpn. Soc. Appl. Phys.*, preprint 881117-01, 175 (1988).
13. A. Furusawa, K. Horie, K. Kuroki, and I. Mita, *J. Appl. Phys*, **66**, 6041 (1989).
14. K. Sakoda, K. Kominami, and M. Iwamoto, *Jpn. J. Appl. Phys.*, **27**, L1304 (1988).
15. S. Volker and R. M. Macfarlane, *J. Chem. Phys.*, **73**, 4476 (1980).

16. K. Horie, K. Kuroki, I. Mita, H. Ono, and S. Okumura, and A. Furusawa, *Polymer*, **31**, (1990) in press.
17. H. P. H. Thijissen, R. E. Van der Berg, and S. Volker, *Chem. Phys. Lett.*, **103**, 23 (1983).
18. L. Kador, G. Schulte, and D. Haarer, *J. Phys. Chem.*, **90**, 1264 (1986).
19. T. Tani, A. Itani, Y. Iino, and M. Sakuda, *Jpn. J. Appl. Phys.*, **26**, Suppl. 26-4, 77 (1987).
20. W. Reese, *J. Macromol. Sci. Chem.*, **A3**, 1257 (1969).
21. T. Kanaya, K. Kaji, S. Ikeda, and K. Inoue, *Chem. Phys. Lett.*, **150**, 334 (1988).
22. H. M. Rosenberg, *Phys. Rev. Lett.*, **54**, 704 (1985).
23. A. Furusawa, K. Horie, and I. Mita, *Chem. Phys. Lett.*, **161**, 227 (1989).
24. S. Saikan, *PSJ Microsymposium on PHB*, preprint (1989), p. 21.
25. K. K. Rebane and L. A. Rebane, in *Persistent Spectral Hole-Burning: Science and Applications*, W. E. Moerner, ed., Springer, Berlin (1988), p. 23.
26. A. Furusawa, K. Horie, and I. Mita, *Jpn. J. Appl. Phys.*, **28**, Suppl. 28-3, 19 (1989).
27. S. G. Johnson and G. J. Small, *Chem. Phys. Lett.*, **155**, 371 (1989).
28. E. Tsuchida, H. Ohno, M. Nishikawa, H. Hiratsuka, K. Arishima, and T. Shimada, *J. Phys. Chem.*, **92**, 4255 (1988).

Ultrafast Dynamics of Macromolecular Complexes

Hiroshi Masuhara and Akira Itaya

Department of Polymer Science and Engineering, Kyoto Institute of Technology, Matsugasaki, Kyoto 606, Japan

4-1. Introduction

Interchromophoric interactions and dynamic behavior of macromolecular complexes have received attention because of their specific and novel functions, which cannot be realized in simple molecules and other molecular assemblies. Although no detailed analysis on electronic processes in macromolecular systems has been given, sequential multiprocesses and geared cycles are considered to be responsible to their characteristic behavior. In order to elucidate these mechanisms and to design new functional macromolecular systems, it is important and indispensable to analyze their elementary steps and to explore new relevant concepts. One approach is to follow the sequential processes and coupled dynamics by using various kinds of time-resolved laser spectroscopy.

Direct measurements of each elementary step in molecules and molecular assemblies have provided a new insight, which has been proved by laser flash photolysis and single photon counting techniques. Through these studies, it has become clear that photoinduced electron transfer is a key process for understanding various relaxation processes and reactions. Not only typical oxidation–reduction reactions, but also hydrogen abstraction and isomerization are initiated by electron transfer. The steric factor of addition reactions is determined by charge-transfer complex formation. The photosynthetic unit in green plants and algae consists of sequential steps of excitation energy and electron transfer. In many solid systems a carrier generation process is due to a photoinduced electron transfer. Catalytic activity and hydrogen evolution are also related to charge-transfer interaction, and polymer-coated electrodes provides a field where photogenerated carriers diffuse. These electron-transfer processes are usually quite rapid and can be followed directly only by time-resolved spectroscopy that has nanosecond and picosecond time resolution. Photoinduced electron transfer forms a very active chemistry field, and detailed mechanisms of how electron transfer polarizes the surrounding environments, induces the solvent reorientation, results in complex formation, and triggers chemical reaction are being clarified.

In macromolecular complexes, charge-transfer interactions, microenvironmental conditions, and molecular motion are quite different from those of small molecular systems. Conformational change is slow even in solution, so that the solvent structure is inhomogeneous even in the nanosecond time scale. Micropolarity and microviscosity may not be averaged completely. In principle dynamic interactions and electronic processes in these macromolecular complexes are statistical and depend on the geometrical and the orientational structures. This is particularly crucial for polymer solids.

In order to elucidate these problems, it is very fruitful to measure electronic processes directly in the nanosecond and picosecond time domains. Fast kinetic spectroscopy and related time-resolved measurement provide a new viewpoint for dynamic interaction and electronic processes of macromolecular complexes. It should also be mentioned that an application of time-resolved spectroscopy is not simple compared to dilute solutions of small molecular systems. Nonlinear photochemical behavior characteristic of polymers excited by an intense pulsed laser is generally induced. We have studied intrapolymer multiphoton charge separation, intrapolymer interactions between excited states, and transient polyelectrolyte formation in detail and have already reviewed the relevant processes.[1,2]

The second difficulty in polymer solids is due to optical conditions. The chromophore concentration of the polymer solids is very high so that the exciting and monitoring lights penetrate only the surface area. To overcome this problem, we had to develop two time-resolved laser spectroscopies under total internal reflection.[3,4] The first is a time-resolved total internal reflection fluorescence spectroscopy, and the second is a dynamic attenuated total reflection UV-visible spectroscopy. Sometimes polymers are insoluble in solvent, so that a transparent film is not obtainable. In this case, diffuse reflectance laser photolysis study on polymer powders is effective, because this gives transient UV absorption spectra of optically inhomogeneous, scattering materials.[2,5]

Using these methodologies we have studied carbazole systems: N-ethylcarbazole, 1,ω-di-N-carbazolylalkanes, $meso$- as well as rac-2,4-di-N-carbazolylpentane, 1,2-$trans$-di-N-carbazolylcyclobutane, poly(N-vinylcarbazole) (abbreviated hereafter as PVCz), poly(N-carbazolylethyl vinyl ether), and polyurethanes having the carbazolyl chromophore. The carbazolyl chromophore is excited by laser pulse and quenched by electron acceptor or electron donor, leading to ionic species. We have revealed a relation among interchromophoric interaction, electronic spectrum, and geometrical structure of bichromophoric molecules in the excited singlet, triplet, cationic, and anionic states.[6] On the basis of these results, electronic structure and dynamic behavior of carbazole polymers in excited and ionic states have been considered.[7] We can say that PVCz constitutes the representative macromolecular electron donor–acceptor complexes whose dynamic interaction and electronic processes have been revealed by ultrafast kinetic spectroscopy.

In the past three years, we have devoted a lot of efforts to developing our instrumental techniques and exploring new dynamics in carbazole macromolecular complexes. First, we summarize here the principle of some time-resolved laser spectroscopic methods and describe their new aspects. Second, current advances in ultrafast dynamics of PVCz macromolecular complexes are described.

4-2. Time-Resolved Laser Spectroscopic Methods

4-2-1. Transmittance Laser Photolysis Method

Transient UV-visible absorption spectroscopy has been used for a long time and its origin comes from flash photolysis proposed by Norrish and Porter.[8] Because this measurement is performed in the transmittance mode, the sample should be optically clear and transparent. In these measurements, dynamic processes have been analyzed primarily by probing a transient absorption at one wavelength. In general, however, absorption spectra of excited states and chemical intermediates overlap with each other. Furthermore, conformational change of the molecule and orientational relaxation of the surrounding environments result in a time-dependent spectral change. Measuring absorption spectra over a wide range of wavelengths for is indispensible elucidating the photophysical and photochemical processes accurately.

From this viewpoint, we have constructed several laser photolysis systems all of which are controlled and processed by a microcomputer. Nanosecond excimer and picosecond Nd^{3+}:YAG transmittance laser photolysis systems are shown in Figs. 4-1 and 4-2, respectively. The monitoring lamp is a 150 W Xe lamp (Wacom) that is operated in the DC and pulsed modes, and is synchronized to an excimer laser (Lumonics 430T2, Hyper 400, or Lambda Physik MSC 101) through a timing circuit (homemade or Leonics). A time profile of the transmitted monitoring light is digitized by a transient memory (Kawasaki Electronica M-50E) or a storage scope (Iwatsu TS-8123), transferred to a microcomputer (NEC PC9801), and processed.

In the picosecond photolysis the monitoring beam is a picosecond continuum produced by focusing the fundamental pulse (1064 nm) into a quartz cell of 10 cm length with a lens of 15 cm focal length. The solution for generating this continuum is ^2H_2O (Merk UVASOL, 99.95%), which makes it possible to measure the full range of the visible region. The excitation pulse is either the second (532 nm), the third (355 nm), or the fourth (266 nm) harmonic of the Nd^{3+}:YAG laser. The spectrum of the picosecond continuum is detected by a multichannel photodiode array (MCPD). Two diode arrays are attached to each spectrograph ($f = 25$ cm, Jovin Yvon, grating 300 lines/mm), covering a wavelength range of 380 nm. The output of each MCPD is sent to a microcomputer (SORD M223 Mark II). A double-beam optical arrangement was adopted, because shot-to-shot fluctuation is large. This one-shot measurement is also possible in nanosecond photolysis by using a gated photodiode array or a streak camera.[9] These optical and electronic arrangements are considered standard at the present stage of investigation.

Transient absorption spectroscopy, of course, can be applied to luminescent as well as nonluminescent species, which makes it possible to investigate more systems compared to fluorescence techniques. This spectroscopy probes several electronic

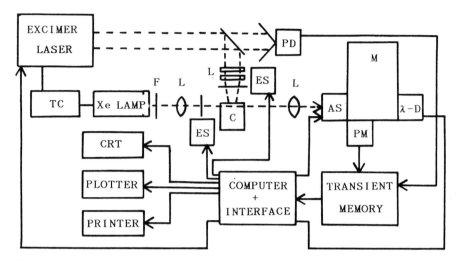

Figure 4-1 ▪ Schematic diagram of the microcomputer-controlled nanosecond excimer laser photolysis system. PD, photodiode; PM, photomultiplier; ES, electromagnetic shutter; AS, automatic slit; λ-D, wavelength driver; L, lens; C, sample cell; F, filter; TC, timing circuit; M, monochromator.

transitions, whereas the fluorescence measurement follows only one. This indicates that more detailed information on electronic structure of the transient species and the factors affecting them is available in absorption spectroscopy. However, two disadvantages have to be pointed out. The first is its small dynamic range: It is rather difficult to obtain an accurate transient absorption spectral shape and rise as well as decay curves. The second is that a high intensity laser has to be used as an excitation light source, because a concentration of 10^{-4}–10^{-5} M is required to detect the excited states or intermediates. This sometimes induces nonlinear photochemical behavior that affects absorption spectroscopy.

The wide applicability and usefulness of transient absorption spectroscopy in solution photochemistry are well recognized, but reports on macromolecular solid films are rare. In the latter case the chromophore concentration is very high, leading to the fact that the excited states are produced only in the surface part of the film. Interactions between excited states are induced efficiently, resulting in rapid local heating. This thermal energy cannot be dissipated as it is in solution and it causes cracking or optical damage of the solid. This is considered to be the main difficulty in performing laser photolysis studies on films. If excited states were formed homogeneously throughout the bulk, rather than densely in the surface, interactions between excited states would be suppressed and no cracking would be induced. This condition is attained simply when the molar extinction coefficient at the excitation wavelength is sufficiently small. Actually, the $T_n \leftarrow T_1$ absorption spectrum of a single crystal of benzophenone was measured by N_2 gas laser photolysis, where the excitation photon (337 nm) is absorbed into the weak n-π^* transition.[10] We came to the conclusion that

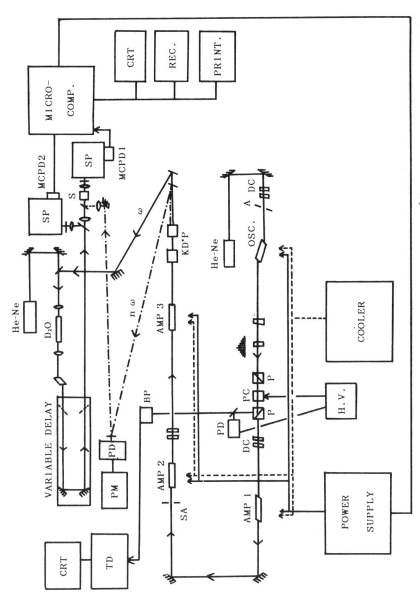

Figure 4-2 ■ Schematic diagram of the microcomputer-controlled picosecond Nd^{3+}:YAG laser photolysis system. DC, dye cell; A, aperture; PC, Pockels cell; PD, photodiode; SA, soft aperture; BP, biplanar photodiode; SP, spectrograph; S, sample; Rec., recorder; PM, power meter; TD, transient digitizer.

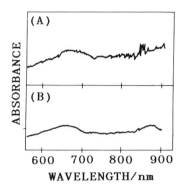

Figure 4-3 ■ The absorption spectrum of the excited singlet state of poly(*N*-vinylcarbazole) at 100 ps. (A) 532 nm two-photon photolysis of the glass-like solution. (B) 355 nm one-photon photolysis of the dilute solution.

a simultaneous two-photon excitation is a useful excitation method for producing the excited states homogeneously in bulk.

We[11] and Hamanoue et al.[12] reported independently that simultaneous two-photon excitation of neat aromatics is so efficient that the absorption spectrum of the excited singlet state can be measured under the normal picosecond photolysis condition. This nonlinear effect provides a new spectroscopic method for studying various kinds of solids, as was first demonstrated for poly(*N*-vinylcarbazole).[1] We applied this to a very viscous glass-like solution of the polymer in THF (this sample was used instead of a block of solid polymer because the latter was very difficult to prepare) and the picosecond 532 nm photolysis gave the absorption spectrum shown in Fig. 4-3. Because the carbazolyl chromophore has no one-photon absorption at this wavelength, the present result is due to the simultaneous two-photon excitation. A comparison with the spectrum of a dilute solution of the same polymer, measured by the normal 355 nm photolysis, is also shown. It is claimed that the electronic structure of the excited singlet state of the polymer glass is very similar to that of the isolated polymer in dilute solution. We consider that the two-photon photolysis is a general method for photophysical and photochemical studies on macromolecular solids. Actually, excitation energy migration in polystyrene film was elucidated in the same way by Miyasaka, Ikejiri, and Mataga.[13]

4-2-2. Diffuse Reflectance Laser Photolysis Method

The transmittance absorption spectroscopy is now being extended to opaque as well as scattering systems. Instead of a transmitted light, a diffuse reflected light from powder samples is used as a monitoring light. A laser-induced change of the latter is due to photoexcited species and chemical intermediates, which provides spectroscopic and kinetic information. This is called a diffuse reflectance laser photolysis method and was first developed by Wilkinson and Kelly[5] and Kessler and Wilkinson.[14]

We have improved its time resolution up to 10 ps using the picosecond continuum as a wide-band monitoring light pulse.[15-17] A microcomputer-controlled diffuse re-

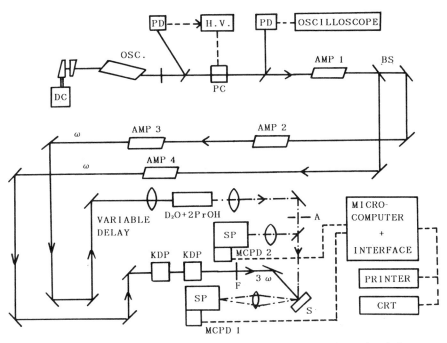

Figure 4-4 ■ Schematic diagram of the picosecond diffuse reflectance laser photolysis system. DC, dye cell; HV, high voltage power supplier; PC, Pockels cell; PD, photodiode; A, aperture; SP, spectrograph; S, sample; BS, beam splitter; F, filter.

flectance laser photolysis system with a repetitive mode-locked Nd^{3+}:YAG laser was set up and used to measure time-resolved absorption spectra. Figure 4-4 shows a schematic diagram where a double-beam optical arrangement was adopted. The sample, contained in a Suprasil cell with 2 mm thickness, was excited with a single 355 nm pulse with 17 ps FWHM and millijoule output power. This excitation pulse was focused onto the sample with 2 mm ϕ spot size. A picosecond continuum was led to the sample as an analyzing light. Only the central part of the excited area was monitored. The picosecond diffuse reflected light from the sample was collected with a lens into a polychromator and detected by a multichannel photodiode array (256 channels; MCPD1). A part of the continuum beam was directed into another polychromator with MCPD2 and was used as the reference for the spectral shape of the analyzing light. A wavelength range of 190 nm was monitored by one-shot laser excitation. Other control and data processing procedures are similar to those of the transmittance laser photolysis system.

Diffuse reflectance spectral data are usually analyzed according to the Kubelka–Munk theory, which assumes that the medium consists of randomly distributed particles whose absorption and scattering coefficients are constant. On the other hand, the concentration of the transient species in diffuse reflectance photolysis has a gradient from the front to the bulk. In this case, it was reported by Wilkinson that the percent absorption is an appropriate measure for absorption:

$$\%\text{abs.} = 100 \times (1 - R/R_0)$$

where R and R_0 represent the intensity of the reflected monitoring light with and without excitation, respectively. A linearity between this value and the concentration of the transient species was theoretically confirmed to hold up to 10%. Actually, transient absorption spectral shape and dynamic behavior analyzed with this unit are reasonable.

4-2-3. Time-of-Flight Photoconductivity Measurement

Photoinduced electron transfer leads to the formation of various kinds of ionic species. One is a contact ion pair composed of the donor cation and the acceptor anion. Sometimes this is in the excited electronic state, and characteristic fluorescence, so-called exciplex emission, is observed. The second is a loose ion pair where interionic distance is longer than that of the contact ion pair. In the solution system some solvent molecules are inserted between the cation and the anion, giving a so-called solvent-shared ion pair. This type of ion pair is also observed in macromolecular solids where the hole or the electron migrates over the chromophores. The last is free ions (or carriers) diffusing independently according to concentration gradient and applied electric field. Absorption spectra of these ionic species are, as the first approximation, a superposition of the bands of the donor cation and the acceptor anion. In the case of the contact ion pair, the interionic interaction is rather strong and modifies the absorption spectral shape. The loose ion pair and free ions give the same absorption spectra, so that a new method for discriminating these species is indispensable. Photoconductivity measurement has been used as a very sensitive and powerful tool for detecting the latter free ions. Therefore, transient absorption, fluorescence, and photoconductivity measurements are complementary and should be used together.

The photoconductivity is governed by both photogeneration yield of the carrier and its drift mobility. By applying a time-of-flight method, it is possible to measure these two values at the same time. We have constructed a pulse-photoconductive measurement system that is controlled and processed by a microcomputer.[18] As shown in Fig. 4-5, an excimer laser is used as an excitation light source, and the intensity of each laser pulse is monitored by a photodiode, the output of which is calibrated with a power meter. Films were prepared on NESA-coated quartz plates by casting polymer solution and drying it in vacuum. A semitransparent gold electrode with a typical area of 1 cm^2 was evaporated on the film. This sandwich-type cell was excited by an excimer laser, and an induced transient photocurrent is detected by a series of load resistors, using a storage oscilloscope. The data are digitized, transferred to a microcomputer, and processed.

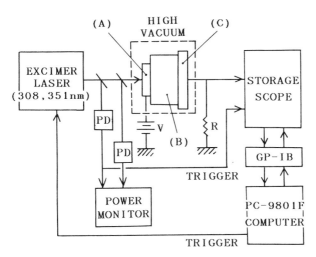

Figure 4-5 ■ Schematic diagram of the microcomputer-controlled time-of-flight photoconductivity measurement system. A, semitransparent Au electrode; B, polymer film; C, nesa-glass; PD, photodiode.

A typical set of transient photocurrent signals for the irradiation of a positively biased gold electrode is shown in Fig. 4-6.[19] The observed curve generally has an initial spike-like signal superimposed on a slow decay component. The former is not due to a loss of carrier by bimolecular recombination in the excitation layer. Rather, it is due to fast carrier transport in a short duration immediately after its generation. The number of collected carriers per unit area N_p is obtained from the integral of the

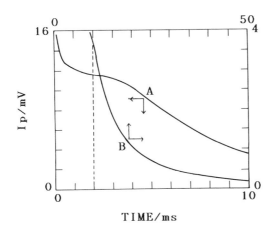

Figure 4-6 ■ A set of typical transient photocurrent signals. (A) The early part and (B) the late part.

photocurrent over the entire drift times, using a set of transient signals with different time regions. The photogeneration yield of the carrier ϕ is obtained by dividing N_p by the excitation light intensity I_{ex} (photons per square centimeter per pulse).

4-3. Current Advances in Photoinduced Electron-Transfer Dynamics in Poly(N-vinylcarbazole) Complexes

The study on photoinduced electron-transfer phenomena has a long history and provides the field where theories and experiments are compared. Most detailed analyses have been given for small molecular systems in solution, although a few groups are studying macromolecular electron donor–acceptor complexes. Concerning solid macromolecular complexes, practical approaches are noticeable and molecular mechanism is not established. We summarize briefly the mechanism of photoinduced electron transfer.

In solution this process is very sensitive to solvent polarity and quite characteristic behavior is observed. In nonpolar environments photoinduced electron transfer results in fluorescence quenching of aromatic hydrocarbons and gives a new emission in the long wavelength region. The broad and structureless band shape indicates that the formed species giving this emission are dissociative in the ground state. This type of emission was first called heteroexcimer by Mataga and Ottolenghi[20] and is now well known as an exciplex. As the dielectric constant increases a little, the fluorescence maximum shifts to the red and the fluorescence yield as well as the lifetime decrease. From analysis of this behavior it has been concluded that the electronic structure of aromatic hydrocarbon–electron donor (or acceptor) pairs is quite polar and is deemed to be a kind of contact ion pair. In polar solvents ionic species, such as free ions and solvent-shared ion pairs of the donor cation and the acceptor anion, are formed instead of exciplexes. These processes are summarized in Scheme 1, where D* and A are the excited state of an aromatic molecule and a quencher (electron acceptor), respectively. Energetic and kinetic aspects of these electron-transfer processes have been elucidated in detail and still form the most active area in chemistry.[21-23]

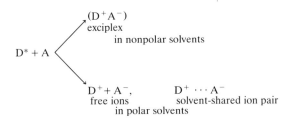

$$D^* + A \Big\langle \begin{array}{l} (D^+A^-) \\ \text{exciplex} \\ \quad \text{in nonpolar solvents} \\[1em] D^+ + A^-, \qquad D^+ \cdots A^- \\ \text{free ions} \qquad \text{solvent-shared ion pair} \\ \quad \text{in polar solvents} \end{array}$$

Scheme 1 ■ Photoinduced electron-transfer mechanism in solution.

According to recent quantitative analyses by time-resolved laser spectroscopy, plural exciplexes and solvent-shared ion pairs are involved in solution. The relative geometry of the donor–acceptor pair in an exciplex and a wide distribution of interionic distance in the ion pairs are confirmed directly in solution.[24] In early delay times of picoseconds and femtoseconds, solvent orientation is not completely averaged and is deemed to have a structure. Electron-transfer dynamics occurs just in the time scale of the solvation relaxation. In the cases of molecular assemblies and macromolecular complexes, an exterplex composed of two donors and one acceptor is preferred to the usual 1:1 pair.[25, 26] A new concept on dynamic interaction and electronic processes is expected.

Concerning macromolecular solids, time-resolved studies are few and direct observation on the elementary process is scarce. The molecular mechanism of photoinduced electron-transfer phenomena has been considered on the basis of steady-state fluorescence and photoconductivity measurements. At the present stage of investigation, the mechanism shown in Scheme 2 was considered the most probable one.[27, 28] A singlet excited state D^* in the film migrates effectively, encounters the doped acceptor during its lifetime, and forms an encounter complex $(D^* \cdots A)$. This complex induces a rapid electron transfer and changes to a nonrelaxed exciplex state $(D^+ - A^-)^{**}$. This state has excess energy and undergoes thermalization, resulting in a loose ion pair state $(D^+ \cdots A^-)$. The relaxation from the ion pair to the relaxed fluorescent exciplex state $(D^+ - A^-)^*$ occurs in competition with the electric field-assisted thermal dissociation into free carriers, $D^+ + A^-$. The interionic distance r_0 in the ion pair is longer than that of the contact fluorescent exciplex. In Scheme 2, only one kind of exciplex usually has been assumed.

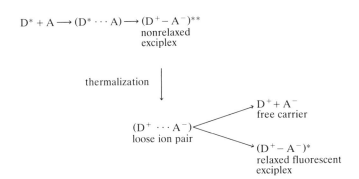

Scheme 2 ■ Photoinduced electron-transfer mechanism in a solid.

At the present stage of investigation, both schemes are not completely consistent with each other, because solution and solid electron-transfer dynamics have been studied independently. We consider, however, that molecular aspects of the macromolecular solid should be clarified and correlated with results on the solution system in order to understand their dynamic interaction and electronic processes. This is strongly required because excitation energy, hole, and electron are, in general, localized in the trap sites and its molecular geometry determines their functionality.

We have performed experimental studies to elucidate electron transfer processes of PVCz solid systems in the same molecular and electronic levels as those of the solution systems. Application of time-resolved and time-of-flight measurements are very fruitful, although a new experimental difficulty due to high intensity of laser pulse is sometimes involved. We next describe the recent results.

4-3-1. Mutual Interactions between Loose Ion Pairs in Film

In general, the carrier-photogeneration efficiency of photoconductive polymer films was obtained by means of a xerographic discharge technique. Although it is convenient and practical, especially for high fields, this technique seems inferior to the time-of-flight method because it is difficult to avoid two effects: trapped space charge due to preexposure and photodegradation of film surface caused by exposing it to corona charge. We used the microcomputer-controlled pulse photoconductive measuring system on the basis of the time-of-flight method, which was previously described.

In the case of undoped PVCz films, the photogeneration efficiency of holes was investigated in detail as a function of applied electric field, temperature, and absorption band excited. The efficiency in PVCz films, which were slightly photooxidized, was also investigated by comparing the results with those obtained for the fresh film. It was shown that the photogeneration of carriers via nonrelaxed exciplexes formed between excited singlet state and acceptor-like impurities or photooxidation products are predominant in PVCz films for all the excitation bands.[19] That is, the electron transfer from the excited carbazolyl chromophore to the photooxidation products, which are produced within the surface layer of the film, is the most probable mechanism, as shown in Scheme 2. This mechanism is considered to be common for photoconductive aromatic vinyl polymers such as poly(5-vinylbenzo[b]carbazole), poly(7-vinylbenzo[c]carbazole), and poly(1-vinylpyrene) films.[29, 30]

As the second step, we have investigated the electron transfer mechanisms of PVCz films doped with electron acceptors with various electron affinity (E_a). The PVCz was prepared by AIBN-initiated free radical polymerization. Dimethyl terephthalate (DMTP), 4-cyanopyridine (CNP), p-dicyanobenzene (DCNB), phthalic anhydride (PA), tetrachlorophthalic anhydride (TCPA), 1,2,4,5-tetracyanobenzene (TCNB), 2,3,7-trinitrofluorenone (TNF), p-chloranil (CA), and tetracyanoquinodimethane (TCNQ) were used as acceptor compounds. DMTP, CNP, and DCNB are weak electron acceptors and do not form the ground state charge transfer complex. Electron-transfer interaction is induced only in the excited state, namely, they are just the exciplex systems. Other acceptors gave charge-transfer bands due to the complex with carbazolyl chromophores.

As previously mentioned, overall carrier-photogeneration efficiency ϕ is obtained by dividing N_p (carriers cm^{-2}) by I_{ex}. Because the ϕ should be independent of light intensity, the measurement is always carried out under the condition that N_p is proportional to I_{ex}. If it is performed at high temperature, under high electric field, or under high excitation light intensity, the condition is not often retained and the

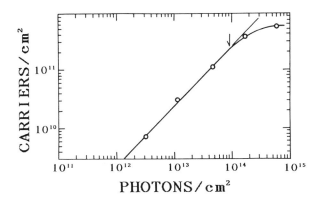

Figure 4-7 ■ Excitation intensity dependence of the carrier number of poly(N-vinylcarbazole) film. The electric field is 1.6×10^5 V cm^{-1}.

induced photocurrent signal is influenced mainly by the space charge effect. When a 3 μs light pulse from an air-gap flash was used as an excitation source for undoped PVCz films, a linear relation between N_p and I_{ex} was confirmed for the excitation light intensity range from 1×10^{12} to 1×10^{14} photons cm^{-2} pulse^{-1}.[19] However, when the dependence of N_p on I_{ex} was measured by using our system in order to check the experimental condition, a very interesting phenomenon was observed. As shown in Figs. 4-7 through 4-9, a saturation tendency was observed for PVCz films undoped and doped with various acceptors at high laser excitation intensity, even if the electric field was not very high and the normal induced transient photocurrent signal was observed. The excitation laser intensity at which the saturation begins (I_{ex}^s) and related parameters are listed in Table 4-1.

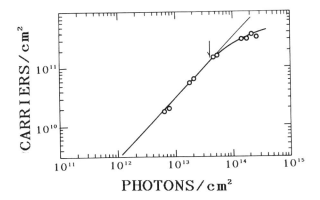

Figure 4-8 ■ Excitation intensity dependence of the carrier number of poly(N-vinylcarbazole) film doped with 2 mol% 4-cyanopyridine. The electric field is 1.0×10^5 V cm^{-1}.

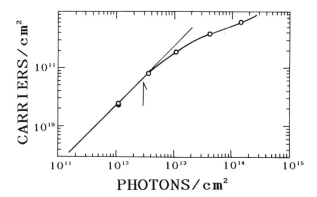

Figure 4-9 ▪ Excitation intensity dependence of the carrier number of poly(N-vinylcarbazole) doped with 2 mol% 1,2,4,5-tetracyanobenzene. The electric field is 1.0×10^5 V cm^{-1}.

Table 4-1 ▪ **Photoconductive parameters of poly(N-vinylcarbazole) films**

Acceptor (concentration, mol%)	E_A (eV)	I_{ex}^s (photons cm^{-2})	r_0 (Å)	ϕ_0	n_{ion}^s
Undoped		9.0×10^{13}	26	0.047	4.2×10^{12}
DMTP (2.0)	1.00	1.0×10^{13}	22	0.90	0.9×10^{13}
CNP (2.0)	1.05	4.0×10^{13}	22	0.50	2.0×10^{13}
DCNB (1.5)	1.10	1.0×10^{13}	23	0.86	1.0×10^{13}
PA (2.0)	1.18	5.0×10^{12}	30	0.35	1.8×10^{12}
TCPA (2.0)	1.63	4.0×10^{12}	30	0.20	0.8×10^{12}
TCNB (2.0)	2.05	3.0×10^{12}	28	0.96	2.9×10^{12}
TNF (2.0)	2.10	2.0×10^{12}	22	1.00	2.0×10^{12}
CA (2.0)	2.48	1.0×10^{13}	26	0.10	1.0×10^{12}
TCNQ (2.0)	2.88	4.0×10^{12}	22	1.00	4.0×10^{12}

This saturation effect with the excitation laser intensity is considered as follows.

1. An S_1-S_1 annihilation between excited singlet states in films results in the saturation of singlet excited state D*. The S_1-S_1 annihilation process is often observed in transmittance laser photolysis studies on polymers even in solution.[1,31] A PVCz film decay time obtained by using an excimer laser with an intensity of 1×10^{16} photons cm^{-2} pulse^{-1} was ca. 20 ns—much shorter than that obtained by the time-correlated single photon counting method (35 ns).[32] Furthermore, the intensity ratio of the sandwich excimer fluorescence to the second excimer was lower than that by a weak excitation.[33] Hence, the presence of the S_1-S_1 annihilation in PVCz films was confirmed, and a threshold is estimated to be 8×10^{13} photons cm^{-2} pulse^{-1} for the XeCl excimer laser (308

nm, 6 ns).[33] Therefore the saturation for the undoped film might be responsible for the S_1-S_1 annihilation. This possibility, however, is excluded for doped films, because values of I_{ex} are smaller than the light intensity of the threshold and different from sample to sample.

2. The saturation in the concentration of carrier-forming sites results in the saturation of N_p at the higher excitation intensity. Although the concentration of acceptor molecules is the same (2 mol%) except for DCNB system, the different values of I_{ex}^s were obtained. Thereby, this possibility is also rejected.

3. The interaction among/between loose ion pairs with the large interionic distance may cause the saturation phenomenon. Under the normal condition that N_p is proportional to I_{ex}, we have measured the electric field dependence of ϕ and analyzed the data according to the Onsager theory.[18] On the basis of the results, we have obtained both the primary quantum yield ϕ_0, of the loose ion pair ($D^{+\cdots}A^-$) and the interionic distance r_0.[18,34] These values are also listed in Table 4-1. The product, n_{ion}^s of ϕ_0 and I_{ex}^s corresponds to the number of ion pairs at the point where the saturation begins. For exciplex systems (DMTP, CNP, DCNB), the values of n_{ion}^s are almost the same $(0.9–2) \times 10^{13}$ pairs cm^{-2}). For CT complex systems (PA, TCPA, TCNB, TNF, CA, TCNQ), the values are between 0.8×10^{12} and 4×10^{12} pairs cm^{-2} and are slightly divergent.

In this discussion, the interionic distance r_0 should be considered also. The r_0 values of the exciplex system are almost the same (22–23 Å). This corresponds to the similarity in the value of n_{ion}^s obtained for the exciplex system. On the other hand, the r_0 values of the CT complex system are between 22 and 40 Å. The n_{ion}^s values of the system with large r_0 (PA, TCPA) are smaller than those of the system with small r_0 (TNF, TCNQ). This relationship between r_0 and n_{ion}^s, which we found here, indicates that the interaction between ion pairs causes the saturation effect with the excitation laser intensity.

An average distance between the centers of the loose ion pairs was evaluated as follows. A half of the incident laser light was absorbed with the layer of 0.27 μm in depth. The average distance between ion pairs in this depth region is calculated to be 140, 175, 240, and 410 Å for 2×10^{13}, 1×10^{13}, 4×10^{12}, and 0.8×10^{12} of n_{ion}^s, respectively. For the high excitation laser intensity region where the saturation was observed, the distance becomes smaller. Thus, the saturation of N_p against the excitation laser intensity is considered to be attributed to the interaction between loose ion pairs.

4-3-2. Charge Recombination Fluorescence in Film

Considerable data on the fluorescence properties of PVCz have been accumulated because of an interest in its photoconductive properties, two spectrally distinct excimer fluorescences, and a clear relation between photophysical properties and tacticity.[32] It was reported that the sandwich excimer fluorescence was quenched more efficiently than the second excimer fluorescence by doping DMTP.[35] Examination of the PVCz-DMTP system in solution by Hoyle and Guillet revealed the presence of both an exciplex and an exterplex. Both are deemed to be complexes composed of donor cation and acceptor anion. The exterplex consists of two car-bazolyl chromophores and one molecule of DMTP and the hole is delocalized over two donor chromophores.[25] As model compounds for exciplexes from isotactic and syndiotactic sequences of PVCz, *meso*- and *rac*-2,4-di(*N*-carbazolyl)pentanes, giving sandwich and partial overlap excimer fluorescence, respectively, are very interesting.[36] Studies on the time-resolved fluorescence spectra of these compounds quenched by *m*-dicyanobenzene (*m*-DCNB) in solution indicated that the special geometrical structure of the two carbazolyl groups leads to two kinds of exterplex where the hole is delocalized in the corresponding sandwich and partial overlap structures.[26] Studies on the dynamics of exciplex fluorescence in film are much less than those in solution, and the mechanism of exciplex formation in films is still beyond our knowledge. We applied for the first time a time-correlated single-photon counting technique to PVCz films doped with DMTP or *p*-DCNB with and without the applied electric field.[37] Time-resolved fluorescence spectra and fluorescence rise and decay curves were measured with a nanosecond single photon counting system with a hydrogen-filled flash lamp. Fluorescence decay curves were measured at some wavelengths. From a series of the decay curves thus obtained, the time-resolved spectra were obtained by plotting the fluorescence intensity at the delay time concerned *versus* wavelength. The wavelength dependence of the instrument response function was not corrected. All the measurements and data processing were fully controlled by microcomputer. Spectra were not corrected for the detector sensitivity.

Fluorescence spectra of PVCz films undoped, doped with DMTP, and doped with DCNB were obtained by integrating the fluorescence intensity at each wavelength up to 168, 4688, and 4668 ns, respectively, and are shown in Fig. 4-10. The spectrum of the undoped film consists of the sandwich excimer fluorescence ($\lambda_{max} = 420$ nm) and

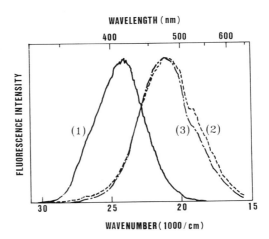

Figure 4-10 ■ Normalized fluorescence spectra of PVCz films. (1) Undoped. (2) Doped with DMTP (3.0 mol%). (3) Doped with *p*-DCNB (1.8 mol%). The spectra were obtained by integrating the fluorescence intensity at each wavelength up to 168, 4688, and 4668 ns after excitation for (1), (2) and (3), respectively, and were not corrected for detector sensitivity.

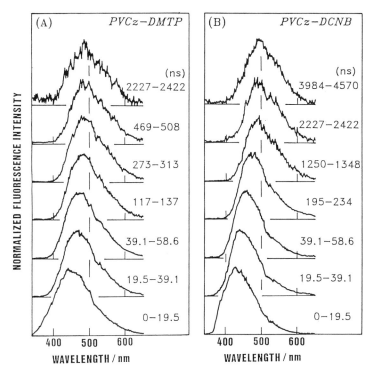

Figure 4-11 ■ Normalized time-resolved fluorescence spectra of PVCz films doped with (A) DMTP (3.0 mol%) and (B) p-DCNB (1.8 mol%) at excitation wavelength of 295 nm. The time window is given in the figure. The spectra were not corrected for detector sensitivity.

the partial overlap one (λ_{max} = 370 nm).[38,39] By doping of DMTP (3.0 mol%) or DCNB (1.8 mol%), the polymer fluorescence is quenched and replaced by the exciplex fluorescence with a peak at 478 and 473 nm, respectively. Time-resolved fluorescence spectra of PVCz films doped with DMTP and DCNB are shown in Fig. 4-11, where each spectrum is normalized at the maximum intensity. In the early-gated spectrum, the host fluorescence was mainly observed with a peak at ca. 430 nm and the exciplex fluorescence in the long wavelength region was weak. The fluorescence peak continuously shifts to the long wavelength with time. At the latest-gated time, the peak is located at 492 and 497 nm for DMTP and DCNB systems, respectively. Their peak wavelength is longer than that of steady-state fluorescence spectra (a sum of time-resolved spectra) for both systems. In the time region after 35 ns, it is difficult to explain the continuous red shift of the spectra only by assuming a superposition of excimer and exciplex fluorescence spectra. This result indicates clearly that the exciplex fluorescence shifted gradually to the long wavelength with time.

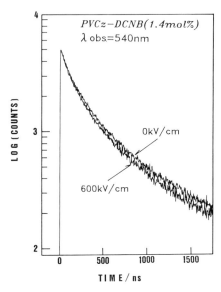

Figure 4-12 ■ Exciplex fluorescence decay curves of PVCz film doped with *p*-DCNB (1.4 mol%) with and without applied electric field.

Yokoyama et al.[27] reported that the steady-state exciplex fluorescence of PVCz films doped with weak electron acceptors, such as DMTP and *p*-DCNB, was quenched by applying an electric field. The decrease of the fluorescence intensity depends on the applied electric field strength, which was interpreted well in terms of the Onsager theory. The effect of electric field on the decay curves of exciplex fluorescence was first examined by us. One example, shown in Fig. 4-12, has a long tail to the microsecond time region and does not obey a multiexponential function, which is quite different from that observed for PVCz solution and is strange as an exciplex. The decay became fast by applying the electric field, but an appreciable change of the rise curve was not observed.

The present result cannot be explained in terms of Scheme **2**. We could not separate a unique exciplex with distinct spectrum and temporal behavior because the fluorescence peak shifts to the long wavelength with time. It is difficult to assume a gradual relaxation of exciplex species, such as reorientation of chromophores, because molecular motion in solid film is very slow. Taking into account the facts that PVCz film shows both sandwich and partial overlap excimer fluorescence and that the structure of the exciplex is usually looser than that of an excimer, we consider that multiple exciplex and exterplex species with various relative configurations between donor and acceptor are responsible for dynamic behavior in the doped film. The continuous shift of the peak of the time-resolved fluorescence spectra can be attributed to a superposition of various exciplex and exterplex species with different temporal behavior. This is the point that we have to modify in the Scheme **2**.

Another result to which we pay attention is the nonexponential decay curve with a long tail up to the several microsecond time region. It is impossible to interpret this behavior only by assuming multiple species, because the lifetime of the singlet exciplex and exterplex is in general shorter than 100 ns. We can explain the slow tail of the decay curve and the electric field effect on the decay as follows: The relaxation from the ion pair state ($D^{+}\cdots A^{-}$) to the relaxed fluorescent exciplex state ($D^{+}-A^{-}$)* as well as free carriers determines the decay of the fluorescent exciplex states. The time constant of this geminate recombination may be around the microsecond order, whereas the exciplex lifetime is still about a few tens of nanosecond. That is, the geminate recombination of the ion pair with an interionic separation r_0 and the dissociation to free carriers are rate-determining steps for the decay process of fluorescent exciplex states. Thus, the formation of fluorescent exciplex states continues even in the microsecond time region. Because this geminate recombination competes with the electric field-assisted thermal dissociation into free carriers, the electric field can affect the apparent decay of the exciplex fluorescence.

Recently, Kim and Webber[40] reported that the PVCz film doped with DMTP showed the exciplex phosphorescence with a peak at ca. 550 nm at room temperature. As the peak of the exciplex fluorescence shifts continuously from nanosecond to microsecond time regions, the present emissions are considered to be due to the singlet state. Furthermore, the electric field effect on the decay curve could not be interpreted in terms of the phosphorescent exciplex. However, we could not exclude a possibility that the slow tail of the decay curve up to the microsecond time region includes phosphorescence to a certain extent.

4-3-3. Excimer and Ionic Species in Solid Powers

As already mentioned, absorption spectral measurement gives more direct information on transient species compared to photoconductivity and fluorescence studies. Detecting ionic species in PVCz film by transmittance laser photolysis has been strongly required. However, an efficient S_1-S_1 annihilation leads to cracking of the sample. Instead, we have applied diffuse reflectance laser photolysis to the solid powders of PVCz macromolecular complexes.

As a monomer model system for PVCz, transient absorption spectra of N-isopropylcarbazole microcrystals were examined. In Fig. 4-13, the absorption spectra are summarized and compared by transmittance photolysis with a spectrum of N-ethylcarbazole in tetrahydrofuran solution. The spectrum at 0 ps is weak and rather broad compared to the spectra at other delay times, which is an artifact due to a distribution of arrival time of the picosecond continuum at each wavelength.[15,41] The spectral shape and the maximum position at 620 nm were independent of the delay time. The absorption maximum is a little shifted to the blue and the spectral width is broader compared to the solution system. Because the spectrum of the solution system is an $S_n \leftarrow S_1$ transition,[42,43] the present band can be ascribed to the fluorescent state. We consider that the above spectral differences are due to a difference of electronic structure between an exciton state in a solid and a molecular electronic state in solution, because N-isopropyl and N-ethyl substituents may not affect π-electronic structure. At 4 ns no appreciable absorption was detected. The fluorescence lifetime of the microcrystals was measured to be 15 ns, so that the present rapid disappearance may be induced by an S_1-S_1 annihilation. We have already confirmed this process for PVCz in film and solution.[31,44]

The transient absorption spectra of PVCz are shown in Fig. 4-14. Immediately after excitation, the peak position was observed around 620 nm; it shifted to 660 nm

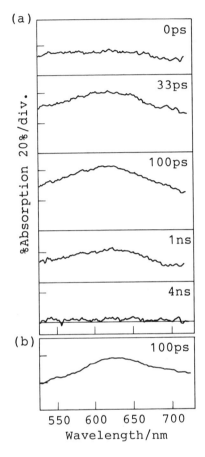

Figure 4-13 ▪ (a) Transient absorption spectra of N-isopropylcarbazole microcrystals using the diffuse reflectance laser photolysis method. (b) Reference spectra of N-ethylcarbazole in tetrahydrofuran using the transmittance laser photolysis method. The delay time is given in the figure.

with the time and became sharp. The former peak position is close to that of the $S_n \leftarrow S_1$ transition of N-ethylcarbazole or the absorption of the partial overlap excimer of the solution system, whereas the latter maximum is similar to that of the sandwich excimer.[7,43] Because the transient absorption spectra usually correspond to the main component as a result of its rather poor S/N value, we consider the present spectral change to be due to a process from the monomer excited state or the partial overlap excimer to the sandwich excimer. Compared to the transient absorption spectra of PVCz in solution by transmittance laser photolysis, the spectral shape is broader in solid powders. This may be due to a different contribution of the sandwich and partial overlap excimers between powder and solution samples.

By solution phase laser photolysis, it was confirmed that the sandwich excimer formation process consists of an instantaneous process and a delayed process with the time constant of 160 ps.[42] The former is the major process, whereas the latter is minor. The present spectral change in solid powders also shows that the sandwich excimer formation is not completed before 1 ns. At the present stage of investigation,

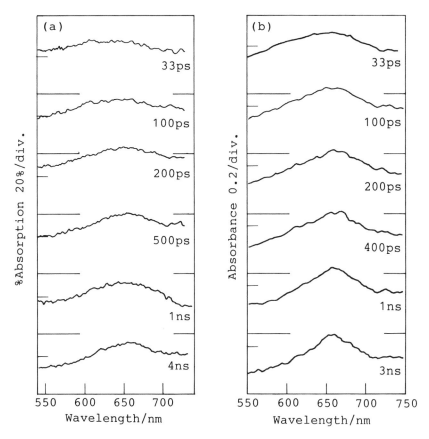

Figure 4-14 ■ (a) Transient absorption spectra of poly(N-vinylcarbazole) powders using the diffuse reflectance laser photolysis method. (b) Reference spectra of the same polymer in tetrahydrofuran using the transmittance laser photolysis method. The delay time is given in the figure.

the difference of excimer formation process between solution and solid phases cannot be interpreted and a more detailed study is in progress.

The absorption intensity of the PVCz system is appreciable even at a few nanoseconds after excitation, which is quite different from N-isopropylcarbazole microcrystalline powders. This is well explained by assuming that the S_1-S_1 annihilation is not as efficient as in the powders of monomer model compounds. Namely, both sandwich and partial overlap excimers do not migrate over chromophores as frequently as those of the monomer system.

Referring to these spectra and their behavior, ionic species in PVCz films doped with electron acceptors were examined by the same photolysis method. The transient absorption spectrum of the PVCz-DMTP system are shown in Fig. 4-15. Absorption bands of the monomer excited state as well as both excimers were not observed at any

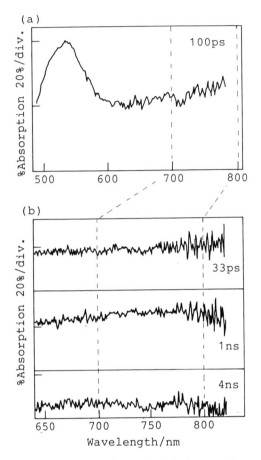

Figure 4-15 ■ Transient absorption spectra of poly(*N*-vinylcarbazole) powders doped with 4.8 mol% dimethyl terephthalate using the diffuse reflectance laser photolysis method. The delay time is given in the figure.

delay times. Instead a distinct band at 535 nm in the shorter wavelength region and a flat band above 600 nm were detected. The former band is the same one reported in solution and can be assigned to the DMTP anion.[7] This is the most direct confirmation of photoinduced charge separation in solid polymer systems.

Concerning carbazole cations, we have seen clear relations among interchromophoric interaction, electronic spectrum, and relative geometrical structure of dicarbazolyl compounds.[7] Compared to these model compounds, absorption spectra of the PVCz cation are broad and independent of the degree of polymerization. The hole in PVCz is delocalized over a number of chromophores and gives a characteristic band,[45] or a superposition of several kinds of the preceding dimer cations corresponds to the present one.[46] At the present stage of investigation, we adopted the latter explanation, because excitation energy and hole are usually stabilized in a dimer site and no direct demonstration for a higher aggregate was presented. This type of explanation may hold for this system.

In order to demonstrate charge separation more clearly, we examined the PVCz-DCNB and PVCz-PA systems where dopants are a little stronger electron acceptor than DMTP. In the case of the DCNB-doped system, a flat band similar to that of the PVCz-DMTP system was observed, although it had a peak in the case of the PVCz-PA system. At early delay times spectral shape was not clear, but a peak around 760 mm at late stages above 100 ps can be seen. We consider that the main component of cations is the sandwich dimer. As pointed out, PA forms a weak CT complex in the ground state that takes a fixed donor–acceptor. This restricts the geometrical structure of the cation, leading to identification of the cation in transient absorption spectroscopy.

In the history of studies on excimer and exciplex dynamics, fluorescence spectral and dynamic measurements first opened the quantitative study from thermodynamic and kinetic viewpoints. By transient absorption spectroscopic measurement, which is rather complementary to the fluorescence technique, electronic as well as geometrical structures of excimers and exciplexes can be clarified and the nature of intermolecular interaction can be elucidated. The present result is the first for transient absorption spectroscopy of excimer and exciplex dynamics in organic solids. A more detailed study is being performed in order to discriminate exciplex, ion pair, and free carriers.

4-3-4. Hole Migration along the Polymer Chain in Solution

Using time-resolved spectroscopic methods, we have confirmed directly ionic species, such as free carriers, ion pairs, and exciplexes, in PVCz film. At the present stage of investigation, our data are the most direct observation in the nanosecond and picosecond time regions. Although more detailed examinations are indispensable and are being performed, it is necessary to measure hole diffusion in film spectroscopically. Whereas an absorption spectroscopic study of polymer solids has just been started (as previously summarized), we have focused our attention on the solution system of the same PVCz.

Systematic laser photolysis studies on PVCz and their bichromophoric model compounds in solution by transmittance laser photolysis method have already been reported and relations among electronic interaction, spectrum, and geometrical structure in the excited singlet, triplet, cationic, and anionic states have been considered.[6,7] Singlet excitation energy and holes are stabilized in dimer sites, such as the excimer and dimer cation, respectively, but no interchromophoric interaction was observed for triplet and anionic states.[6] This also holds for polymers, and some dimers with specific structures, such as sandwich, partial overlap, and open geometry are responsible for localization of singlet excitation energy and hole.[46] Thus, if the hole transferred from one dimer cation site to the other, it would be spectroscopically detected.

One of the reasons why PVCz has been used so frequently is the important relation between the fluorescence spectrum and the tacticity of this polymer. We actually proved a tacticity effect upon fluorescence dynamics.[47] In the case of the polymer prepared by radical polymerization [PVCz (r)], the ratio of syndiotactic to

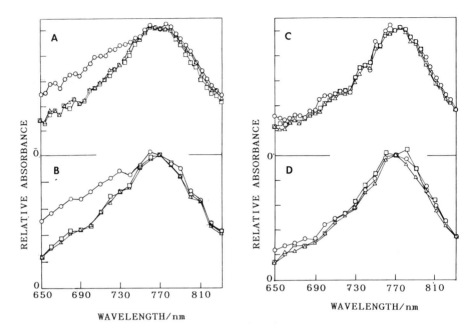

Figure 4-16 ■ Normalized absorption spectra of poly(*N*-vinylcarbazole) cation in *N,N*-dimethylformamide at −55°C. (A) PVCz(r)-84; (B) PVCz(r)-2400; (C) PVCz(c)-70; (D) PVCz(c)-2400. Delay times are 0.1–0.5 μs (○), 5.0–5.5 μs (□), and 10.0–10.5 μs (△).

isotactic sequences is 3:1, whereas it is 1:1 for the polymer prepared by cationic polymerization [PVCz (c)].[35] The relative contribution of the partial overlap and sandwich excimers in fluorescence spectra is determined by this sequence ratio. We consider that the similar relation between spectrum and tacticity is observed also for absorption spectra of polymer cations. Furthermore, we expect that the hole transfer along the polymer chain results in an absorption spectral change of the polymer cation.

Here we report a nanosecond laser photolysis study on PVCz with different tacticity distribution. At room temperature no appreciable difference in absorption spectrum was observed between PVCz(r) and PVCz(c) cations, although it was detected by lowering the temperature. We also examined the effect of degree of polymerization and considered dynamic aspects of the polymer cation in solution. The polymers used are summarized, where the number after the abbreviation represents the mean degree of polymerization as follows: PVCz(r)-20, 84, 140, 500, and 2400; PVCz(c)-8, 16, 70, 140, 500, and 2400. CNP was used as an electron acceptor and the solvent was *N,N*-dimethylformamide.

Some transient absorption spectra at different gate times were normalized at 700 nm and they are summarized in Fig. 4-16. According to our previous results,[7,46] these spectra can be ascribed to the polymer cation that was produced by photoinduced electron transfer of the PVCz-CNP electron donor–acceptor system in a polar solvent. At room temperature absorption spectra of PVCz(r) and PVCz(c) showed no

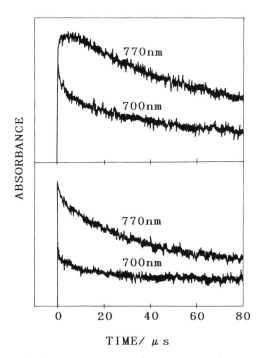

Figure 4-17 ■ Rise and decay curves of poly(N-vinylcarbazole) cation in N,N-dimethyl-formamide at −55°C. (A) PVCz(r)-140 and (B) PVCz(c)-140.

time-dependent spectral change in the submicrosecond domains and are similar to each other. On the other hand, a very broad absorption spectrum of PVCz(r) changes at low temperature and approaches that of PVCz(c) at several microseconds after excitation.

As previously mentioned, absorption spectra of the PVCz cation consists of partial overlap, sandwich, and open dimer cations having peaks at 700, 770, and 800 nm, respectively.[7,46] According to this explanation, the present spectral change of PVCz(r) at low temperature can be ascribed to a decrease of the relative contribution of the partial overlap dimer cation.

The rise and decay curves of PVCz(r)-140 and PVCz(c)-140 at −55°C are shown in Fig. 4-17. In the former case, a decay component at 700 nm and the corresponding rise component at 770 nm were observed, which indicates that a decrease of the partial overlap dimer cation results in an increase of the sandwich dimer cation. Here interconversion between dimer cations in polymers is measured directly by transient UV-visible absorption spectroscopy. The PVCz(c) cation radical did not show such behavior, which is reasonable because the isotactic sequence that gives the sandwich excimer site is rather rich and the hole is rapidly trapped in it.

A dependence of the interconversion rate on the mean degree of polymerization was examined. The spectral shapes of PVCz(r)-84, 140, 500, and 2400 at −55°C are identical at 0.1–0.5 and 5.0–5.5 μs. Namely, the interconversion dynamics of cationic species is independent of the degree of polymerization in this region of the molecular weight. The absorption spectra of PVCz(c) also have nothing to do with the degree of polymerization.

Because the present polymer cations are formed by electron-transfer quenching of the polymer fluorescent state by CNP, the initial concentration of each dimer cation radical is determined by the distribution of excimers with corresponding geometries. Here we describe absorption as well as fluorescence results. Because the ground state absorption spectra are almost common to all PVCz(r) and PVCz(c) with different degrees of polymerization, interchromophoric interaction in the ground state is independent of tacticity. Fluorescence spectra of PVCz consist of two kinds of excimer: the partial overlap around 370 nm and the sandwich at about 420 nm. It is already established that the relative concentration of both excimers is a function of tacticity[35] and is affected by excitation intensity,[31] because the S_1-S_1 annihilation competes with excimer formation and decay processes under laser excitation. Therefore, a comparative discussion of fluorescent and cationic states of PVCz is possible by measuring fluorescence just under the same condition as that of laser photolysis. Furthermore, it should be noted that the relative distribution of excimers just at quenching is responsible for the relative formation yield of dimer cations.

We measured time-resolved fluorescence spectra of PVCz with a nanosecond streak camera system. The results on PVCz(r)-84 and PVCz(c)-70 are summarized in Figs. 4-18 and 4-19, respectively, where the gate width was set to 1.7 ns. Just during laser excitation, namely at 0 ns, no clear difference was observed for PVCz(r) and PVCz(c) at room temperature. The quencher concentration of 0.25 M and room temperature measurement mean that a diffusion-controlled electron transfer occurs in the subnanosecond time region. Similar results were also observed for polymers with high degree of polymerization. At $-55°C$ the rate constant of diffusion-limited processes is reduced to one-tenth that at 20°C, so that the suitable delay time is 10 or 15 ns. No distinct sandwich excimer emission around 420 nm was observed for PVCz(r)-84, although its formation was detected at late stages. On the other hand, a contribution of the sandwich excimer in PVCz(c)-70 was appreciable at 10 or 15 ns. Similar results were also confirmed for PVCz(r)-2400 and PVCz(c)-2400.

These time-resolved fluorescence spectra can be related to the spectral change of polymer cations. At room temperature the main excited state quenched by electron acceptor is the partial overlap excimer for both PVCz(r) and PVCz(c), whereas the observed absorption spectra of cations consist mainly of the sandwich dimer cation; the partial overlap cation is a minor constituent. This means that an interconversion from the partial overlap to the sandwich dimer cations is rapid and cannot be detected in this time range. In the case of PVCz(c) at low temperature, the sandwich excimer site is quenched, leading to formation of the sandwich dimer acation. The preceding interconversion, however, should be involved because the main component is still the partial overlap excimer. The latter excimer is the unique component of PVCz(r) just upon quenching at low temperature, so that the polymer cation observed at the early stage takes a distribution where the partial overlap dimer cation is appreciable and the sandwich is the minor. At late stages the distribution shifts to the sandwich dimer cation and the observed spectrum is the same as that of PVCz(c). Namely, the interconversion dynamics in polymer cations was directly confirmed by time-resolved spectroscopy.

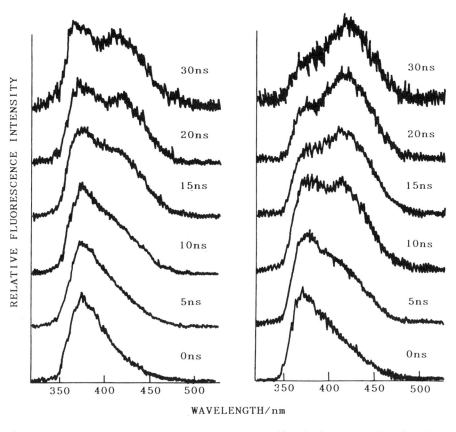

Figure 4-18 ■ Time-resolved fluorescence spectra of PVCz(r)-84 (left) and PVCz(c)-70 (right) in *N,N*-dimethylformamide at 20°C.

Now we consider what mechanism is responsible for the present dynamics. The spectral shape of the polymer cation at room temperature is independent of the delay time and identical with the low temperature result at late stages. This corresponds to the relaxed and the most stable state, which is in an equilibrium among some dimer cations. The interconversion from the partial overlap to the sandwich dimer cations in the polymer is due to the conformational change of the polymer chain or to hole transfer between the corresponding dimer sites.

Concerning the conformational dynamics, bichromophoric compounds, such as 1,3-di-*N*-carbazolylpropane and *meso-* as well as *rac*-2,4-di-*N*-carbazolylpentane in the excited singlet state, have been examined in detail as a model study.[6] Similar experiments have been recently conducted in our laboratory for the cationic state of the same model compounds.[48] At −55°C in the same *N,N*-dimethylformamide solvent as used here, only 1,3-di-*N*-carbazolylpropane showed a conformational change from an extended form to the folded sandwich structure. No other compounds in the

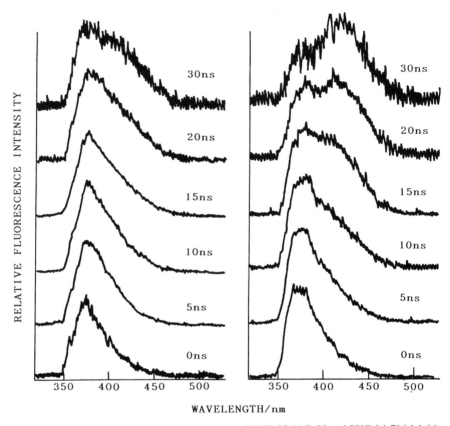

Figure 4-19 ■ Time-resolved fluorescence spectra of PVCz(r)-84 (left) and PVCz(c)-70 (right) in *N,N*-dimethylformamide at −55°C.

cationic state showed spectral change, which means conformational change is slower than the recombination process between the cation and the counteranion. It is reasonable to assume that conformation dynamics of the present polymer is slower than that of 1,3-di-*N*-carbazolylpropane. Therefore, we conclude that the hole transfer results in the present spectral change observed at low temperature. Because an interpolymer transfer process is neglected due to the low polymer concentration, this occurs along one polymer chain.

Here we mention the tacticity of PVCz. The ratio of syndiotactic to isotactic sequences in PVCz(r) and PVCz(c) is 3:1 and 1:1, respectively. However, this sequence does not take a completely statistical distribution, and it is usually accepted that the sequence forms a stereoblock structure, namely, some of the sandwich and partial overlap structures are incorporated as a group in the polymer chain. Therefore the hole migrates in the sequence of the partial overlap geometry and transfers to the sandwich. The migration in the latter structure is also possible, and sometimes goes back to the partial overlap sites. An open dimer structure that we confirmed by measuring its cation[7] may be in the kink and contribute to this dynamics. The hole produced by photoinduced electron transfer migrates along the polymer chain and an equilibrium state where at least three dimer cation sites are involved is attained. Immediately after quenching the relative contribution of the partial overlap and the sandwich dimer cations is determined by the excimer distribution. During tens of microseconds the hole is redistributed, which was first directly followed by transient absorption spectroscopy.

Acknowledgment. The present work is partly supported by a grant-in-aid for scientific research on Priority Area for Macromolecular Complexes (63612510) from the Japanese Ministry of Education, Science, and Culture. The authors wish to express their sincere thanks to Professor K. Yoshihara (Institute for Molecular Science), Dr. N. Ikeda (Osaka University), H. Yoshida, M. Koshioka, H. Sakai, and T. Kuroda.

References

1. H. Masuhara, in *Photophysical and Photochemical Tools in Polymer Science*, M. A. Winnik, ed., Reidel, Dordrechet (1986), pp. 43–63.
2. H. Masuhara, in *Lasers in Polymer Science and Technology*; *Application*, Vol. II, J. P. Fouassier and J. F. Rabek, eds., CRC Press, Boca Raton, FL (1989).
3. M. Masuhara, in *Photochemistry on Solid Surface*, M. Anpo and T. Matsuura, eds., Elsevier, Amsterdam (1989), pp. 15–29.
4. H. Masuhara and A. Itaya, in *Lasers in Polymer Science and Technology: Application*, Vol. II, J. P. Fouassier, and J. F. Rabek, eds., CRC Press, Boca Raton, FL (1989), pp. 217–234.
5. F. Wilkinson and G. P. Kelly, in *Photochemistry on Solid Surface*, M. Anpo and T. Matsuura, eds., Elsevier, Amsterdam (1989), pp. 30–40.
6. H. Masuhara, *J. Mol. Struct.*, **126**, 145 (1985).
7. H. Masuhara, *Makromol. Chem. Suppl.*, **13**, 75 (1985); in *Photophysical and Photochemical Tools in Polymer Science*, M. A. Winnik, ed., Reidel: Dordrecht (1986), pp. 65–84.
8. S. Claesson, *Fast Reactions and Primary Processes in Chemical Kinetics*; Almqvist and Wiskell, Uppsala (1967).
9. T. Kuroda, Thesis, Kyoto Institute of Technology (1988).
10. J. M. Morris and K. Yoshihara, *Mol. Phys.*, **36**, 993 (1978).
11. H. Masuhara, H. Miyasaka, N. Ikeda, and N. Mataga, *Chem. Phys. Lett.*, **82**, 55 (1981).
12. K. Hamanoue, T. Hidaka, T. Nakayama, and H. Teranishi, *Chem. Phys. Lett.*, **156**, 446 (1989).
13. H. Miyasaka, F. Ikejiri, and N. Mataga, *J. Phys. Chem.*, **92**, 249 (1988).
14. R. Kessler and F. Wilkinson, *J. Chem. Soc. Faraday Trans. 1*, **77**, 309 (1981).
15. N. Ikeda, K. Imagi, H. Masuhara, N. Nakashima, and K. Yoshihara, *Chem. Phys. Lett.*, **40**, 281 (1987).
16. N. Ikeda, M. Koshioka, H. Masuhara, and K. Yoshihara, *Chem. Phys. Lett.*, **150**, 452 (1988).
17. N. Ikeda, M. Koshioka, H. Masuhara, N. Nakashima, and K. Yoshihara, in *Ultrafast Phenomena VI*, T. Yajima, K. Yoshihara, C. B. Harris, and S. Shionoya, eds., Springer, Berlin (1988), p. 428.
18. H. Yoshida, Thesis, Kyoto Institute of Technology (1989).
19. K. Okamoto and A. Itaya, *Bull. Chem. Soc. Jpn.*, **57**, 1626 (1984).
20. N. Mataga and M. Ottolenghi, in *Molecular Association*, Vol. II, R. Foster, ed., Academic, London (1979), pp. 1–78.

21. N. Mataga, in *Molecular Interactions*, Vol. II, H. Ratajczak and W. J. Orville-Thomas, eds., Wiley, New York (1980), pp. 509–570.
22. H. Masuhara and N. Mataga, *Accounts Chem. Res.*, **14**, 312 (1981).
23. N. Mataga, *Pure Appl. Chem.*, **56**, 1255 (1984); *Acta Phys. Polonica*, **A71**, 767 (1987).
24. N. Mataga, private communication.
25. C. E. Hoyle and J. E. Guillet, *Macromolecules*, **11**, 221 (1978); **12**, 956 (1979).
26. H. Masuhara, J. Vandendriessche, K. Demeyer, N. Boens, and F. C. De Schryver, *Macromolecules*, **15**, 1471 (1982).
27. M. Yokoyama, Y. Endo, A. Matsubara, and H. Mikawa, *J. Chem. Phys.*, **75**, 3006 (1981).
28. K. Okamoto, A. Oda, A. Itaya, and S. Kusabayashi, *Chem. Phys. Lett.*, **35**, 483 (1975).
29. A. Itaya, K. Okamoto, and S. Kusabayashi, *Bull. Chem. Soc. Jpn.*, **58**, 2078 (1985).
30. A. Itaya and K. Okamoto, *Bull. Chem. Soc. Jpn.*, **60**, 83 (1987).
31. H. Masuhara, S. Ohwada, N. Mataga, A. Itaya, K. Okamoto, and S. Kusabayashi, *J. Phys. Chem.*, **84**, 2363 (1980).
32. A. Itaya, H. Sakai, and H. Masuhara, *Chem. Phys. Lett.*, **138**, 231 (1987) and references cited therein.
33. H. Masuhara, S. Eura, H. Fukumura, and A. Itaya, *Chem. Phys. Lett.*, **156**, 446 (1989).
34. A. Itaya, H. Yoshida, and H. Masuhara, unpublished.
35. A. Itaya, K. Okamoto, and S. Kusabayashi, *Bull. Chem. Soc. Jpn.*, **49**, 2082 (1976).
36. J. Vandendriessche, P. Palmans, S. Toppet, N. Boens, F. C. De Schryver, and H. Masuhara, *J. Amer. Chem. Soc.*, **106**, 8057 (1984).
37. H. Sakai, A. Itaya, and H. Masuhara, *J. Phys. Chem.*, **93**, 5351 (1989).
38. A. Itaya, K. Okamoto, and S. Kusabayashi, *Bull. Chem. Soc. Jpn.*, **50**, 22 (1977).
39. G. Rippen and W. Klopffer, *Ber. Bunsenges. Phys. Chem.*, **83**, 437 (1979).
40. N. Kim and S. E. Webber, *Macromolecules*, **18**, 741 (1985); H. Masuhara, N. Tamai, and N. Mataga, *Chem. Phys. Lett.*, **91**, 209 (1982).
41. H. Masuhara, H. Miyasaka, A. Karen, T. Uemiya, N. Mataga, M. Koishi, A. Takeshima, and Y. Tsuchiya, *Opt. Commun.*, **44**, 426 (1983).
42. H. Masuhara, N. Tamai, N. Mataga, F. C. De Schryver, J. Vandendriessche, and N. Boens, *Chem. Phys. Lett.*, **95**, 471 (1983).
43. H. Masuhara, N. Tamai, N. Mataga, F. C. De Schryver, and J. Vandendriessche, *J. Am. Chem. Soc.*, **105**, 7256 (1983).
44. H. Masuhara, N. Tamai, K. Inoue, N. Mataga, A. Itaya, K. Okamoto, and S. Kusabayashi, *Chem. Phys. Lett.*, **91**, 113 (1982); M. Yamamoto and A. Tsuchida, private communication.
45. M. Yamamoto, Y. Tsujii, and A. Tsuchida, *Chem. Phys. Lett.*, **154**, 559 (1989).
46. H. Masuhara, K. Yamamoto, N. Tamai, K. Inoue, and N. Mataga, *J. Phys. Chem.*, **88**, 3971 (1984).
47. A. Itaya, H. Sakai, and H. Masuhara, *Chem. Phys. Lett.*, **146**, 570 (1988).
48. H. Baba, Thesis, Tokyo Institute of Technology (1988).

III Soft Interaction and Multiple Interaction

Multistep Complexation of Macromolecules

Y. Kurimura

Department of Chemistry, Ibaraki University, Mito, Ibaraki 310, Japan

Macromolecule-metal complexes display characteristic reaction behaviors that are not observed in the reactions of ordinary inorganic complexes. Such unique properties of the macromolecule-metal complexes are mainly because the reactions of macro-molecule-metal complexes proceed only in a microheterogenous region occupied by the polymer chain, which is a so-called domain: In most cases, the physicochemical environment of the domain differs from that of the bulk solution. In addition, the nature of the domain is altered with the progress of the reactions and is sensitive to the environment.

Soft interactions, such as electrostatic, hydrophobic, π-π, and hydrogen bonding, are susceptible to the microenvironment in a dynamic way. Multiple interactions due to such soft interactions are, of course, very important because macromolecule-metal complexes have a great number of functional groups on their backbones. Thus the macromolecule-metal complexes often display cooperative and/or allosteric effects or stereoselective reactivity as a result of the multiple interactions.

Now, we consider the two limiting cases for the reactions of macromolecule-metal complexes: One case is the reaction of the macromolecule-metal complexes having a high degree of coordination, which is defined by [mole fraction of complex residue]/[mole fraction of total monomer units]. The other case is that which has an extremely low degree of coordination. In the former case, the nature of the domain is the most important factor that governs the reaction rate and, in the latter case, conformation of the polymer chain is probably the most important factor.

In this chapter, we consider multiple complexation due to soft or weak interaction as well as that due to coordination bonds.

5-1. Complexation of Polymer Ligands with Metal Ions and Metal Complex Ions

A typical example of the complexation of polymer ligands with metal ions is that of polyacrylic acid (PAA) with Cu^{2+} ions. Detailed discussion on the equilibrium study of the complexation was first reported by Gold and Gregor.[1]

Complexation of PAA with Cu^{2+} proceeds by the two step processes

$$RCOO^- + Cu^{2+} \rightleftharpoons RCOOCu^+ \qquad (5\text{-}1)$$

$$RCOO^- + RCOOCu^+ \rightleftharpoons (RCOO)_2Cu \qquad (5\text{-}2)$$

where R represents the polymer chain of PAA. For polyelectrolytes such as PAA, charges on the polyelectrolyte chain are varied with the progress of the reaction in a neutral pH region, leading to variation of the formation constants of Eqs. (5-1) and (5-2), which are defined respectively, by

$$K_1 = \frac{[RCOOCu^+]}{[RCOO^-][Cu^{2+}]} \qquad (5\text{-}3)$$

$$K_2 = \frac{[(RCOO)_2Cu]}{[RCOO^-][RCOOCu^+]} \qquad (5\text{-}4)$$

The formation constants K_1 and K_2 are not constant because charges on the polymer chains and, thus, the effect of neighboring groups on dissociation of the protons vary with the progress of the reaction.

If we consider the equilibrium reaction

$$2RCOOH + Cu^{2+} \longrightarrow (RCOO)_2Cu + 2H^+ \qquad (5\text{-}5)$$

the modified formation constant B_2, defined by

$$B_2 = \frac{[(RCOO)_2Cu][H^+]^2}{[RCOOH]^2[Cu^{2+}]} \qquad (5\text{-}6)$$

should be constant because no charged polymer species are included in these equations. The values of B_2 for PAA and the corresponding low molecular weight ligand, glutaric acid, are summarized in Table 5-1. The table shows that the value of B_2 for the Cu^{2+}-PAA complex is considerably greater than that of the low molecular weight system due to the electrostatic attraction force between the anionic PAA chains and Cu^{2+} cations.

Successive complexation of poly-4-vinylpyridine (PVP) and those of the corresponding low molecular weight analogue with M^{2+} ions are expressed by

$$M^{2+} + L \underset{}{\overset{K_1}{\rightleftharpoons}} ML^{2+} \qquad (5\text{-}7)$$

$$ML^{2+} + L \underset{}{\overset{K_2}{\rightleftharpoons}} ML_2^{2+} \qquad (5\text{-}8)$$

$$ML_2^{2+} + L \underset{}{\overset{K_3}{\rightleftharpoons}} ML_3^{2+} \qquad (5\text{-}9)$$

$$ML_3^{2+} + L \underset{}{\overset{K_4}{\rightleftharpoons}} ML_4^{2+} \qquad (5\text{-}10)$$

where L is PVP and pyridine (Py) and K_1–K_4 are the formation constants for Eqs. (5-7)–(5-10), respectively.

The formation constants of Cu^{2+}-PVP and Cu^{2+}-Py complexes are shown in Table 5-2. Remarkable differences in the formation constants for the polymer system (L = PVP) and those for the low molecular weight system (L = Py) are ascribed to the polymeric nature of PVP. One of the most interesting features of the complexation of the polymer ligand is that the successive formation constants for the polymer system increase slightly in the order K_1, K_2, K_3, K_4, whereas those for the low molecular weight system decrease in the same order. The characteristic feature of such multi-step complexation of the polymer ligand is clearly indicated in Fig. 5-1 where the relative concentrations of the dissolved chemical species in the Cu^{2+}/L system

Table 5-1 ■ Formation Constants for the PAA-Cu²⁺ and Glutaric Acid-Cu²⁺ Complexes

Acid	Neutral Salt	B_2
PAA (0.01 M)	—	9.1×10^9
PAA (0.01 M)	0.2 M NaNO₃	3.0×10^7
PAA (0.01 M)	2 M NaNO₃	9.6×10^6
PAA (0.01 M)	1 M KCl	6.9×10^5
Glutaric acid	—	7.2×10^3
Glutaric acid	0.2 M NaNO₃	4.4×10^3
Glutaric acid	2 M NaNO₃	6.9×10^2

Table 5-2 ■ Successive Formation Constants of Cu²⁺-L Complexes at $I = 0.10$ and 20°C[a]

L	$\log K_1$	$\log K_2$	$\log K_3$	$\log K_4$
PVP[b]	2.2	2.3	3.0	3.1
Py	2.5	1.9	1.3	0.8

[a]Concentrations of L and Cu²⁺ are 1.0×10^{-2} M and 2.0×10^{-3} M, respectively.
[b]The degree of polymerization of PVP is 19.
Source: N. Nishikawa and E. Tsuchida, *J. Phys. Chem.*, **79**, 2027 (1975).

are plotted as a function of log[L]. For the low molecular weight system, relative concentrations of the dissolved species increase in the order $Cu^{2+}, Cu(Py)^{2+}, Cu(Py)_2^{2+}, Cu(Py)_3^{2+}, Cu(Py)_4^{2+}$ with an increase in the concentration of pyridine, and the concentration of $Cu(Py)_4^{2+}$ becomes predominant at about several molar concentrations of pyrdine. On the other hand, for the polymer system, predominant species are Cu^{2+} at $[PVP] < 10^{-3}$ M and $Cu(PVP)_4^{2+}$ at $[PVP] > 5 \times 10^{-2}$ M and concentrations of other species are extremely low compared to that of Cu^{2+} or $Cu(PVP)_4^{2+}$ at any concentration of the polymer ligand. The results show that the tetradentate ligand complex $Cu(PVP)_4^{2+}$ is thermodynamically the most stable compared to other Cu^{2+}-PVP complexes. When the monodentate complex of $Cu(PVP)^{2+}$ is formed, subsequent coordinations of the neighboring pyridine moieties bound to the same polymer chain are promoted, leading to preferential formation of the tetradentate ligand complex. Once the first Py of PVP is bound, the local concentration of Py ligands near Cu^{2+} becomes very large. Thus the actual K value may not be that different but the Py concentration is about 1 M rather than 5×10^{-2} M.

The rates of the rapid complex formation between poly-4-vinylpyridine (PVP) with Ni^{2+} ions, represented by

$$Ni^{2+} + L \underset{k_b}{\overset{k_f}{\rightleftharpoons}} NiL^{2+} \tag{5-11}$$

$$Ni^2 + HL^+ \underset{k_{b'}}{\overset{k_{f'}}{\rightleftharpoons}} NiL^{2+} + H^+ \tag{5-12}$$

are determined by means of the conductance stopped flow (CSF) and conductance

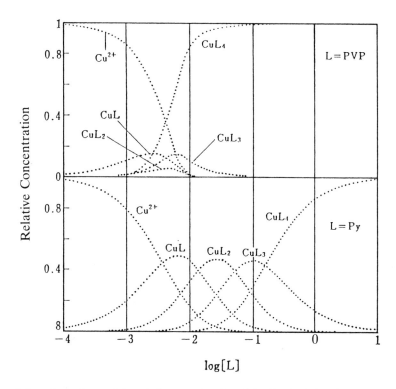

Figure 5-1 ■ Dependence of the relative concentration of the dissolved chemical species on the logarithmic concentration of the ligand [L].

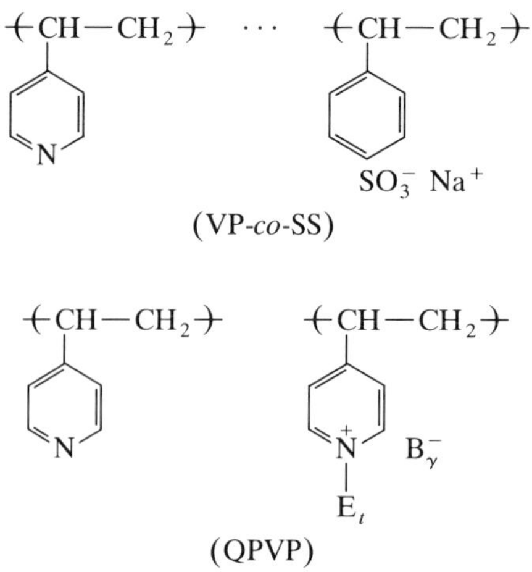

(VP-*co*-SS)

(QPVP)

Figure 5-2 ■ Chemical structures of VP-*co*-? and QPVP. [From Y. Kurimura and K. Taka_ *J. Chem. Soc. Faraday Trans. 1*, **84**, 841 (198_

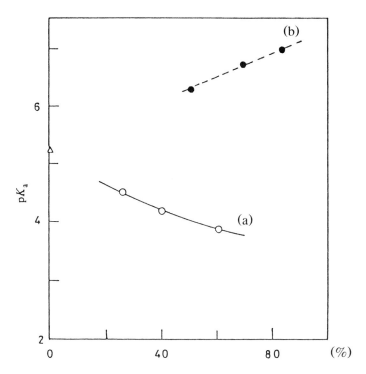

Figure 5-3 ▪ (a) Dependence of pK_a on the mole fraction of vinyl-pyridine groups in VP-*co*-SS and (b) degree of quarternization of the QPVP.

pressure-jump (CPJ) techniques.[3] In these equations, L and HL$^+$ are the nonprotonated and protonated pyridine moieties on PVP, respectively. The observed rate and formation constants at 25°C are $k_f = 4.6 \times 10^3$ M^{-1} s^{-1}, $k_b = 6.7$ s^{-1}, and $K(= k_f/k_b) = 690$ M^{-1} for Eq. (5-11) and $k'_f = 10$ M^{-1} s^{-1}, $k'_b = 10^3$ M^{-1} s^{-1}, and $K'(= k'_f/k'_b) = 10^{-2}$ M^{-1} for Eq. (5-12). The values of k_f, k_b, and K for the Ni^{2+}/PVP system are quite similar to those of the corresponding low molecular weight system Ni^{2+}/imidazole. The results indicate that the effect of the polymer backbone on the rate is not as important a factor for the first-step complexation as in the Cu^{2+}/PVP system.

The acid dissociation constants of the polymeric ligand, such as PAA, PVP, and polyvinylimidazole (PVIm), are known to be dependent on the charges on the polymer chains. Dependence of the pK_a of vinylpyridine moieties on VP-*co*-SS copolymer (VP = vinylpyridine, SS = styrenesulfonate) and vinylpyridine partially quarternized by ethyl bromide (QPVP) on the chemical compositions of the copolymers are shown in Fig. 5-3. Figure 5-3 shows that, for VP-*co*-SS, the pK_a value increases with an increase in the mole fraction of SS. On the other hand, for QPVP, the pK_a value decreases with increasing degree of quarternization. The results demonstrate that the dissociation of protons from the pyridine rings on VP-*co*-SS is

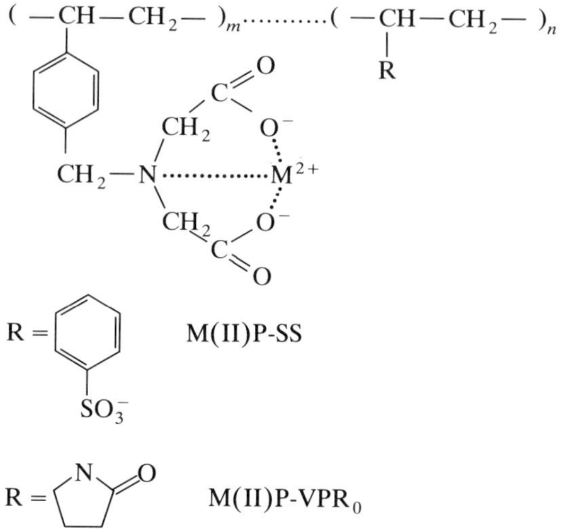

Figure 5-4 ■ Chemical structures of M(II)P-SS and M(II)P-VPRo.

suppressed by increasing the negative charge density on the polymer chain due to electrostatic interactions between protons and the negatively charged groups on the polymers. The protonation is reduced with increasing the positive charge density on QPVP due to the electrostatic repulsion force between the protons and the positively charged groups on the polymer backbone.

Reactions of the polymer-bound chelating agents, such as 4-vinylbenzylamine-N,N-diacetate-*co*-styrenesulfonate (P-SS) and 4-vinylbenzylamine-N,N-diacetate-*co*-vinylpyrrolidone (P-VPRo) (Fig. 5-4) with transition metal ions have been investigated in aqueous solution.[4] The acid dissociation constants of P-SS and P-VPRo are determined by means of potentiometric titration using modified Henderson-Hasselbalck equations[5]:

$$pH = pK_{a1} - n_1 \log\{(1 - \alpha)/\alpha\} \tag{5-13}$$

$$pH = pK_{a2} - n_2 \log\{(2 - \alpha)/(\alpha - 1)\} \tag{5-14}$$

where α is degree of neutralization and n_1 and n_2, which vary in value from somewhat above 1 to about 2, measure the extent of the effect of the neighboring groups on dissociation of the first and second acid groups in the chelating agent moieties.

Plots of pH versus $\log\{(1 - \alpha)/\alpha\}$ for the first step dissociation and that of pH versus $\log\{(2 - \alpha)/(\alpha - 1)\}$ for the second step are shown in Figs. 5-5 and 5-6, respectively. Estimated values of pK_{a1}, pK_{a2}, n_1, and n_2 are summarized in Table 5-3 along with those of pK_{a1} and pK_{a2} of the low molecular weight analogue, ben-zylamine-N,N-diacetate (BDA). This table shows that the values of pK_{a1} and pK_{a2} of the polymer-bound chelating agents are greater than those of the corresponding pK_a values of the low molecular weight analogue by about 1.2–1.5 times, indicating that

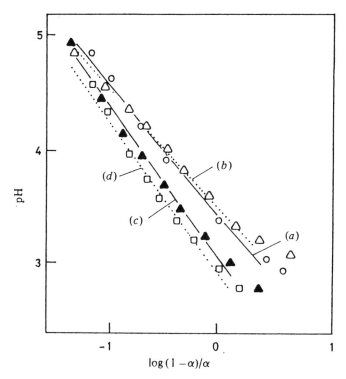

Figure 5-5 ■ Plots of pH versus log{$(1 - \alpha)/\alpha$} for the first-step proton dissociation of P-SS and P-VPRo. a; P-VPRo-1 (○), b; P-$\bar{V}$PRo-2 (△), c; P-SS-1 (▲), d; P-SS-2 (□) (see footnote of Table 5-3).

high negative charge density on the polymer backbones of P-SS and P-VPRo suppresses the dissociation of protons from the acid moieties on the polymer chain relative to the low molecular weight analogue. Thus, an increasing order of pK_a becomes BDA < P-VPRo < P-SS. Furthermore, the values of pK_{a1} for P-SS are larger than those of P-VPRo, reflecting the larger negative charge density on the domain of the former. Relatively smaller values of n_1 and n_2 (1.1–1.3) for both polymer ligands suggest a weak effect of the neighboring groups on dissociation of protons on the acid groups.

Formation constants (K) of P-SS, P-VPRo, and BDA with several transition metal ions (K) are included in the equation

$$\log\left(1 + \frac{[H^+]}{K_{a2}} + \frac{[H^+]^2}{K_{a1}K_{a2}}\right)$$

$$= \log K + \log\frac{([M]_T - [MA])([A]_T - [MA])}{[MA]} \tag{5-15}$$

where [MA], $[M]_T$, and $[A]_T$ are the concentration of metal chelate, total concentra-

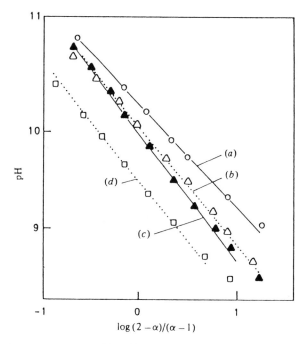

Figure 5-6 ■ Plots of pH versus $\log\{(2-\alpha)/(\alpha-1)\}$ for the second-step proton dissociation of P-SS and P-VPRo. a; P-VPRo-2 (○), b; P-SS-2 (△), c; P-SS-1 (▲), d; P-VPRo-1 (□) (see footnote of Table 5-3).

tion of metal ion, and total concentration of chelating agent, respectively. Equation (5-15) can be used to determine the values of chelate formation constants: the value of $\log K$ is obtained from the intercept of the straight line of the plot of $\log(1 + [H^+]/K_{a1} + [H^+]^2/K_{a1}K_{a2})$ versus $\log(([M]_T + [MA])([A]_T - [MA])/[MA])$. The values of the formation constants obtained are summarized in Table 5-4, which results show an enhancement of the chelate formation in the polymer system over that of

Table 5-3 ■ Acid Dissociation Constants and Values of n_1 and n_2 for the Polymer-Bound Chelating Agent at $I = 0.1$ and 25°C

Chelating Agent[a]	pK_{a1}	n_1	pK_{a2}	n_2
P-SS-1	3.32	1.1	10.27	1.1
P-SS-2	3.38	1.1	10.06	1.2
P-VPRo-1	2.94	1.2	9.97	1.3
P-VPRo-2	2.78	1.2	9.52	1.2
BDA	2.24	—	8.90	—

[a]Chemical compositions of the polymer-bound chelating agents are: P-SS-1, [VBDA]:[SS] = 0.12:0.88; P-SS-2, [VBDA]:[SS] = 0.49:0.51; P-VPRo-1 = [VBDA]:[VPRo] = 0.09:0.91; P-VPRo-2, [VBDA]:[VPRo] = 0.25:0.75.

Table 5-4 ■ **Chelate Formation Constants of Co^{2+}, Ni^{2+}, and Cu^{2+} with P-SS or BDA at $I=0.1$ and 25°C[a]**

Chelating Agent	log K		
	Co^{2+}	Ni^{2+}	Cu^{2+}
P-SS-1	8.68	9.74	11.50
P-SS-2	8.96	10.07	11.76
BDA	6.59	8.07	10.70

[a] 1.00×10^{-3} M M^{2+} and 1.00×10^{-2} M chelating agent.
Source: Y. Kurimura and K. Takato, *J. Chem. Soc. Faraday Trans. 1*, **84**, 841 (1988).

the corresponding low molecular weight system. The values of the formation constants for all the polymer-bound chelating agents are 1–2 orders of magnitude greater than those of the corresponding low molecular weight system. Such an enhancement of the chelate formation is ascribed to the multi-step complexation, that is, complexation processes for the polyelectrolyte-bound chelating agents proceed by multistep complexation process (a) accumulation of M^{2+} ions into the negative potential field of the P-SS domain (preequilibrium process) and (b) complexation of M^{2+} ions with

Class I

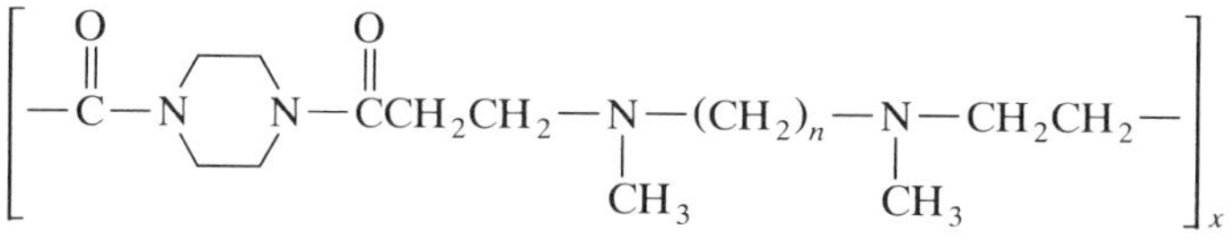

$$n = 2, L_1$$
$$n = 3, L_2$$
$$n = 4, L_3$$

Class II

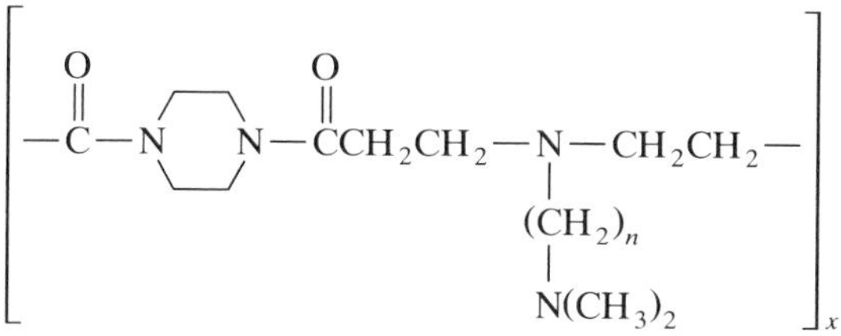

$$n = 2, L_4$$
$$n = 3, L_5$$
$$n = 4, L_6$$

Figure 5-7 ■ Chemical structures of poly(amido amines).

the chelating agent moieties (chelation process) in the domain. A higher electrostatic potential on the polymer-bound chelating agents would attract a larger number of divalent metal ions, leading to high local distribution of M^{2+} in the P-SS domains.

Enhancement factor F, defined by $F = K_p/K_m$, where K_p and K_m are the formation constants of $(\text{P-SS-1})\text{-}M^{2+}$ and $\text{BDA-}M^{2+}$ (see Table 5-4), respectively, are ca. 120 for Co^{2+}, 50 for Ni^{2+}, and 15 for Cu^{2+} under the conditions employed: The enhancement factor decreases as the corresponding K_m increases. Conversely, metals that have little affinity for small ligand molecules in solution may bind to the polymer with significant K values. For metal complexes having larger K_m values for the corresponding low molecular weight complex, the degree of electrostatic accumulation of metal ions into the polymer domains is smaller than those having small K_m values, because relatively small amounts of the metal ions remain in the bulk solution whereas the larger part of the metal ions are already combined with the chelate groups under the conditions employed. This is the reason for such dependence of F on K_m as shown in Table 5-4.

Poly(amido amines) are unusual polyelectrolytes, because their repeating units behave independently toward protonation and complex formation with transition metal ions. Thermodynamic parameters for the complexation of some of the poly(amido amines) belonging to two different classes, class I and class II (Fig. 5-7), are summarized in Table 5-5. This table shows that CuL complexes are more stable for class I polymer (L_1) than for class II polymer (L_4), the difference being due mainly to the enthalpy terms. This effect is probably due to the involvement of the carbonyl groups in class I complexes, although this is not the case for unsymmetrically substituted compounds of class II, due to high steric hindrance. The entropy term is not strongly influenced because the mobility of the amidic group in the CuL complexes is very low for steric reasons, even if not directly involved in the complex. The decrease of the complexing ability with lengthening of the aliphatic chain is related to the increased steric strain when going from five- to six-membered chelate rings; with seven-membered rings binding does not occur at all. In this case the decrease of stability is mainly due to the decrease of enthalpy.

The rates of the ligand substitution reactions of macromolecule-metal complexes also depend on the properties of the polymer domains.[7] The acid and base hydrolysis reactions of cis-$Co(en)_2PVPCl^{2+}$ (en = ethylenediamine) are expressed respectively,

Table 5-5 ▪ Thermodynamic Functions for the Complexation of L with Cu^{2+} in Aqueous 0.1 M NaCl Solution at 25°C[a]

	$\log K$	$-\Delta G$	$-\Delta H$	ΔS
CuL_1	8.96	12.22	8.04	14.8
CuL_2	5.36	7.31	7.1	0.7
CuL_4	8.47	11.55	7.57	13.4

[a]L_3, L_5, and L_6 do not form stable complexes.
Source: R. Burbucci, M. Casolaro, V. Barone, P. Ferruti, and M. Tramontini, *Macromolecules*, **16**, 1159 (1983).

Table 5-6 ■ Rate Constants for the Acid and Base Hydrolyses
of cis-Co(en)$_2$LCl^{2+} (L =PVP and EtPy) at 25°C

L	P_n^c	X^d	k_H^a (s^{-1})	k_{OH}^b (M^{-1} s^{-1})
PVP	19	0.60	—	4.3×10^2
PVP	49	0.62	—	7.2×10^2
PVP	100	0.58	8.1×10^{-7}	8.7×10^2
EtPy	—	—	2.4×10^{-6}	3.2×10^2

^{a}Acid hydrolysis rate constant at $I = 0.02$ and 40°C.
^{b}Base hydrolysis rate constant at $I = 0.08$ and 25°C.
^{c}Degree of polymerization of PVP.
^{d}Degree of coordination.

by

$$cis\text{-Co(en)}_2\text{PVPCl}^{2+} + \text{H}_2\text{O} \longrightarrow cis\text{-Co(en)}_2\text{PVP(OH}_2)^{3+} \text{Cl}^- \quad (5\text{-}16)$$

$$cis\text{-Co(en)}_2\overline{\text{PVP}}\text{Cl}^{2+} + \text{OH}^- \longrightarrow cis\text{-Co(en)}_2\text{PVP(OH)}^{2+} + \text{Cl}^- \quad (5\text{-}17)$$

The rate constants for the acid and base hydrolysis reactions of *cis*-Co(en)$_2$PVPCl^{2+} and the corresponding low molecular weight analogue, *cis*-Co(en)$_2$EtPyCl^{2+} (EtPy = ethylpyridine), are shown in Table 5-6.

The smaller acid hydrolysis rate constant for the polymer complex is attributable to the existence of the bulky polymer backbone attached to the Co(III) complex moieties: The polymer chains present in the neighborhood of the complex moieties tend to break up the solvation shell of the complex, whereas the complex makes greater demands on solvation when passing through the process of charge separation. Therefore, factors producing poorer overall solvation lead to lower acid hydrolysis rates.[8]

On the other hand, the rate of base hydrolysis of *cis*-Co(en)$_2$PVPCl^{2+} is greater than that of *cis*-Co(en)$_2$(EtPy)Cl^{2+} by a factor of about 3 for PVP with degree of polymerization $P_n = 100$. This is ascribed to a greater local concentration of hydroxide ions in the polymer domains. The higher positive potential field in the polymer domains attracts large amounts of hydroxide ions at lower ionic strength, leading to high local concentration of hydroxide ions in the domain.

5-2. Multiple Interactions of Polyelectrolytes with Metal Ions and Metal Complex Ions

In aqueous solutions, simple polyelectrolytes, such as sodium salts of polyacrylic acid, polymethacrylic acid, polyvinylsulfate (PVS), and polystyrenesulfonic acid (PSS), concentrate the counterions in their domains. The fraction of bound Na$^+$ ions into the domains of PAA and poly-D-glutamic acid (PDG) is determined by means of electrolysis in aqueous solution: Plots of the degree of binding of Na$^+$ ions (f) for PAA and

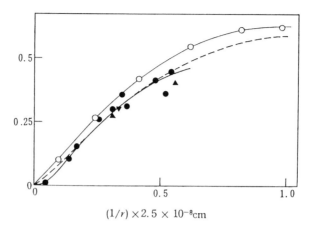

$(1/r) \times 2.5 \times 10^{-8}$cm

Figure 5-8 ■ Plot of the fraction of the bound Na$^+$ (F) versus $1/r$ for sodium acrylate solution. ○, PAA; _____, copolymer of maleic acid; ●, PDG; △, PDLG. [From K. Nitta and S. Sugai, *Ann. Rep. Res. Group Biophys. Japan*, **5**, 83 (1965).]

PMA, and that of PDG versus density of dissociated acid groups $(1/r)$ on unit length of the polymer chain give similar curves as shown in Fig. 5-8. The results seem to indicate that interactions between polyelectrolytes, such as PAA, PMA, and polyglutamic acid (random coil), with alkaline metal ions are the same in nature and essentially electrostatic rather than a coordination bonds.[9]

Weak and soft interactions in the polymer domains can solubilize some kinds of polynuclear metal complex species without precipitation of the hydrolysis species: The polynuclear complex of $Cu(OH)_2$ (oleated Cu^{2+} species) is solubilized by polyvinylalcohol (PVA) at pH > 6. The solubilization mechanism is of an inclusion type due to multisite hydrophobic interactions, where the polynuclear complex of $Cu(OH)_2$ is included by PVA chains with hydrophobic backbones toward the inside and with hydrophilic OH groups toward the outside bulk water.[10]

The binding between chiral polymers and some metal or metal complex ions has been studied in order to clarify the nature of the binding force, selective binding ability, binding site of the polymer chain, recognition or resolution of optical isomers, and possibility of the metal or metal complex ions as a chiral probe.[11,12]

The results of the investigation of the nature of Eu(III)-poly(L-glutamate) interactions show that intimate binding with loss of waters of hydration is possible only for the helical configuration. Free glutamate anions or anion residues in the random polymer do not induce a loss of Eu(III) waters of hydration.[13]

In aqueous solution, Tb(III) ions are bound strongly to polyelectrolytes or to polysaccharides, such as carboxymethyl cellulose or heparin, and some or all of the coordinated water molecules are expelled upon binding.[14-17]

The fluorescence intensity and lifetime of polyelectrolyte-Tb(III) complexes have been measured in aqueous solutions of PAA, copolymers of maleic acid with ethylene (MA-E), MA-isobutene (MA-iBu), and 2,4,4-trimethylpentene (MA-3MPe). The fluorescence intensity decreases in the order MA-3MPe > MA-iBu > Ma-E > PAA and

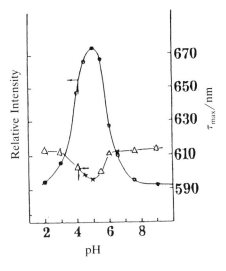

Figure 5-9 ■ Relative intensity ($\bigcirc$) and the wavelength of the maximum ($\triangle$) of luminescence of Ru(bpy)$_3^{2+}$ in PMA aqueous solution as a function of pH. [From D. Y. Chen and J. K. Thomas, *J. Phys. Chem.*, **89**, 4065 (1985).]

the lifetime is in the reverse order. The results indicate that the binding ability of metal ions increases with increasing size of the alkyl group on the copolymer chains. The number of water molecules coordinated to the Tb(III) ion in PAA and MA-E is about 3.5. On the other hand, in the corresponding monomeric carboxylates, such as propionate and 2,3-dimethylsuccinate, this number is 5.8–6.0.[18]

Polyelectrolytes such as PMA and PAA often provide unique environments for guest molecules. It is well known that the probe molecule tris(2,2'-bipyridine)ruthenium(II), Ru(bpy)$_3^{2+}$, is an excellent monitor of the environment where it resides. In aqueous solution of PMA, the luminescence intensity of Ru(bpy)$_3^{2+}$ is markedly affected by the pH. Dependence of the luminescence intensity on pH and that of the maximum wavelength of the luminescence spectra are shown in Fig. 5-9. The relative intensity of the luminescence exhibits a marked enhancement with pH increase from 2–5, followed by a decrease to the original spectrum with pH increase above 5. The luminescence intensity reaches a maximum and, in addition, the maximum wavelength exhibits a blue shift at pH 5. Increasing the pH of the PMA solution from 2 through 5–9 results in an increased degree of ionization of the acid. At pH 2, the polymer is in a nondissociated condition and tight coils of PMA are formed. At pH 5 the polymer is partially neutralized and the degree of ionization is about 24%. At pH > 9, the polymer becomes a long flexible molecule, because it is strongly negatively charged, and then Ru(bpy)$_3^{2+}$ strongly binds to the anionic sites on the polymer. At pH 2 no binding of Ru(bpy)$_3^{2+}$ is observed and it is in the aqueous phase. The increased luminescence lifetime and the blue spectral shift are indicative of a restricted environment for Ru(bpy)$_3^{2+}$ at pH 5: The ligands are restricted around the metal.

Quenching studies also follow this behavior over the full pH range and are interpreted according to whether the probe is in the water phase, pH 2, on a stretched polymer, pH 9, or in the interior of a polymer at pH 5. At pH 5, Ru(bpy)$_3^{2+}$ is shielded from attack by molecules from the aqueous phase. Other molecules that bind

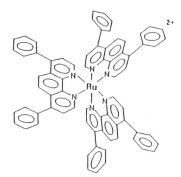

Figure 5-10 ■ Chemical structure of Ru(DIP)$_3^{2+}$.

to the polymer, such as cupric and chromic ions, also quench the Ru(bpy)$_3^{2+}$, and the kinetics follow a Poisson distribution.[19]

The chiral ruthenium complexes, such as Ru(bpy)$_3^{2+}$, Ru(phen)$_3^{2+}$ (phen-1,10-phenanthroline), and Ru(DIP)$_3^{2+}$ (DIP = 4,7,-diphenyl-1,10-phenanthroline) (Fig. 5-10), with luminescence characteristics indicative of the binding mode and stereoselectivities that may be tuned to the helix topology, may be useful molecular probes in solution for secondary structures of nucleic acid.[20]

In the aqueous solution, increasing luminescence intensity is seen for Ru(phen)$_3^{2+}$ and Ru(DIP)$_3^{2+}$ with DNA addition: A two-exponential decay in luminescence is found for Ru(phen)$_3^{2+}$ and Ru(DIP)$_3^{2+}$ with luminescence lifetimes of the complexes bound to DNA appearing 3–5 times larger than those of the free complexes. On the other hand, no enhancement in luminescence intensity is observed for Ru(bpy)$_3^{2+}$.[21]

For quenching studies, biphasic Stern–Volmer plots are found for Ru(phen)$_3^{2+}$ and Ru(DIP)$_3^{2+}$ in aqueous solution in the presence of ferrocyanide, indicating extensive protection of their excited state complexes in the presence of DNA from ferrocyanide. Here luminescence quenching is completely static as a result of counterion condensation at the DNA polyion. The results of luminescence polarization measurements for Ru(phen)$_3^{2+}$ are interpreted in terms of two binding modes: *electrostatic*, which is essentially quenched by ferrocyanide and contributes no polarization in luminescence, and *intercalative*, which is protected from cyanide quenching and, because it is rigidly bound, retains luminescence polarization. The distinction becomes more apparent for Ru(DIP)$_3^{2+}$ where significant enantiometric selectivity is observed on binding to DNA. The Δ isomer binds to DNA both electrostatically and by intercalation: Extensive curvature in the Stern–Volmer plots and increases in polarization are observed. On the other hand, the Λ isomer, which gives strictly linear Stern–Volmer plots, binds only electrostatically.[21]

It is known that unstable metal complex species are often stabilized in the polymer matrices due to the effect of the polymer backbone. Stabilization of unstable species is mostly accomplished due to multistep complexation, soft or weak multisite interactions, and restrictions of the translational and rotational motions of the molecules in the network of the polymer chains. In polymer matrices several metal complexes that are unstable under usual conditions are found to be affected by the environment

Table 5-7 ▪ Stabilization of Unstable Complex Species in the Polymer Matrices

System	Stabilized Species	Process
$CO/CuCl_2/AlCl_3/PS^a$	Cu-Co	Selective adsorption of NO[42]
N_2/P-MeCpMn(CO)$_3^b$	P-MeCpMn(CO)$_2$(N$_2$)	Reversible coordination of dinitrogen[43]
C_2H_2/P-MeCpMn(CO)$_3^b$	P-MeCpMn(CO)$_2$(C$_2$H$_2$)	Reversible coordination of acetylene[44]
O_2/Fe(II)(TPP)/lipz/H$_2$O^c	Fe(II)-O$_2$	Reversible adsorption of dioxygen under physiological conditions[45]
NO/CR/Fe(II)^d	Fe(II)-NO	Selective adsorption of NO[46]

^{a}PS = polystyrene bead.
^{b}P-MeCpMn(CO)$_3$ = polymer-bound methylcyclopentadienyltricarbonylmanganese.
^{c}TPP = tetraphenylporphyrin; lipz = lipozome.
^{d}CR = chelate resin.

where they reside. Several examples 42–46 of such stabilization of complex species in polymer matrices are summarized in Table 5-7.

Stereoselective reactions can be achieved due to multisite interactions using structurally regulated functional polymers. One example is a model of DNA cleavage restriction enzymes.

Attachment of Fe(II)EDTA analogues to DNA binding molecules creates a DNA cleaving molecule due to multiple and soft interactions between them. For example, sequence-specific double-strand cleavage of DNA can occur using bis(EDTA-dis-tamycine-Fe(II)) [BED.Fe(II)] (Fig. 5-11) in the presence of oxygen molecules and dithiothreitol.[22] A histogram of the electrophoresis of the cleavage products of the DNA cleavage pattern reveals a major cleavage site contiguous to the eight base pair sequence 5'-TTTTTATA-3' and a minor site contiguous to the five base pair sequence 5'-AATAA-3'. Schematic representation for the interactions of BED.Fe(II) with double helical DNA is also shown in Fig. 5-11. The cleavage products are consistent with oxidative cleavage of the deoxyribose ring affording a 5'-phosphate DNA terminus and approximately equal proportions of 3'-phosphate and 3'-phosphoglycolic acid termini.[22] DNA cleavage occurs by attack of superoxide ions, which are produced by the reaction of O_2 with the terminal Fe(II)EDTA moieties on BED.Fe(II) bound to DNA.

5-3. Multistep Complexation in the Electron-Transfer Process

For the electron-transfer reactions of macromolecule-metal complexes and metalloproteins, electron transfer often proceeds through a preequilibrium step (formation of intermolecular association complex). For example, electron-transfer reactions of polyanion-bound metal chelates, such as Fe(II)P-SS (P-SS = vinylbenzylamine-N,N'-diacetate-co-styrene-sulfonate) (Fig. 5-4), with cationic Co(III) complexes proceed in the domain of the intermolecular association complex that formed by the electrostatic

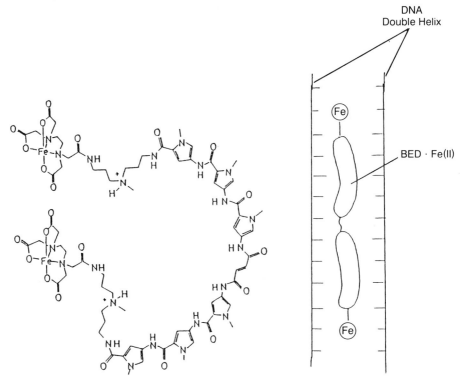

Figure 5-11 ■ Schematic representation of the interaction of DNA with BDA · Fe(II). (Modified from ref. 22.)

interaction between the polyanion-bound metal complex and the low molecular weight metal complex. Schematic representation of the formation of an association complex between polyanion-bound metal complex (P-M$_1$X) and cationic metal complex (M$_2$Y^{n+}) is shown in Fig. 5-12. Formation of such an association complex is caused by partial exchange of the countercations, such as Na$^+$, in the Fe(II)P-SS domains by positively charged complex ions M$_2$Y^{n+}. As the result of such exchange equilibrium, the local concentration of M$_2$Y^{n+} becomes greater than that in the bulk of the solution, leading to enhancement of the reaction rate.

An example of the rate enhancement by the electrostatic interactions between the reactant ions is shown in Fig. 5-13. The enhancement factor is defined by k_p/k_m where k_p and k_m are the rate constants for the Fe(II)P-SS reduction and that for the Fe(II)BDA (BDA = benzylamine-N,N-diacetic acid) reduction of the given Co(III) complex, respectively. Figure 5-13 shows that the enhancement factor increases exponentially with increasing positive charges on the Co(III) complexes.

Another characteristic behavior in the electron-transfer reaction of macromolecule-metal complexes caused by change in the microenvironment in the domain is observed for the reduction of *cis*-azidobis(ethylenediamine)poly-4-vinyl-pyridine-cobalt(III), *cis*-Co(en)$_2$PVP(N$_3$)$^{2+}$ (PVP = poly-4-vinylpyridine; en = ethyl-

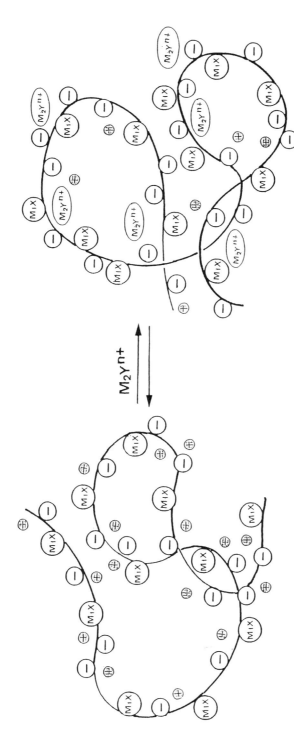

Figure 5-12 ■ Schematic representation of the association of polyanionic metal complex with cationic metal complex.

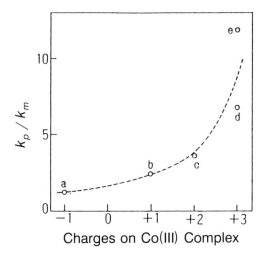

Charges on Co(III) Complex

Figure 5-13 ▪ Dependence of the enhancement factor (k_p/k_m) on the charges of the Co(III) complexes. (a) $[\text{CoCl(Hedta)}]^-$; (b) $[\text{Co(ox)(en)}_2]^+$; ox = oxalate; (c) $[\text{CoCl(NH}_3)_5]^{2+}$; (d) $[\text{Co(en)}_3]^{3+}$; (e) $[\text{Co(NH}_3)_6]^{3+}$. [From Y. Kurimura, K. Takato, and N. Suzuki, *Bull. Chem. Soc. Jpn.*, **59**, 2356 (1986).]

ene-diamine) (Fig. 5-14), by hexaamineruthenium(II), $\text{Ru(NH}_3)_6^{2+}$. Domains of this polymer bound complex have large positive charge density because the polymer-bound Co(III) complex has a great number of positively charged metal complex moieties on a polymer backbone. Dependence of the rate constant on the concentration of added salt for the reduction of *cis*-$\text{Co(en)}_2\text{L(N}_3)^{2+}$ (L = PVP, Py) by $\text{Ru(NH}_3)_6^{2+}$ is shown in Fig. 5-15:

$$\textit{cis}\text{-Co(en)}_2(\text{N}_3)\text{PVP}^{2+} + \text{Ru(NH}_3)_6^{2+} \longrightarrow \text{Co(II)} + 2\text{en} + \text{PVP} + \text{N}_3^- + \text{Ru(NH}_3)_6^{3+}$$

$$(5\text{-}18)$$

At given concentration of the added salt, the rate constant for *cis*-$\text{Co(en)}_2(\text{PVP})(\text{N}_3)^{2+}$ is considerably greater than that of the corresponding low

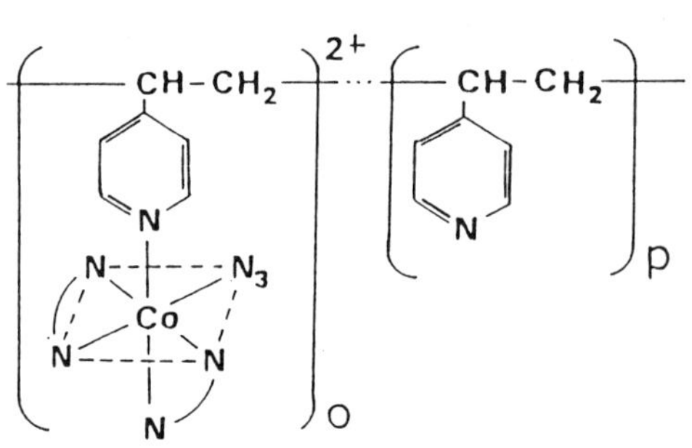

Figure 5-14 ▪ Chemical structure of *cis*-$\text{Co(en)}_2(\text{N}_3)\text{PVP}^{2+}$.

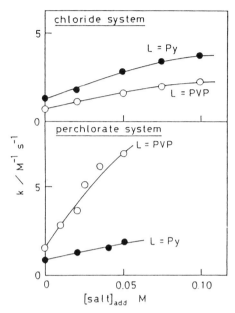

Figure 5-15 ▪ Dependence of the rate constant for the reductions of *cis*-Co(en)$_2$L(N$_3$)$^{2+}$ (L = PVP and Py) on the concentration of added salt at pH 2.0 and 25°C. The ionic strength in the absence of the added salt was 0.04. [From Y. Kurimura, T. Kikuchi, and E. Tsuchida, *Macromolecules*, **22**, 1645 (1986).]

molecular weight Co(III) complex, *cis*-Co(en)$_2$(EtPy)(N$_3$)$^{2+}$ (EtPy = ethylpyridine), in perchlorate solution; the rate constant for the polymer system is smaller than that for the low molecular weight system in chloride solution.

The rate enhancement in the perchlorate solution is mainly ascribed to the difference in the physicochemical property of the microenvironment occupied by the polymer chain and the environment of the bulk of the solution. The results of the kinetic analysis suggested that, in the perchlorate solution, a great number of perchlorate anions are concentrated in the polycationic domain, leading to partial elimination of the water molecules attached to the complex moieties and, probably, the Ru(NH$_3$)$_6^{2+}$ localized in the domain. For the polymer complex, a distinct decrease in the activation enthalpy occurs with an increase in the concentration of added sodium perchlorate, whereas no apparent decrease in the activation enthalpy is observed for the polymer complex in the chloride system or for the low molecular weight analogue in perchlorate and chloride systems (Fig. 5-16). A tentative explanation for the smaller activation enthalpy for the reduction of the polymer complex in perchlorate solution is that greater destabilization of the ground state of the Co(III) and, probably, the Ru(II) dications in the domain occurs than in the activated complex, which has four positive charges, due to different degrees of dehydration of these species.[24]

Studies of the reactions between synthetic macromolecule-metal complexes and other synthetic macromolecule-metal complexes would give useful basic information

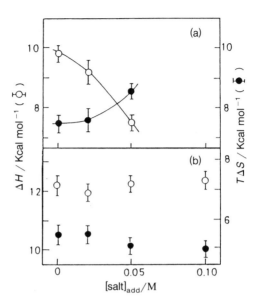

Figure 5-16 ■ Dependences of ΔH and $-T\Delta S$ on the concentration of added salt for the reduction of cis-Co(en)$_2$PVP(N$_3$)$^{2+}$ in (a) the perchlorate system (b) chloride system at pH 2.0 and 25°C.

about the interactions between polymers, effects of steric factors, conformations of the macromolecules, structure of the activated complex, and important factors that govern the rate of interpolymer reaction. However, only a few studies have been carried out, probably because difficulties arose in the design of macromolecule-metal complexes suitable for such reactions and arose in analysis of the kinetic data.

cis-Co(en)$_2$(OH$_2$)(P-VIm)$^{3+}$ (P-VIm = 1-vinylimidazole-co-1-vinyl-2-pyrrolidone) interacts with Fe(II)P-VPRo to give an intermolecular complex through electrostatic interactions. Rates of the electron transfer between the Co(III) and the Fe(II) complex moieties in the interpolymer complex are determined by means of spectrophotometry. The kinetic parameters for the electron-transfer reactions are summarized in Table 5-8 along with those of the corresponding polymer–low molecular weight system and that of the low molecular weight–low molecular weight system. In the Co(III)P-Im/Fe(II)P-VPRo system, the reactions proceed via two parallel second-order processes. The rate constants for the fast process at the initial stage and the subsequent slow process are $k_f = 1.3 \times 10^5$ M^{-1} s^{-1} and $k_s = 6.9 \times 10^2$ M^{-1} s^{-1} at 15°C, respectively. The fast reaction is ascribed to the electron transfer between the Co(III) and Fe(II) moieties, both of which are located on the surfaces of the Co(III) and the Fe(II) polymers, respectively. The slow reaction reflects the conformational change of the interpolymer complex.[25]

Reactions of Fe(II)P-SS and Fe(II)P-VPRo (P-SS = vinylbenzylaminediacetate-co-styrenesulfonate and VPRo = vinylbenzylaminediacetate-co-vinylpyrrolidone) with Co(NH$_3$)$_5$PAA [charges on the Co(III) complex are omitted] were investigated in

Table 5-8 ■ **Rate Constants and Activation Parameters for the Reduction of the Co(III) Complexes by Fe(II)EDTA at 25°C**

Kinetic	System[a]			
	I			
Parameter	Fast	Slow	II	II
k (M^{-1} s^{-1})	1.3×10^5	6.9×10^2	1.3×10^4	3.1×10^3
ΔH (Kcal mol^{-1})	38	15	20	16
ΔS (eu)	95	6.6	29	12

[a] I: Co(III)P-Im/Fe(II)P-VPRo (P-Im = VIm-co-VPRo, VBDA-co-VIm); II: Co(III)P-VPRo/Fe(II)EDTA; III: Co(III)N-EtIm/Fe(II)EDTA (N-EtIm = N-ethylimidazole).

aqueous solutions.[26] For Fe(II)P-SS, Fe(II)P-VPRo, and their low molecular weight analogue Fe(II)BDA (BDA = benzylamine-N,N-diacetate), the Co(III) complex moieties on the polymer chains could not be reduced completely or a part of the Co(III) is reduced too slowly to determine the rate. There are two kinds of Co(III) species on the Co(NH$_3$)$_5$PAA polymer chains: One is a reactive species and the other is inert. This may be ascribed to the conformational effect of the polymer chain: The reactive species is probably in close contact with the aqueous phase, whereas the inert species is protected by the polymer chains from attack of the ferrous chelate from the bulk solution. The fraction of the reactive Co(III) species defined by the ratio of the concentration of the reactive species to that of the total Co(III), which is abbreviated as R_{co}, depends on the degree of coordination X. The dependence of the rate constant k and R_{co} on degree of coordination is shown in Fig. 5-17.

Detailed discussion on the effect of the degree of coordination X on the rate is, however, difficult because the accuracy of the rate constants is not very good. Nevertheless, it is clear that the rate constants of the Fe(II)P-SS and Fe(II)P-VPRo reductions are not very different from those of the corresponding low molecular weight system, indicating that there is no appreciable steric hindrance due to the polymer chains of the polymer-bound ferrous chelates on the rate. Furthermore, it is noteworthy that no appreciable effect of the coulombic repulsion force between the polyanion chains is observed for the Fe(II)P-SS reduction. For the Fe(II)P-SS/Co(NH$_3$)$_5$PAA system, the polyanionic domains of Co(NH$_3$)$_5$PAA and Fe(II)P-SS are almost shielded by the counterions (Na$^+$). Therefore, the repulsive force between the two polyanions is not so strong that no appreciable contribution of such long range electrostatic repulsion force on the free energy for the formation of the activated complex would be very small.

Figure 5-17 indicates that the value of R_{co} increases with increasing degree of coordination. Under the conditions employed, about 40% of the deprotonated carboxylic acid groups bound to the Co(NH$_3$)$_5$PAA ($X = 0.010$) chains are presented in the reaction mixture. It seems likely that a large number of carboxylate anions and nonprotonated carboxylic acid groups are necessary to shield a Co(III) center from attack by Fe(II)L (L = P-SS, P-VPRo, and BDA) from the aqueous phase. Concentra-

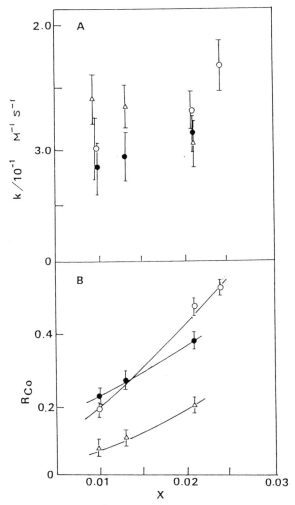

Figure 5-17 ■ Dependence of (A) the rate constant k and (B) R_{co} on degree of coordination X for the reduction of $Co(NH_3)_5PAA$ by $Fe(II)EDTA$ at pH 6.5 and 25°C. [From Y. Kurimura, *3rd International Conference on Macromolecule-Metal Complexes, Rider College, Lawrenceville, NJ, 1989,* preprint.]

tion of the shielded Co(III) complex moiety, that is, the value of R_{co}, decreases with an increase in the degree of coordination.

Ru(bpy)$_3^{2+}$ photosensitized reduction of methyl viologen (MV^{2+}) occurs in the networks of chelate resin beads[27] and chelate filter paper[28]; that of a Co(III)-Schiff base complex, $Co(dop)(OH_2)_2^{2+}$ (dop = 1 − diacetylmonoximeimino-3-diacetyl-monoximatoiminopropane monoanion), proceeds in gelatin hydrogel.[29] In the net-

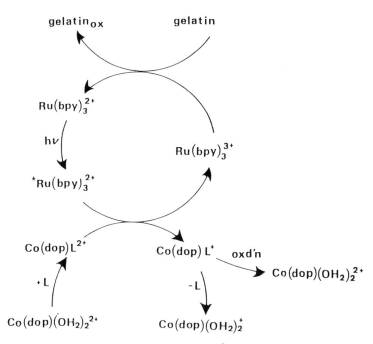

Figure 5-18 ■ Schematic representation of the $Ru(bpy)_3^{2+}$ photosensitized charge separation in a gelatin hydrogel matrix containing $Ru(bpy)_3^{2+}$ and $Co(dop)(OH_2)_2^{2+}$. L; $(OH_2)R$ or R_2 (R = amino acid residue on gelatin); gelatin$_{ox}$; oxidized species of gelatin. [From Y. Kurimura et al., *J. Chem. Soc. Faraday Trans. 1*, **86**, xxx (1990).]

works of such media, soft and multiple interactions and geometrical configuration of the reactive species are of considerable significance for effective charge separation.

For example, on irradiation of gelatin hydrogel containing $Ru(bpy)_3^{2+}$ and $Co(dop)(OH_2)^{2+}$, the reduced intermediate species of the Co(III) having its absorption maximum at 450 nm, which coordinates to an amino acid residue of a gelatin molecule at the axial site of the Co(II) center (R_{450}), is rapidly generated under anaerobic conditions. After the irradiation stops, this intermediate species turns slowly to the diaqua Co(II)-Schiff base complex, $Co(dop)(OH_2)_2^{2+}$ (R_{507}). Therefore, visible-light–induced charge separation is accomplished as a prototype model for solar energy conversion. In this case, a gelatin molecule acts as the ligand coordinated to the Co(III) and Co(II) centers and the electron donor as well as a medium of the photoinduced reaction as shown in Fig. 5-18.

Electron transfer in biological systems takes place through the mediation of a number of proteins, which contain a variety of active sites such as heme, Fe-S, and flavin. These active sites are protected from the solvent by a hydrophobic environment created by the protein chain.[30, 31]

Characteristics of the electron transfer in biological systems are long-distance and directional electron mediation and regulation of the rate. The direction from the

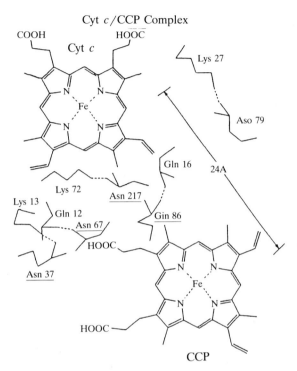

Figure 5-19 ■ A model of the complex formed between cyt c and ccp based on the model of Poulos.[33, 34, 47]

source of electron to sink can be understood by the scheme: electron source → way in → trap → way out → sink.[32]

Recently, direct measurements of long-distance electron-transfer reactions were carried out using several types of modified metalloproteins, such as $Ru(NH_3)_5^{3+}$ labeled metalloproteins. Flash photolysis reduction of the His-83 bound Ru(II) of blue copper (azurin) followed by electron transfer from the Ru(II) to Cu^{2+} is observed. The result shows that intramolecular long-distance (ca. 10 Å) electron transfer takes place rapidly.[37, 38]

Electron transfer between metalloproteins and those between metalloproteins and other molecules, such as inorganic complexes are, in most cases, known to occur through association complexes or electron-transfer complexes in which soft interactions exist between the particular sites of the proteins and those of their partners. For the electron-transfer reaction between metalloproteins, prior to the electron transfer, both reacting proteins bind to form an electron-transfer complex, 42–46 which is stabilized by both electrostatic and/or hydrophobic interaction (Fig. 5-19).

For example, the binding constant of cyt c/ccp (cyt c = cytochrome c and ccp = cyt c peroxidase) is 10^7 M^{-1} at $I = 1$ mM. At low ionic strength, the rate constant for the Fe(II)(cyt c)/Fe(IV)(ccp) reaction is 150 s^{-1}.[35]

Table 5-9 ▪ **Rate Constants for the Intramolecular and Intermolecular Electron-Transfer Reaction**

Substance	Redox Couple	d^a (Å)	k (s^{-1})	Ref.
LRu(His-83)azurin(Cu^{2+})^b	Ru(II)-Cu^{2+}	ca. 10	1.9	36, 37
LRu(His-48)Mb(Fe^{3+})^b	Ru(II)-Fe^{3+}	-13^c	0.019	38
LRu(His-33)cyt c(Fe^{3+})^b	Ru(II)-Fe^{3+}	12–16	82	39
cyt c(Fe^{3+})/ccp(Fe^{2+})	Fe^{2+}-Fe^{3+}	24	0.23	41, 48
α(FeH$_2$O)β(Zn)	^{3}ZnP-Fe(III)	20	10^2	40
	Fe(II)-ZnP$^+$	20	10^3	40
cyt c/cyt b$_5$	Fe(II)/Fe(III)	8	10^5	47

^{a}Distance between redox centers.
^{b}L = (NH$_3$)$_5$.
^{c}Distance from the His-48 imidazole ring to the edge of the heme.

Rate constants for the intermolecular electron transfer within the associated complexes formed from the reacting modified metalloproteins are summarized in Table 5-9 where abbreviations of the modified metalloproteins are: LRu(His-83)azurin(Cu^{2+}), Ru(NH$_3$)$_5$ modified azurin His-33; LRu(His-48)Mb(Fe^{3+}), Ru(NH$_3$)$_5$ modified myoglobin His-48; LRu(His-33)cyt c(Fe^{3+}), Ru(NH$_3$)$_5$ modified cytochrome c; cyt c(Fe^{3+}), cytochrome c; ccp(Fe(IV)), cytochrome c peroxidase.

In this chapter, several examples of multiple complexation and soft interaction in macromolecule-metal complexes have been illustrated. However, the chemistry of macromolecule-metal complexes is a novel field connecting the chemistries of polymers and metal complexes and metalloproteins and inorganic metal complexes. Therefore, basic research on the effect of the polymer chains on the reactivities of macromolecule-metal complexes is necessary to obtain more knowledge about the chemistry of macromolecule-metal complexes as well as of macromolecules, metal complexes, and metalloproteins.

References

1. D. H. Gold and H. P. Gregor, *J. Phys. Chem.*, **64**, 1468 (1960).
2. N. Nishikawa and E. Tsuchida, *J. Phys. Chem.*, **79**, 2027 (1975).
3. T. Ohkubo, K. Hongo, and A. Enokida, *J. Chem. Soc. Faraday Trans. 1*, **80**, 2087 (1984).
4. Y. Kurimura and K. Takato, *J. Chem. Soc. Faraday Trans 1*, **84**, 841 (1988).
5. H. Morawetz, *Macromolecules in Solution*, Wiley, New York (1965).
6. R. Burbucci, M. Casolaro, V. Barone, P. Ferruti, and M. Tramontini, *Macromolecules*, **16**, 1159 (1983).
7. Y. Kurimura, Y. Takagi, and M. Saito, *J. Chem. Soc. Faraday Trans. 1*, **84**, 1025 (1988).
8. F. Basolo and R. G. Pearson, *Mechanism of Inorganic Reactions in Solution*, 2nd ed., Wiley, New York (1967).
9. K. Nitta and S. Sugai, *Ann. Rep. Res. Group Biophys. Japan*, **5**, 83 (1965); P. Doty, A. Wada, J. T. Yong, and E. R. Blout, *J. Polym. Sci.*, **23**, 851 (1957).
10. H. Yokoi, S. Kuwana, and M. Iwaizumi, *J. Am. Chem. Soc.*, **108**, 3358 (1986).

11. H-Y. Mai and J. K. Barton, *J. Am. Chem. Soc.*, **108**, 7414 (1986).
12. Y. Kurimura, *Kobunshi Kako*, **30**, 37 (1987).
13. S. Starzak and M. Cohen, *Biopolymers*, **23**, 847 (1984).
14. B. K. Gallagher, *J. Chem. Phys.*, **41**, 3061 (1964).
15. I. Nagata and Y. Okamoto, *Macromolecules*, **77**, 773 (1977).
16. N. Yoshino, S. Paoletti, J. Kido, and Y. Okamoto, *Macromolecules*, **16**, 749 (1983).
17. V. Crescenzi, H. G. Briten, N. Yoshino, and Y. Okamoto, *J. Polym. Sci.*, *Phys. Ed.*, **23**, 437 (1985).
18. Y. Okamoto, J. Kido, H. G. Britain, and S. Paoletti, *J. Macromol. Sci. Chem.*, **A25**, 1385 (1988).
19. D. Y. Chen and J. K. Thomas, *J. Phys. Chem.*, **89**, 4065 (1985).
20. H-Y. Mai and J. K. Barton, *J. Am. Chem. Soc.*, **108**, 7414 (1986).
21. C. V. Kumar, J. K. Barton, and N. J. Turro, *J. Am. Chem. Soc.*, **107**, 5518 (1985).
22. P. G. Schultz and P. B. Dervan, *J. Am. Chem. Soc.*, **105**, 7748 (1983).
23. Y. Kurimura, K. Takato, and N. Suzuki, *Bull. Chem. Soc. Jpn.*, **59**, 2356 (1986).
24. Y. Kurimura, T. Kikuchi, and E. Tsuchida, *Macromolecules*, **22**, 1645 (1989).
25. Y. Takano, H. Ohno, and E. Tsuchida, *Waseda Daigaku Rikougaku Kenkyusho Houkoku*, **117**, 62 (1987).
26. Y. Kurimura, *3rd International Conference on Macromolecule-Metal Complexes, Rider College, Lawrenceville, NJ, 1989*, preprint.
27. Y. Kurimura, M. Nagashima, K. Takato, E. Tsuchida, M. Kaneko, and A. Yamada, *J. Phys. Chem.*, **86**, 2432 (1982).
28. Y. Kurimura, N. Matsuo, E. Kokuta, Y. Takagi, and Y. Usui, *J. Chem. Res. (Suppl.)*, 238 (1986).
29. Y. Kurimura, K. Hiraizumi, T. Harakawa, M. Yamashita, Y. Osada, K. Shigehara, and A. Yamada, *J. Chem. Soc. Faraday Trans. 1*, **86**, 609 (1990).
30. S. S. Isied and A. Vassilian, *J. Am. Chem. Soc.*, **106**, 1726 (1984).
31. Y. Kurimura, *Advances in Polymer Science*, Vol. **90**, Springer, Berlin (1989).
32. G. R. Moore and R. J. P. Williams, *Coord. Chem. Rev.*, **18**, 125 (1976).
33. T. Poulos and B. Pept, *Protein Rev.*, **4**, 115 (1984).
34. M. Mauk and A. G. Mauk, *Biochemistry*, **21**, 1843 (1982).
35. G. McLendon, *Acc. Chem. Res.*, **21**, 160 (1988).
36. R. Margalit, N. M. Kostic, C-M. Cho, D. F. Blair, H-J. Chiang, I. Pecht, J. R. Shelton, W. A. Schroeder, and H. B. Gray, *Proc. Nat. Acad. Sci. USA*, **105**, 7765 (1983).
37. N. M. Kostic, R. Margalit, C-M. Cho, and H. B. Gray, *J. Am. Chem. Soc.*, **105**, 7765 (1983).
38. R. J. Crutchley, W. R. Ellis, Jr., and H. B. Gray, *J. Am. chem. Soc.*, **107**, 5002 (1985).
39. S. S. Isied, G. Worosila, and S. J. Atherton, *J. Am. Chem. Soc.*, **104**, 7659 (1982).
40. S. E. Peterson-Kennedy, J. L. McGourty, and B. M. Hoffman, *J. Am. Chem. Soc.*, **106**, 5010 (1984).
41. E. Cheung, K. Taylar, J. A. Kornlatt, A. M. English, G. McLendon, and J. R. Miller, *Proc. Nat. Acad. Sci. USA*, **83**, 1330 (1986).
42. H. Hirai, S. Hara, and M. Komiyama, *Bull. Chem. Soc. Jpn.*, **59**, 1051 (1986).
43. Y. Kurimura, Y. Takagi, and E. Tsuchida, *Macromol. Sci.-Chem.*, **A24**, 419 (1987).
44. Y. Kurimura, F. Ohta, N. Ohtsuka, and E. Tsuchida, *Chem. Lett.*, **1987**, 1787 (1987).
45. E. Tsuchida, E. Hasegawa, Y. Matsushima, K. Eshima, M. Yuasa, and H. Nishide, *Chem. Lett.*, **1985**, 969 (1985); E. Tsuchida, *Macromol. Chem. Suppl.*, **12**, 239 (1985).
46. H. Hirai, N. Toshima, and H. Asanuma, *Chem. Lett.*, **1985**, 655 (1985).
47. G. McLendon, K. Simole, and A. G. Mauk, *J. Am. Chem. Soc.*, **106**, 5012 (1984).
48. G. McLendon and J. R. Miller, *J. Am. Chem. Soc.*, **107**, 7811 (1985).

Multiple Interaction in Molecule Transport through Macromolecular Complexes

Hiroyuki Nishide and Eishun Tsuchida

Department of Polymer Chemistry, Waseda University, Tokyo 169, Japan

6-1. Introduction

One of the characteristic chemical functions of metal complexes is specific and reversible coordination of gaseous molecules. A typical example is the efficient oxygen transport by hemoglobin, which is a conjugated protein that has an iron-porphyrin complex as the oxygen-coordinating site. Hemoglobin uptakes oxygen rapidly and selectively at the lungs and efficiently releases oxygen in response to partial oxygen pressure at terminal tissue. 100 ml of human blood absorbs 23 ml of oxygen when exposed to an air atmosphere. This absorption rate is about 60 times the volume physically dissolved in water. The extraordinarily high capacity and efficiency in oxygen transport by hemoglobin are attributed to the 1:1 coordination of molecular oxygen to the iron-porphyrin complex combined with macromolecular protein.

The metal complexes bound to synthetic polymer matrices often show specific interactions in coordination reactions of small molecules and ions, because the reactions are affected by the polymers that exist outside the coordination sphere and surround the metal complex moieties.[1] Immobilizing, orienting, and coordinating the structure of the complex moieties and physical properties of the polymer matrices strongly reflect on the kinetic and equilibrium profile of the molecule-coordination reactions, which are multiplied or enhanced by the integrated structure of the macromolecular complexes.

It has been shown that synthetic and macromolecular metalloporphyrin complexes can specifically, rapidly, and reversibly bind molecular oxygen from air.[2] For solid-state membranes of these macromolecular complexes, oxygen-transport across the membrane is expected to be strongly affected by the oxygen-coordinating reactivity of metalloporphyrin complex, concentration and orienting degree of the complex moiety,

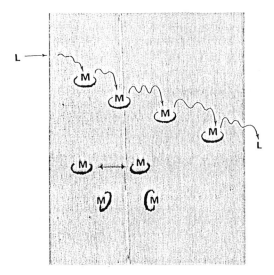

Figure 6-1 ▪ Scheme for oxygen coordination and transport through the membrane of macromolecular metalloporphyrin complexes.

and solid properties of the polymer matrix, as illustrated in Fig. 6-1. In this chapter these specific and multiple interactions observed for the macromolecular complexes are described, by discussing experimentally and theoretically the coordination reaction of simple gaseous molecules, such as oxygen and nitrogen, rapid coordination reactions of solid-state complexes, diffusion of small molecules in polymer matrices, and molecule-transport phenomena via the coordination reactions. The multiple interactions will be characterized as specifically observed dynamic interactions in the macromolecular complexes.

6-2. Oxygen Coordination to Macromolecular Cobalt Complexes in the Solid State

One of the major difficulties encountered in attempts to prepare 1:1 oxygen–metal coordinated complexes that can reversibly bind and dissociate molecular oxygen is the strong driving force toward the irreversible formation of the stable μ-oxo M(III)dimer [M(III) = higher valence metal ion] as represented by

$$M(II) + O_2 \longrightarrow M\text{-}O_2 \longrightarrow M\text{-}O_2\text{-}M \longrightarrow M(III)\text{-}O\text{-}M(III) \qquad (6\text{-}1)$$

The oxygen-coordinated complex rapidly reacts with another deoxy M(II) complex [M(II) = lower valence metal ion] and the binuclear dioxygen-bridged complex is formed. This binuclear complex is irreversibly oxidized to produce the oxo-bridged

M(III) dimer. In other words, the first problem of reversible oxygen coordination is how to inhibit the dimerization.[2]

Much work has been aimed at overcoming this problem, and two approaches have been partially successful.[3] The first is an elegant steric approach: Planar ligands of the complexes have been substituted in a fashion that inhibits dimerization. The second approach is to attach the complexes to a rigid polymer chain so as to prevent two complexes from approaching each other closely enough to lead to dimerization. By these means a reversible oxygen coordination to metalloporphyrins and cobalt Schiff base complexes has been achieved under limited conditions.

Immobilization of the complex in a solid-state polymer considerably inhibited its irreversible dimerization and prolonged the lifetime of the complex for reversible oxygen coordination.[4, 5] However, the rate of oxygen coordination to these solid-state macromolecular complexes was much lower than with the complexes in homogeneous solutions, because ligand-exchange reactions of the metal complexes in the solid state occur very slowly. By combining the preceding two approaches, that is, polymer attachment and chemical modification of the complex, we recently succeeded in preparing dry or solvent-free solid membranes of macromolecular metalloporphyrin complexes that had the sixth coordination site vacant even in the solid state and were, therefore, able to bind oxygen very rapidly.

Macromolecular complexes with oxygen-coordinating ability are shown in Schemes 1–6. A series of cobalt- and iron-porphyrin derivatives, tetrakis(alkylamido-phenyl)porphinatometals (1–5), were synthesized as the oxygen-coordinating site to be fixed in the polymer membranes.[6–9] Their stereostructure was confirmed by NMR and other types of molecular spectroscopy. The polymer-combined Schiff-base chelate, salicylaldehydeethylenediimine cobalt (6), was also an effective oxygen-coordinating site and is discussed in comparison with the results for the metalloporphyrin complexes.[10–12] These modified cobalt-porphyrin complexes were successfully fixed in polymer matrices with their oxygen-coordinating ability by physically dispersing (1), coordinate bonding (2, 4, 5), and covalent bonding (3), because the complexes have a steric pocket-like structure on the porphyrin plane to keep their oxygen-coordinating site vacant before oxygen coordination even in solid or solvent-free states. This stereostructure of the porphyrin complex is one of the key points in preparing an active macromolecular complex in the solid state. For example, a solution of 2 was carefully cast under an absolutely oxygen-free atmosphere, followed by drying in vacuo, to yield a transparent wine-red colored membrane.[7]

Selective and reversible oxygen-coordination to the fixed complexes was first confirmed with microgravimetric measurement (Fig. 6-2). For example, the 2 membrane containing 20% complex moiety sorbed ca. 0.7 ml oxygen per gram, which is more than 70 times larger than that of physically dissolved nitrogen.[7] This extraordinarily large amount of dissolved oxygen in the polymer is based on the specific and reversible oxygen coordination to the cobalt-porphyrin moiety in the polymer. Figure 6-2 also indicates that the oxygen sorption is in response to an atmospheric oxygen pressure, and according to Langmuir's isotherm, leads to the description in Section 6-5, where the contribution of the complex moiety to an oxygen-transport phenomenon is represented as the Langmuir term.

Oxygen coordination to the cobalt-porphyrin complex fixed in the membrane was also confirmed by IR spectroscopy.[7] With an increase of partial oxygen pressure the

1

2

3

4

5

6

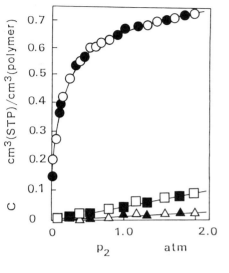

Figure 6-2 ▪ Oxygen sorption to the membrane of the macromolecular cobalt-porphyrin complex (**2**) (microgravimetric measurement). p_2, atmospheric pressure; C, sorption amount; ○, oxygen sorption; □, oxygen sorption only into the polymer matrix; △, nitrogen sorption; closed plots, repeated data.

polymer membrane shows strong IR absorption at 1150 cm^{-1} for $^{16}O_2$ and 1060 cm^{-1} for $^{18}O_2$ (Fig. 6-3), which are attributed to an end-on type coordination of dioxygen to the metal ion. Because concentration of the metal complex moiety in the polymer domain of the solid-state membrane is much higher (> 50 mmol l^{-1}) than that of a homogeneously solubilized complex in solution (1 mmol l^{-1}), even a simple spectrophotometer is effective for observing coordination reactions and ligand-bonding characteristics of complexes.

Exposure of the cobalt-porphyrin membrane to an oxygen atmosphere causes the color of the membrane to change from dark red like venous blood to brilliant red like

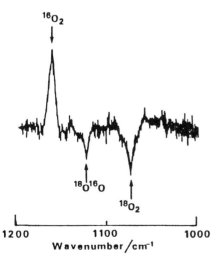

Figure 6-3 ▪ Differential IR absorption of the membrane of the macromolecular cobalt-porphyrin complex (**2**) under oxygen atmosphere.

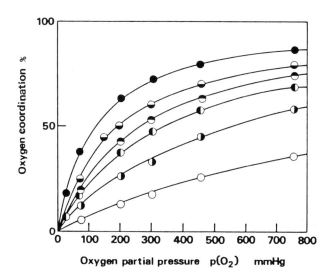

Figure 6-4 ■ Oxygen-coordination equilibrium curves for the cobalt-porphyrin complexes fixed in the membrane (**1**) (pK_a of the axial ligand). ●, 1-methylimidazole; ◓, dimethylaminopyridine(9.5); ◑, aminopyridine(9.0); ◐, 1,2-dimethylimidazole(6.8); ◑, pyridine(5.2); ○, cyanopyridine(1.8).

arterial blood. This color change is attributed to the oxygen coordination and can be monitored easily with visible absorption spectrometry. This oxy–deoxy spectral change was reversible in response to partial oxygen pressure with isosbestic points. The visible absorption spectra for the various cobalt-porphyrin complexes agreed with those for the corresponding complex in toulene solution. This result also means that the cobalt complex acts like an effective oxygen coordination site from the equilibrium viewpoint, even after fixing in the solid-state membrane.

The oxygen-coordination equilibrium constant (K),

$$CoP + O_2 \underset{k_{off}}{\overset{k_{on}}{\rightleftharpoons}} CoP\text{-}O_2 \qquad k = k_{on}/k_{off} \tag{6-2}$$

was determined[13] from the oxygen-coordination equilibrium curves (Fig. 6-4). Figure 6-4 shows that the oxygen-coordinating affinity of the cobalt-porphyrin complex can be controlled by variation of complexed ligand species and the porphyrin structure. For example, a linear relationship between K and the basicity (pK_a) of the axial ligand is noticed in Fig. 6-4 (the same relationship has been summarized for cobalt porphyrin complexed with various nitrogenous ligands in toulene solution[14]). Study on the coordination behavior of the complex fixed in the membrane is an effective procedure for estimating coordination parameters of complexes.

Photodissociation and recombination of the coordinated oxygen from and to the cobalt-porphyrin complex (CoP) in the membrane was successfully observed by pulse and laser flash spectroscopy.[6] It is known that the metalloporphyrin-coordinated oxygen is photodissociated under a flash irradiation,

$$CoP\text{-}O_2 \xrightarrow{h\nu} CoP + O_2 \longrightarrow CoP\text{-}O_2 \tag{6-3}$$

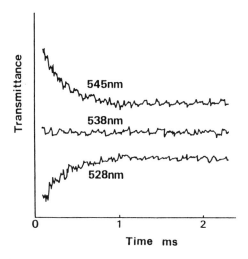

Figure 6-5 ■ Flash photolysis of coordinated oxygen to the cobalt-porphyrin complex (4) fixed in the poly(butylmethacrylate) membrane under exposure to air.

and that the rapid oxygen-coordinated process can be analyzed.[15] Improving the laser flash spectroscopic technique allowed flash photolysis to be applied to be solid-state complex in the membrane. The laser flash was applied perpendicularly to the light path of the spectrophotometer, and the membrane was placed at the intersection of the laser flash and the light path and at 45° to both. The rapid absorption change was recorded with a contact-type photomultiplier to cancel the noise caused by scattered light.

An example[9] of the recombination time curve of oxygen is shown in Fig. 6-5. Changing the monitoring wavelength in the visible region allowed measurement of a differential spectrum before and after flash photolysis. The positive and negative extremes in the differential spectrum, 528 and 545 nm, were selected as the monitoring wavelengths as shown in Fig. 6-5. These wavelengths agreed with the absorption maxima of the oxygen-coordinated and the deoxy complexes. The validity of this measurement is supported further by the following results: Absorbance of 538 nm is constant before and after the flash photolysis. This wavelength agrees with the isosbestic point of the oxygen coordination. The absorption changes followed at 528 and 545 nm are symmetric (Fig. 6-5).

The k_{on} and k_{off} [Eq. (6-2)] were estimated[6-9] by pseudo-first-order kinetics and are given in Table 6-1. The k_{on} and k_{off} values of the complex in the membrane were a little faster than or similar to those of the complex in toulene solution. This means that the cobalt-porphyrin complex is kinetically active for oxygen coordination even in the membrane. The equilibrium constant K as well as the kinetic constant k is much affected or controlled by the structure of the complex in Table 6-1. The details will be discussed in Section 6-6.

Table 6-1 ▪ Oxygen-Coordination Rate and Equilibrium Constants (25°C) in the
Solid-State Membranes of Macromolecular Cobalt-Porphyrin Complexes

Complex Moiety	$10^{-2}K$ (M^{-1})	$10^{-5}k_{on}$ $(M^{-1} s^{-1})$	$10^{-2}k_{off}$ (s^{-1})
1	5.7	1.6	2.8
1[a]	10	1.2	1.2
2	19	5.2	2.8
3	25	13	5.2
4	31	12	3.9
5	28	39	14

[a] Toluene solution.

The advantages of the membrane of solid-state macromolecular complexes are that the fixed complex maintains its rapid and reversible penetrant coordinating capability even after fixation in the membrane and that the coordination rate and equilibrium constants of the penetrant at the complex moiety can be evaluated by spectroscopy in situ.

6-3. Effects of Polymer Matrices on Oxygen Coordination to a Cobalt Complex

The complex moiety fixed in polymers is an effective probe for estimating the physical properties of the polymers. A membrane sample was stuck on the cell wall of a spectrophotometer and a gaseous mixture of oxygen and nitrogen was introduced into the cell through gas inlet and outlet tubes with a standing pressure such that the total pressure in the cell was 1 atm. The apparent oxygen-coordination equilibrium constant (K_{app} per centimeter of mercury) of the cobalt porphyrin complex (1) fixed in various polymers was measured by changing atmospheric oxygen partial pressure (centimeters of mercury) and is given in Table 6-2. The K_{app} values for the poly(dimethylsiloxane) and poly(phosphazene) membranes are larger than those for the methacrylate polymers. A large amount of oxygen dissolves in the rubbery poly(dimethylsiloxane) and poly(phosphazene) matrix (glass transition temperature $T_g < 0°C$) with a larger oxygen solubility coefficient k_D: This brings about the larger K_{app}. That is, K_{app} is a good indication of oxygen solubility or free volume of polymer matrices, which is quantitatively supported by the following calculation.[16]

Apparent thermodynamic parameters for the oxygen coordination were estimated from the temperature dependence of K_{app} and are also given in Table 6-2. These apparent thermodynamic parameters for the oxygen-coordination equilibrium can be corrected by including those for oxygen dissolution in the polymer matrix, as

$$K = [\text{CoP-O}_2]/[\text{CoP}][\text{O}_2] \qquad (6\text{-}4)$$

where $[O_2] = k_D p(O_2)$ is the oxygen concentration around the complex moiety and $p(O_2)$ is atmospheric oxygen pressure to which the membrane is exposed. $K' =$

Table 6-2 ■ Oxygen Solubility Parameters, Oxygen-Coordination EquilibriumConstants, and Thermodynamic Parameters of the Membranes of Macromolecular Complexes (1) (25°C)

Polymer Matrix[a]	T_g (°C)	k_D $\left(\dfrac{cm^3(STP)}{cm^3\,cmHg}\right)$	ΔH_d $\left(\dfrac{kcal}{mol}\right)$	K_{app} $\dfrac{10^2}{\left(cmHg\right)}$	K' $\left(\dfrac{10^2\,cm^3}{cm^3(STP)}\right)$	ΔH_{app} $\left(\dfrac{kcal}{mol}\right)$	ΔH $\left(\dfrac{kcal}{mol}\right)$	ΔS_{app} (eu)	ΔS (eu)
PDMS	−120	3.8×10^{-3}	−2.8	38	1.0	−16	−13	−55	−35
PEPZ	−67	2.0×10^{-3}	—	25	—	−13	—	−37	—
PBMA	20	9.6×10^{-4}	−1.2	5.8	0.51	−14	−13	−54	−36
PMMA	105	5.9×10^{-4}	−1.2	5.2	—	−13	—	−49	—

[a]PDMS = poly(dimethylsiloxane), PEPZ = poly(ethoxyphosphazene), PBMA = poly(bumtylethacrylate), PMMA = poly(methylmethacrylate).

K_{app}/k_D can be written. Accordingly, the thermodynamic parameters from van't Hoff plots of K_{app} are just affected by the k_D value and its heat of dissolution (ΔH_d):

$$\Delta H = \Delta H_{app} - \Delta H_d \qquad (6\text{-}5)$$

Here, ΔH is the enthalpy change for oxygen coordination in which the effect of the polymer matrix is reduced. Similar reduction was available also for ΔS and is given in Table 6-2. Both the values of ΔH and ΔS in the poly(dimmethylsiloxane) membrane are equal to those in the poly(butylmethacrylate) membrane. Although the apparent equilibrium and thermodynamic parameters of the oxygen coordination of the complex fixed in polymer membranes are found to be significantly affected by the oxygen solubility in the polymer matrices, the parameters corresponding to the oxygen coordination to the complex itself are not essentially dependent on the species composing the polymer matrix.

The reversible oxygen coordination occurred very rapidly. For example, for a 50 μm thick poly(laurylmethacrylate) membrane the coordination equilibrium was established within a minute after exposure to oxygen or nitrogen. The time course of the degree of oxygen coordination to the complex in the membrane (ϕ) was easily measured by spectroscopy after exposure of the membrane to a certain oxygen pressure (p_2), which gave apparent oxygen-coordination rate constants (k_{on}^{app} and k_{off}^{app}) according to pseudo-first-order kinetics[17] (Table 6-3). The apparent rate constants in Table 6-3 are greater for the membrane composed of the methacrylate copolymer with a longer alkyl side chain, which shows lower T_g. Oxygen diffusivity in the polymer increases with decreasing T_g of the polymer, which clearly is reflected in the apparent oxygen-coordination rate constants to the complex moiety fixed in the polymer membrane.

Combination of the ϕ value and the oxygen-coordination equilibrium curve (e.g., Schemes 1–6) yields the oxygen pressure, which gives the ϕ value. The product of this oxygen pressure and k_D gives the amount of physically dissolved oxygen. The total absorbed amount of oxygen is the sum of the physically dissolved oxygen and the coordinated oxygen. This treatment is based on the assumption that the equilibrium between the physical absorption mode and the chemically coordinated mode is attained fairly rapidly. Flash photolysis measurements of the oxygen-coordination rate to and from the complex in the membrane mentioned in Section 6-2 supports the

Table 6-3 ▪ Apparent Oxygen-Coordination Rate Constants for Poly(alkylmethacrylate-co-vinylimidazole)-Cobalt-Porphyrin(2) Membranes

Alkyl Group of the Copolymer	T_g (°C)	$10^5 k_{on}^{app}$ $(M^{-1} s^{-1})$	$10^3 k_{off}^{app}$ (s^{-1})	$10^7 D_D$ $(cm^2 s^{-1})$	$10^9 D_C$ $(cm^2 s^{-1})$
Methyl	120	1.6	1.5	1.1	2.8
Octyl-	15	4.7	2.5	7.0	7.0
Lauryl-	−5	16	8.0	12	12
Stearyl-	−20	46	23	17	27

validity of this assumption. Absorption kinetic runs monitored spectroscopically fitted the Fickian diffusion model. Based on the Fickian diffusion equations, oxygen diffusion coefficients[18] through the polymer matrix (D_D) and via the fixed complex moieties (D_C) were calculated, respectively, as given in Table 6-3. The results show that the complex-coordinated oxygen has a 10^2 smaller mobility than that of the physically dissolved oxygen.

The oxygen coordination to the complex moiety fixed in polymer membranes is a convenient probe for estimating dissolution and diffusion of oxygen in polymers and physical properties of polymers.

6-4. Nitrogen Coordination in Macromolecular Manganese Complexes

A macromolecular metal complex to which molecular nitrogen coordinates rapidly and reversibly is expected to adsorb and/or transport nitrogen selectively and to offer the possibility of a material for enriching nitrogen from air. A large number of transition-metal complexes coordinated with molecular nitrogen have been synthesized.[19] However, these studies are aimed at nitrogen fixation through reduction of the nitrogen coordinated to a metal ion, and the nitrogen complexes themselves often undergo degradation under an air atmosphere. Therefore, kinetic and equilibrium profiles of the nitrogen coordination to the metal ion have received surprisingly little attention.

Two effects of a polymer matrix in reducing the degradation of a metal complex have been reviewed[1]: an immobilization effect to inhibit a dimerization of metal complexes and an environmental effect to suppress a redox reaction of metal complexes. Cyclopentadienylcarbonylmanganese and benzenecarbonylchromium complexes were successfully introduced in a polymer chain,[20, 21] which provides a transparent, flexible, and solvent-free membrane. UV irradiation of the membrane in an argon atmosphere, converted them to the corresponding dicarbonyl metals[22] (**7–9**), which have one unsaturated coordination site even in the solid membrane state to coordinate nitrogen rapidly and reversibly.

On exposure of the **7** membrane to a nitrogen atmosphere, the polymer membrane showed a strong IR absorption peaks (ν_{N_2}) at 2160 cm^{-1} for $^{14}N_2$ and 2090 cm^{-1} for $^{15}N_2$, assigned to an end-on-type coordinated dinitrogen.[23] ν_{N_2} for **7–9** are listed in Table 6-4. The intensity of ν_{N_2} reversibly increased/decreased in response to the partial pressure of nitrogen. This ν_{N_2} intensity change occurred very rapidly: For a 10 μm thick **7** membrane containing 6.4 mol% cyclopentadienylmanganese moiety, the nitrogen-coordination equilibrium was established within 1 min after exposure of the membrane to nitrogen (saturation) or in vacuo[23] (degassing).

The intensity of the ν_{N_2} peak in air decreased slowly with time, corresponding to decomposition of the dinitrogen complex. Its degradation obeyed first-order kinetics from which the lifetime (half-life period) of the dinitrogen complex could be calculated,[20, 21] as in Table 6-4. It has been reported that the corresponding cyclopentadienyldicarbonylnitrogenmanganese and benzenedicarbonylnitrogenchromium complexes are decomposed immediately after exposure to open air. The polymer

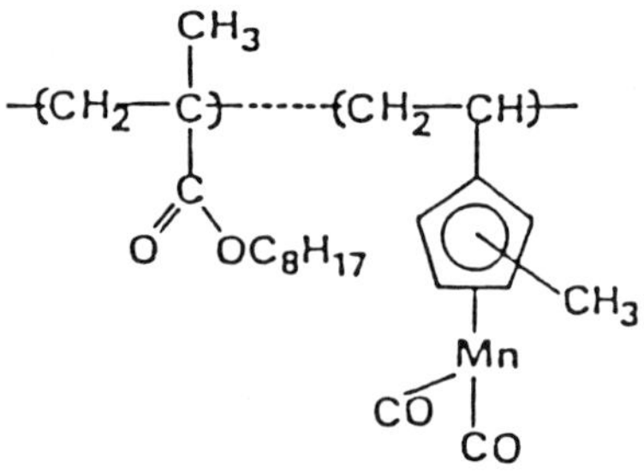

7

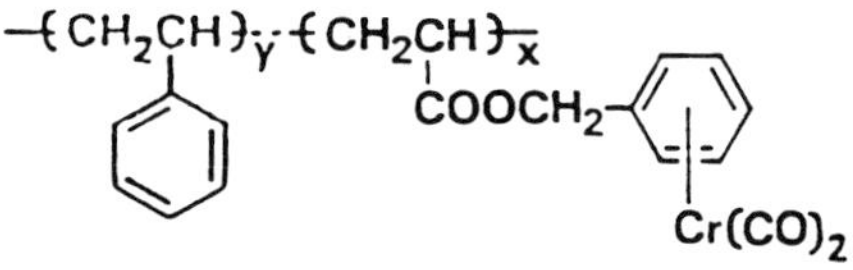

8

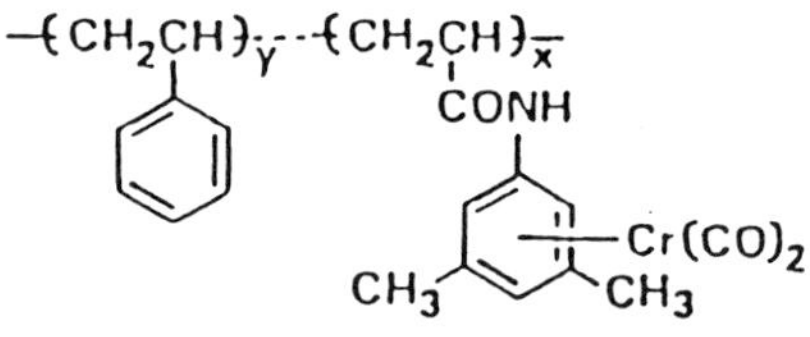

9

Table 6-4 ▪ IR Absorption Peak and Lifetime of the Nitrogen-Coordinated Complexes under Open Air Atmosphere

Complex	ν_{N_2} (cm^{-1})	Lifetime (day)
7	2160	1.8
8	2140	0.38
9	2120	0.88

matrix protects the dinitrogen complex even in air, probably because the polymer matrix retards diffusion of moisture to the complex and fixes the complex to suppress irreversible dimerization.

The nitrogen coordination in response to partial nitrogen pressure was also observed by the color change from brown to orange or the spectral change in the UV-visible absorption, which exhibited good isosbestic behavior. The nitrogen-coordination equilibrium curve was drawn from the spectral change of the membrane. The curve obeyed a typical Langmuir isotherm for which the equilibrium constant ($K = $ [CpMnN$_2$]/[CpMn][N$_2$]) for the nitrogen coordination was obtained. The K (per centimeter of mercury) obtained was converted to K per mode per liter by substituting the solubility coefficient of nitrogen into the membrane determined by the nitrogen absorption measurement and is given in Table 6-5. Thermodynamic parameters for the nitrogen coordination were estimated from the temperature dependence of K and are given also in Table 6-5. Control data for the tetrahydrofuran (THF) coordination to the monomeric and polymeric complexes were measured in the cooled THF solution (Table 6-5): They support the validity of the nitrogen-coordination measurement for the cyclopentadienylmanganese complex moiety in the polymer. The K value for the nitrogen coordination is comparable with the K values for the THF coordination. K and thermodynamic parameters for oxygen coordination to a macromolecular iron-porphyrin complex are listed in Table 6-5 as references. The K value for nitrogen coordination is a little smaller than that for oxygen coordination. Both the enthalpy and the entropy changes for the nitrogen coordination to the cyclopentadienylmanganese complex are more positive than those for the oxygen coordination to the iron-porphyrin complex. In comparison with the oxygen coordination, the smaller enthalpy gain in the nitrogen coordination is compensated with the smaller entropy decrease in the immobilization of nitrogen, which provides the relatively large nitrogen-coordination equilibrium constant.

Rapid and reversible nitrogen coordination to the cyclopentadienylmanganese residue in the membrane was also confirmed by laser flash photolysis of the nitrogen complex. Photodissociation and recombination of the coordinated nitrogen to and from the manganese in the membrane was successfully observed: The reaction was completed within 10 ms. It is a surprisingly rapid reaction. The nitrogen coordination and dissociation rate constants (k_{on} and k_{off}) were estimated by pseudo-first-order

Table 6-5 ■ Ligand-Coordination Rate, Equilibrium Constants, and Thermodynamic Parameters

Complex	Ligand	Physical State	$10^{-2}K$ (M^{-1})	$10^{-5}k_{on}$ ($M^{-1} s^{-1}$)	$10^{-3}k_{off}$ (s^{-1})	ΔH (kcal mol^{-1})	ΔS (eu)
7	THF	THF solution, 0°C	8.3	61	7.4		
7	N$_2$	Membrane, 20°C	9.8	2.9	0.3	11	− 35
7	C$_2$H$_2$	Membrane, 25°C	0.26	0.00024	0.00094	14	− 60
C$_P$Mn(CO)$_2$	N$_2$	Cyclohexane, 22°C	—	3.7	—	—	—
FeP	O$_2$	Toluene, 25°C	23	1060	46	14	− 42

kinetics and are given also in Table 6-5 with kinetic data for the oxygen coordination of the iron porphyrin.

The k_{on} value with a 10^5 M^{-1} s^{-1} order for the nitrogen coordination means that the organometallic nitrogen coordination is also a rapid reaction, although the rate constants are 10^2 times smaller than those for the oxygen coordination. This k_{on} value determined for the macromolecular manganese complex in the membrane state is consistent with the k_{on} value recently reported for the nitrogen-coordinated intermediate of the corresponding monomeric manganese complex (η^5-$C_5H_5)Mn(CO)_2$ with time-resolved spectroscopy.[24] One of the advantages of our system is that kinetic and equilibrium constants of the nitrogen coordination can be evaluated in situ spectroscopically: The spectroscopy of organometallic species in a polymer membrane can extend the information about intermediates beyond that already available from low-temperature and/or time-resolved experiments.

Ultimately, the complex maintains its rapid and reversible coordinating capability for molecular nitrogen even after immobilization in the dry membrane.

This macromolecular manganese complex 7 selectively coordinates with acetylene in the presence of ethane and ethylene. IR absorption at 1740 cm^{-1} was assigned to a side-on type coordination of acetylene to the metal ion. The acetylene coordination in response to acetylene partial pressure also obeyed Langumir's isotherm, giving K for the acetylene coordination as shown in Table 6-5. The K value for the acetylene coordination to the manganese complex is ca. 50 times smaller than K for the nitrogen coordination. Although the larger enthalpy gain (negative ΔH value in Table 6-5) suggests a stronger bond between C_2H_2 and Mn in comparison with the $N_2 - Mn$ bond, much more negative entropy change through the acetylene coordination contributes to the smaller K value. A larger entropy decrease in the coordination step for the bulkier acetylene molecule reduces the coordination equilibrium constant, as compared with the coordination of the smaller nitrogen molecule.

The flash photolysis procedure was also adequate to estimate k_{on} and k_{off} for the acetylene coordination (Table 6-5). The k_{on} value for acetylene is 10^4 times smaller than that for the nitrogen coordination: This means more organometallic character for the acetylene coordination.

6-5. Facilitated Transport of Oxygen and Nitrogen in the Membranes of Macromolecular Complexes

The membranes of macromolecular complexes were set under pressure gradient. Oxygen permeation coefficients for various upstream gas pressures were measured with a low-vacuum permeation apparatus.[25] The pressure on the upstream side was maintained essentially constant and the pressures on the upstream and downstream sides were detected by using an absolute pressure gauge. The permeation coefficient was calculated from the slope of the steady-state straight line of the permeation curve.

The oxygen permeation coefficient (P_{O_2}) for the membrane containing macromolecular cobalt-porphyrin complex 4 is given in Fig 6-6. P_{O_2} is larger than the nitrogen permeation coefficient (P_{N_2}) and enhanced with the decrease in oxygen upstream pressure $[p_2(O_2)]$. On the other hand, P_{N_2} is small and independent of nitrogen upstream pressure $[p_2(N_2)]$ because the fixed complex does not interact with nitrogen. That is, the cobalt-porphyrin complex fixed in the membrane facilitates oxygen transport in the membrane and enhances the oxygen permeation additionally, as represented by the shadowed area in Fig. 6-6.

The time course of the permeation of gaseous molecules through membranes shows an induction period followed by permeation with a constant slope (steady state). The induction period (θ) for the oxygen permeation is longer than θ for

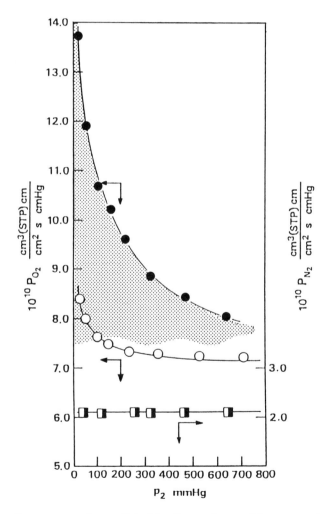

Figure 6-6 ■ Oxygen permeation coefficient for the membrane of the macromolecular cobalt-porphyrin complex (**4**) at 30°C. ●, oxygen (for the membrane containing 4.5 wt% cobalt-porphyrin); ○, oxygen (1.2 wt%); ▯, nitrogen.

nitrogen and prolonged with the decrease in $p_2(O_2)$ (Fig. 6-7). On the other hand, θ for nitrogen permeation is short and independent of $p_2(N_2)$. This behavior indicates that oxygen clearly interacts with the complex moiety in the membrane and its diffusivity in the membrane is reduced by repeatedly coordinating to the complex. In Fig. 6-7 one also notices that θ_{O_2} and the $p_2(O_2)$ dependence of θ_{O_2} decrease with temperature. θ_{O_2} and the $p_2(O_2)$ dependence are based on oxygen coordination to the fixed complex and enhanced at lower temperatures because the oxygen-coordination kinetic constant decreases with decreasing temperature.

These results suggest that a dual-mode-transport theory[26] is mathematically applicable to the oxygen transport in the membrane of a macromolecular complex. Figure 6-8 shows the oxygen permeability in the membrane of a macromolecular complex that is governed by two modes. That is, the oxygen permeation coefficient is equal to the sum of a first term that represents the physical Henry mode permeation and a second term that represents the chemical Langmuir mode permeation. For the Henry mode, oxygen physically dissolves in a polymer membrane according to Henry's law and the dissolved oxygen diffuses physically. For the Langmuir mode, oxygen selectively coordinates to the complex moiety fixed in a membrane and diffuses via the

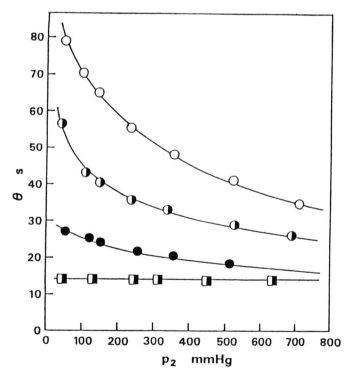

Figure 6-7 ▪ Induction period for the oxygen permeation through the membrane of **2**. Oxygen, at 25 (○), 35 (◐), 45°C (●); nitrogen (◼) at 35°C.

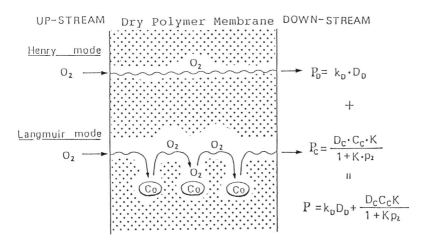

Figure 6-8 ■ Oxygen transport in the membrane of a macromolecular complex.

fixed complex. The oxygen transport is accelerated by this Langmuir mode in addition to the physical Henry mode:

$$P = k_D D_D + D_C C_C K/(1 + K p_2) \tag{6-6}$$

Here, P is the permeability coefficient, k_D is the solubility coefficient for Henry's law, D_D and D_C are the diffusion coefficients for Henry- and Langmuir-type diffusion, respectively, C_C is the saturated amount of oxygen reversibly coordinated to the fixed complex, K is the oxygen-coordination equilibrium constant, and p_2 is the upstream gas pressure. Equation (6-6) is a function of p_2, and P increases with decreasing p_2 (see Fig. 6-6).

The induction period for the membrane of a macromolecular complex is also governed by both the Henry and Langmuir modes:

$$6D_D\theta/l^2 = \left[1 + R\left[f_0(y) + FRf_1(y) + (FR)^2 f_2(y)\right]\right.$$

$$\left. + FRf_3(y) + (FR)^2 f_4(y)\right]\Big/\left[1 + FR/(1+y)\right]^3 \tag{6-7}$$

Here, $F = D_C/D_D$, and $R = KC_C/k_D$, $y = Kp_2$, and l is the thickness of the membrane. θ also depends on p_2. The combination of Eqs. (6-6) and (6-7) gives R/D_D and $1/D_D$, and the permeation parameters D_C, K_D, and C_C can be estimated.

The effect of $p_2(O_2)$ on P_{O_2} (Fig. 6-6) was analyzed by using Eq. (6-6); that is, P_{O_2} was plotted against $1/(1 + Kp_2)$. The plots showed a linear relationship: The oxygen permeability in the complex membrane can be explained in terms of the dual-mode model. The effect of $p_2(O_2)$ on the induction period (Fig. 6-7) was analyzed by using Eq. (6-7), which data also support the dual-mode oxygen transport in the membrane and a pathway of oxygen permeation via the fixed complex.

Table 6-6 ■ Parameters of Oxygen Coordination and Transport for the Membranes of Macromolecular Cobalt-Porphyrin Complex (1) (25°C)[a]

5th Ligand of 1	$10^2 K$ (cmHg^{-1})	k_{off} (s^{-1})	$10^{10} P$ $\left(\dfrac{cm^3 (STP)\ cm}{cm^2\ s\ cmHg} \right)$	$10^8 D_C$ (cm^2 s^{-1})	D_C/D_D
Im	5.6	0.28	12	1.4	0.02
MIm	2.6	0.48	9.2	2.0	0.03
Py	1.4	1.1	10	3.9	0.06
MAPy	3.5	0.21	8.8	0.65	0.01
APy	2.5	0.15	8.5	0.35	0.005
(Fe)Im	12	0.026	4.3	0.71	0.001

[a] $C_C = 0.2$ cm^3 (STP) cm^{-3}, $D_D = 7 \times 10^{-7}$ cm^{2}s^{-1}, $k_D = 1 \times 10^{-3}$ cm^3 (STP) cm^{-3} cmHg^{-1}.
[b] Im = 1-methylimidazole, MIm = 1,2-dimethylimidazole, Py = pyridine, MAPy = dimethylaminopyridine, Apy = aminopyridine, (Fe)Im = iron-porphyrin–imidazole.

Parameters calculated by this procedure are summarized in Table 6-6. For example, the C_C value is independent of the complex species and is not affected by temperature. This also supports the validity of the mathematical treatment by the dual-mode transport model. A rapid oxygen exchange between two transport modes is established, as projected by the kinetic constants determined by flash photolysis spectroscopy on the complex fixed in the membrane.

The permeability ratio (P_{O_2}/P_{N_2}) was 3.2, 5.7, and 12 for the membrane containing 0, 2.5, and 4.5 wt% cobalt-porphyrin complex (4) moiety, respectively. This result indicates the possibility of high permselectivity with a membrane containing a macromolecular complex.

For the macromolecular cyclopentadienylmanganese complex 7, nitrogen transport through the membrane was selectively augmented due to the rapid and reversible coordination of nitrogen to the fixed cyclopentadienylmanganese moiety. The nitrogen permeation coefficient (P_{N_2}) and induction period (θ) for nitrogen permeation are shown in Fig. 6-9. P_{N_2} increases with a decrease in $p_2(N_2)$, whereas oxygen permeability is independent of $p_2(O_2)$. $\theta(N_2)$ also depends on $p_2(N_2)$. These results indicate that the manganese complex in the membrane interacts specifically with nitrogen and not with oxygen. Nitrogen transport through the membrane is facilitated by the manganese complex.

Acetylene also coordinates to the macromolecular manganese complex 7 reversibly, as previously mentioned. The acetylene permeation coefficient ($P_{C_2H_2}$) and the induction period for the acetylene permeation [$\theta(C_2H_2)$] are also plotted in Fig. 6-9. $\theta(C_2H_2)$ is long and increases with a decrease in $p_2(C_2H_2)$, which indicates that acetylene clearly interacts with the manganese complex in the membrane, but $P_{C_2H_2}$ enhancement and $p_2(C_2H_2)$ dependency of $P_{C_2H_2}$ are not observed. The acetylene transport through the membrane is not facilitated, that is, the acetylene is totally immobilized to the manganese moiety in the polymer. In Table 6-6, the acetylene-coordination rate constants, especially the k_{off} value, were much smaller than those of the oxygen and nitrogen coordination. This kinetically inactive coordination in acetylene hardly contributes to acetylene transport in the macromolecular complex.

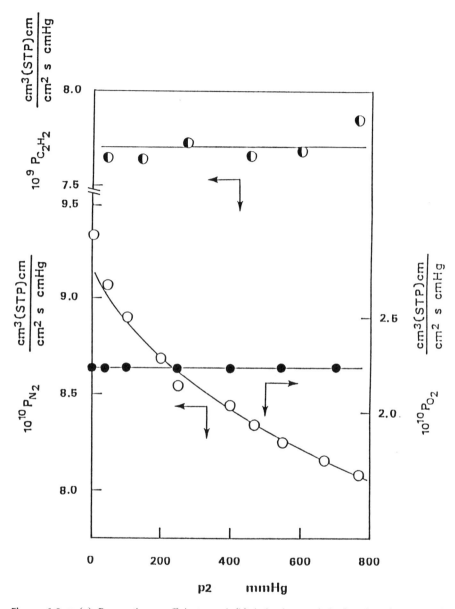

Figure 6-9 ▪ (a) Permeation coefficients and (b) induction periods for the nitrogen and acetylene permeation through the membrane of the macromolecular cyclopentadienylmanganese complex (7). ○, nitrogen; ◐, acetylene; ●, oxygen.

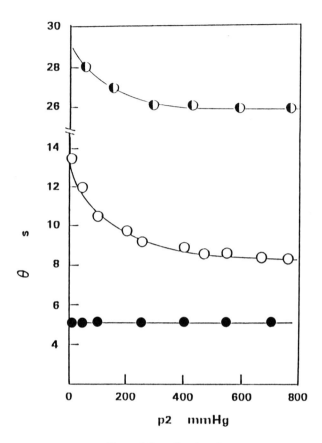

Figure 6-9 ■ Continued

6-6. Multiple Interaction in Facilitated Molecule Transport

Facilitated oxygen transport is much affected by the complex species. For the example of cobalt porphyrin (1) ligated with various nitrogenous ligands, P_{O_2} is relatively large and drastically increases with a decrease in $p_2(O_2)$ for the membrane of the cobalt-porphyrin–methylimidazole complex, whereas it is small and hardly influenced by $p_2(O_2)$ for the membrane of the cyanopyridine complex. The transport parameters are summarized in Table 6-6. The D_D, k_d, and C_C values for the membranes composed of various complexes coincide with each other, which strongly supports the validity of this study or that only the oxygen-coordinating ability of the fixed complex is variable without changing physical properties of the membrane. Equation (6-6) suggests that an increase in the oxygen-coordination equilibrium constant K brings about an enhanced facilitation. In Table 6-6, P_{O_2} is largest for the membrane of the cobalt porphyrin ligated with methylimidazole (Im), which has the largest K value.

The effect of the metal species of the porphyrin complex on facilitated oxygen transport is also given in Table 6-6. Although the oxygen-coordination equilibrium constant of the iron porphyrin is larger than that of the cobalt porphyrin, the facilitation is lower because of its smaller oxygen-coordination kinetic constants. In order to enhance the facilitated transport, the complex has to combine oxygen strongly as well as has to release the coordinated oxygen rapidly.

The oxygen diffusion coefficient via the complex fixed in the polymer matrix (D_C) is plotted against the dissociation rate constant of the coordinated oxygen from the fixed complex in Fig. 6-10. The logarithmic plots of D_C versus k_{off} show a good linear relationship. D_C increases with k_{off} and is independent of K. This result means that the penetrant dissociation kinetic constant from the fixed complex is clearly reflected on the penetrant diffusion coefficient via the fixed complex in the membrane.

For the cobalt porphyrin ligated with pyridines in Table 6-6, the k_{off} value decreases in the order pyridine, dimethylaminopyridine, and aminopyridine. The D_C value decreases in the same order, as does P_{O_2} also.

One can discuss also the effect of the structure of porphyrin in Table 6-7. The porphyrin structure was chemically modified by changing the substituent groups on the porphyrin plane with pivalamide and methacrylamide groups. The pocket-like structure constructed by the bulky substituent groups guarantees that the oxygen-coordination site is vacant before the oxygen coordinating occurs in the solid state. On the other hand, this steric structure sterically hinders oxygen-coordination and -dissociation steps. The porphyrin derivative with methacrylamide substituents pro-

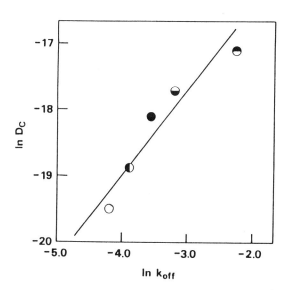

Figure 6-10 ■ Correlation of the oxygen-dissociation rate constant with the diffusion coefficient via the cobalt-porphyrin complex for the membrane of **1**. ●, 1-methylimidazole; ◐, dimethylaminopyridine; ◓, aminopyridine; ◖, 1,2-dimethylimidazole; ◑, pyridine.

duces both large K and k_{off}, or a good balance of coordination kinetics and equilibrium. The complex with larger K better enhances oxygen solubility in the polymer matrix and gives a larger oxygen permeability coefficient. On the other hand, the complex with larger oxygen-coordination rate constants gives a larger oxygen diffusion coefficient via the fixed complex, which reflects also on the higher permeability coefficient. It is concluded that a complex that has both larger kinetic and equilibrium parameters is requisite for a much improved efficiency in the facilitated transport.

The facilitated transport of nitrogen in Fig. 6-9 was mathematically analyzed by the dual-mode transport model[27, 28]

$$P = k_D D_{DD} + \frac{C_C K D_{CC}}{1 + Kp_2} + \frac{C_C K D_{CD} - k_D D_{DC}}{1 + Kp_2}$$

$$+ \frac{2k_D D_{DC} \ln(1 + Kp_2)}{Kp_2} \tag{6-8}$$

That is, P is equal to the sum of the first term, which represents the Henry mode attributed to a physical permeation through a polymer matrix, the second term, which represents the Langmuir mode attributed to a specific coordination and diffusion of nitrogen to and through the complex, and the third term, which represents the additive Langmuir mode attributed to an exchanging term between the first and second terms. Here, D_{DD}, D_{DC}, D_{CD}, and D_{CC} are the diffusion coefficients for the Henry-type physical permeation, for the polymer matrix to the fixed complex, for the fixed complex to the polymer matrix, and for the hopping between the fixed complexes. The K, C_C, and k_D values determined spectroscopically and gravimetrically are substituted in Eq. (6-8) and by using these parameters, we can calculate the transport parameters as given in Table 6-8. The P_{N_2} versus $p_2(N_2)$ curve was calculated and drawn as the solid line in Fig. 6-9. The experimental plots agree with the solid line, which supports the facilitated and dual-mode transport of nitrogen in the membrane and a pathway of nitrogen permeation via the fixed complex, represented by Eq. (6-8).

Table 6-7 ▪ **Structure Effect of the Cobalt-Porphyrin Complex on Parameters of Oxygen Coordination and Transport**

Complex	$10^2 K$ (cmHg^{-1})	$10^{-2} k_{\text{off}}$ (s^{-1})	$10^{10} P$ $\left(\dfrac{\text{cm}^3 \text{ (STP) cm}}{\text{cm}^2 \text{ s cmHg}} \right)$	$10^8 D_C$ (cm^2 s^{-1})	D_C/D_D
2	2.8	2.8	20	3.1	0.01
4	3.8	3.9	32	7.3	0.03
5	3.5	14	40	9.0	0.05

Table 6-8 ▪ Nitrogen Diffusion Constants in the Membrane of Macromolecular Manganese Complex 7

C_C (cm^3 cm^{-3})	$10^7 D_{DD}$ (cm^2 s^{-1})	$10^7 D_{DC}$ (cm^2 s^{-1})	$10^7 D_{CD}$ (cm^2 s^{-1})	$10^8 D_{CC}$ (cm^2 s^{-1})
0.05	11	2.2	1.5	8.8
0.08	7.4	1.9	1.2	6.3
0.12	5.1	1.7	0.69	5.8

^{a}cm^3 cm^{-3} = cm^3 N$_2$ cm^{-3} membrane.

Nitrogen diffusion coefficients in Table 6-8 suggest the following aspects of the facilitated transport.

A rapid molecule exchange between two transport modes is established, as had been expected by the kinetic constants determined spectroscopically for nitrogen coordination to the complex fixed in the membrane.

The diffusion coefficient via the complex has one-quarter of the mobility of that due to the physical permeation mode.

D_{CC}/D_{DD} increases with fixed complex concentration, which suggests a penetrant molecule hopping accompanied by the decrease in the distance between the fixed complexes.

6-7. Conclusion

Specific molecule-transport processes can be analyzed for macromolecular complexes because they are prepared as a thin membrane. In macromolecular complexes, coordination of a small molecule is characterized as a kinetic and continuous reaction system. The kinetic reaction process or dynamic interaction can be quantitatively discussed by modelling the interaction mathematically. Multiple interaction is caused by integration of the kinetically active reaction and is clearly affected by the chemical characteristics of the metal complex and properties of the polymer matrix. Macromolecular complexes are a good model system to understand this dynamic multiple interaction.

References

1. E. Tsuchida and H. Nishide, *Adv. Poly. Sci.*, **24**, 1 (1977).
2. E. Tsuchida and H. Nishide, *Topics Curr. Chem.*, **132**, 64 (1986).
3. E. Tsuchida, *J. Macromol. Sci.-Chem.* **A13**, 545 (1979).
4. D. Woehrle and H. Bohlen, *Makromol. Chem.*, **187**, 2081 (1986).
5. J. N. Gillis and R. E. Sievers, *Anal. Chem.*, **57**, 1572 (1985).

6. H. Nishide, M. Ohyanagi, O. Okada, and E. Tsuchida, *Macromolecules*, **20**, 417 (1987).
7. H. Nishide, H. Kawakami, S. Toda, Y. Kamiya and E. Tsuchida, *Macromolecules*, **23**, 4325 (1990).
8. H. Nishide, M. Ohyanagi, O. Okada, and E. Tsuchida, *Macromolecules*, **21**, 2910 (1988).
9. H. Nishide, H. Kawakami, T. Suzuki, Y. Azechi and E. Tsuchida, *Macromolecules*, **23**, 3715 (1990).
10. H. Nishide, M. Kuwahara, M. Ohyanagi, Y. Funada, H. Kawakami, and E. Tsuchida, *Chem. Lett.*, **1986**, 43 (1986).
11. H. Nishide, M. Ohyanagi, H. Kawakami, and E. Tsuchida, *Bull. Chem. Soc. Jpn.*, **59**, 3213 (1986).
12. E. Tsuchida, H. Nishide, M. Ohyanagi, and H. Kawakami, *Macromolecules*, **20**, 1970 (1987).
13. E. Tsuchida, H. Nishide, and M. Ohyanagi, *J. Phys. Chem.*, **92**, 641 (1988).
14. R. D. Jones, D. A. Summerville, and F. Basolo, Chem. Rev., **79**, 319 (1979)
15. E. Antonini and M. Brunori, *Hemoglobin and Myoglobin in their Reaction with Ligands*, North-Holland, Amsterdam (1971).
16. M. Ohyanagi, H. Nishide, K. Suenaga, T. Nakamura, and E. Tsuchida, *Bull. Chem. Soc. Jpn.*, **60**, 3045 (1987).
17. H. Nishide, K. Suenaga, K. Murata, H. Kawakami, and E. Tsuchida, *Kobunshi Ronbunshu*, **46**, 375 (1989).
18. H. Nishide, M. Ohyanagi, O. Okada, and E. Tsuchida, *Polymer J.*, **19**, 839 (1987).
19. R. A. Henderson, G. F. Leigh, and C. J. Piekett, *Adv. Inorg. Chem.* **27**, 198 (1983).
20. Y. Kurimura, F. Ohta, J. Gohta, H. Nishide, and E. Tsuchida, *Makromol. Chem.*, **183**, 2889 (1982).
21. H. Nishide, H. Kawakami, E. Tsuchida, and Y. Kurimura, *J. Macromol. Sci.-Chem.* **A25**, 1339 (1988).
22. C. V. Pittman, Jr., G. V. Marlin, and T. D. Rounsefell, *Macromolecules*, **6**, 1 (1973).
23. H. Nishide, H. Kawakami, Y. Kurimura, and E. Tsuchida, *J. Am. Chem. Soc.*, **111**, 7175 (1989).
24. B. S. Creaven, A. J. Dixon, J. M. Kelly, C. Long, and M. Poliakofs, *Organometallics*, **6**, 2600 (1987).
25. H. Nishide, M. Ohyanagi, O. Okada, and E. Tsuchida, *Macromolecules*, **19**, 494 (1986).
26. D. R. Paul and W. J. Koros, *J. Polymer Sci.*, *Phys. Ed.*, **14**, 675 (1976).
27. R. M. Barrer, *J. Memb. Sci.*, **18**, 25 (1984).
28. G. H. Fredrickson and E. Helfond, *Macromolecules*, **18**, 2201 (1985).

Soft Interaction of Macromolecule-Lanthanide Complexes

Yoshiyuki Okamoto and Junji Kido

Polytechnic University, Department of Chemistry, Brooklyn, New York and Yamagata University, Department of Polymer Chemistry, Yonezawa, Yamagata, Japan

7-1. Introduction

The lanthanides, symbolically Ln, are the fourteen elements following lanthanum in the Periodic Table, and have fourteen $4f$ electrons successively added to the lanthanum configuration.

Because these $4f$ electrons are relatively uninvolved in bonding, all these highly electropositive elements have, as their prime oxidation state, the Ln^{3+} ion. The radius of these ions decreases with increasing atomic number from La, which is known as the lanthanide contraction.

The term "lanthanide" is used to indicate that these elements form a closely allied group, having lanthanum as the prototype. The term usually includes lanthanum itself. Table 7-1 gives some properties of the atoms and ions. The electronic configurations are not known with complete certainty due to the great complexity of the electronic spectra of the atoms and ions and the attendant difficulty of analysis.

The lanthanide contraction has certain important effects on the properties of these elements. There is a significant and steady decrease in the size of the atoms and ions with increasing atomic number: La has the greatest radius and Lu the smallest; the radius of La^{3+} is about 0.21 Å larger than that of Lu^{3+} (Table 7-1).

The contraction is caused by the imperfect shielding of one electron by another in the same subshell. Going from La to Lu, the nuclear charge and the number of $4f$ electrons increase by 1 at each step. The shielding of one $4f$ electron by another is very imperfect due to the complex shapes of the $4f$ orbitals. Therefore, each $4f$ electron experiences an increase in the effective nuclear charge at each increase in

Table 7-1 ■ Some Properties of Lanthanide Atoms and Ions[a]

Atomic Number	Name	Symbol	Electronic Configuration Atom	Electronic Configuration M^{3+}	Radius M^{3+} (Å)
57	Lanthanum	La	$5d6s^2$	[Xe]	1.061
58	Cerium	Ce	$4f^1 5d^1 6s^2$	$4f^1$	1.034
59	Praseodymium	Pr	$4f^3 6s^2$	$4f^2$	1.013
60	Neodymium	Nd	$4f^4 6s^2$	$4f^3$	0.995
61	Promethium	Pm	$4f^5 6s^2$	$4f^4$	0.979
62	Samarium	Sm	$4f^6 6s^2$	$4f^5$	0.964
63	Europium	Eu	$4f^7 6s^2$	$4f^6$	0.950
64	Gadolinium	Gd	$4f^7 5d6s^2$	$4f^7$	0.938
65	Terbium	Tb	$4f^9 6s^2$	$4f^8$	0.923
66	Dysprosium	Dy	$4f^{10} 6s^2$	$4f^9$	0.908
67	Holmium	Ho	$4f^{11} 6s^2$	$4f^{10}$	0.894
68	Erbium	Er	$4f^{12} 6s^2$	$4f^{11}$	0.881
69	Thulium	Tm	$4f^{13} 6s^2$	$4f^{12}$	0.869
70	Ytterbium	Yb	$4f^{14} 6s^2$	$4f^{13}$	0.858
71	Lutetium	Lu	$4f^{14} 5d6s^2$	$4f^{14}$	0.848

[a]Taken from ref. 1.

atomic number, which causes a shrinkage in size of the entire $4f^n$ shell. The total lanthanide contraction is the result of the accumulation of these successive contractions.

It should be noted also that the decrease, though steady, is not quite regular. The biggest decrease occurs with the first f electrons added; there also appears to be a larger decrease after f^7 (between Tb and Gd). Certain chemical properties of lanthanide compounds show corresponding divergences from regularity as a consequence of the ionic size.

Because the sum of the first three ionization enthalpies is relatively low, the elements are highly electropositive. They easily form $+3$ ions in solids like oxides and in aqueous ions, $[Ln(H_2O)_n]^{3+}$, and complexes. However, cerium can give Ce^{4+}, and Sm, Eu, and Yb form the Ln^{2+} ions in aqueous solutions and in solids. Other elements can give $+4$ states in solids.

Comparison of the coordination chemistry of the lanthanide metal ions with that of the d-type transition metal ions may be profitable. Appreciably stable lanthanide complexes, other than the hydrated cations themselves, are obtained only when strongly chelating ligands are used and, in particular, when these ligands contain highly electronegative donor atoms (e.g., oxygen). Both the electronic configuration of the cation and its radius are important to understanding both the phenomenon and the properties of lanthanide complexes in general.

The electronic configuration has a significant effect on the stability of the metal ion complexes. In the case of d-type transition metal ions, their complex stabilities are the result of participation of the d electrons in the metal-ligand bond through hybridization of metal electronic orbitals and overlap of these hybrid orbitals with appropriate ligand orbitals. In the case of lanthanide metal ions, they have $4f$ orbitals that are effectively shielded from interaction with ligand orbitals by $5s$ and $5p$ electrons. If hybridization occurs, it must involve normally unoccupied higher-energy orbitals (e.g., $5d, 6s, 6p$), and hybridization of this type can be expected only with the most strongly coordinating ligands. Thus complex formation is entirely electrostatic in character, and the lanthanide complexes compare more closely with the complexes derived from calcium, strontium, and barium ions than with those derived from the d-type transition metal ions.[2-4] These cations are thus classified as class a[5,6] or hard[7,8] acceptors.

It is expected, and observed, that the Ln ions prefer donor atoms with the preference O > N > S and F > Cl. Complexes with pure nitrogen or sulfur donors are not stable in aqueous solutions due to hydrolysis.[9] The high degree of ionicity in the Ln complexes can be confirmed by the observed patterns of bond distances. It was observed that Ln — N bonds are generally longer than Ln — O bonds, in some cases, by as much as 0.30 Å.

The coordination numbers reported for the Ln ions range from 6–10; sometimes a maximum value of 12 is known.[9] Among them, coordination numbers of 8 and 9 are the most common, although Ca^{2+} exhibits numbers of 6, 7, and 8.

The spectral behavior of the lanthanides is also different fundamentally from that of the d-block transition elements. The differences are based on the fact that the electrons responsible for the properties of lanthanide ions are $4f$ electrons, and the $4f$ orbitals are very effectively shielded from the influence of external forces by the overlying $5s^2$ and $5p^6$ shells. Therefore, the electronic states arising from the various $4f^n$ configurations are only slightly affected by the surroundings of the ions and remain practically invariant for a given ion in all its compounds.

Because the f orbitals are so well shielded from the surroundings of the ions, the various states arising from the f^n configurations are split by external fields only to the extent of -100 cm^{-1}. Thus when electronic transitions occur from one J state of an f^n configuration to another J state of this configuration (f-f transition), absorption bands are extremely sharp. They are similar to those for free atoms and are quite unlike the broad bands observed for the d-d transitions in transition metal ions. Actually all the absorption and emission bands found in the visible and near-ultra-violet spectra of the lanthanide $+3$ ions have line-like character.

Most metal ion complexes absorb visible and/or UV irradiation; however, very few reemit even a small fraction of the absorbed energy in the form of UV or visible photons. This is due to facile nonradiative deexcitation pathways that are much more efficient than radiative modes. Because the d-electron excited states of the transition metal complexes are strongly coupled to the environment, the excited energy is efficiently degraded by the ligand field. Therefore, luminescent complexes of the transition metal ion complexes are very rare. Contrarily, in the case of the Ln^{3+} ions, their lowest lying excited states are comprised of $4f^n$ configurations. Due to the effective shielding of the $4f$ orbitals from the environment, the $4f$ orbitals minimally participate in bonding; therefore, the total spread of ligand-field splitting of an f-electron term is rarely more than a few hundred per centimeter (cf., more than 25,000 cm^{-1} is known for the d-electron term of the transition metal complexes). As a result, radiationless deexcitation processes in Ln^{3+} complexes are relatively inefficient and the emission of radiation as luminescence is able to compete in many instances. The f-f transitions are electric dipole forbidden, which results in weak absorption and emission. The probability of such a transition is very low, with molar extinction coefficients of less than 10 M^{-1} cm^{-1} and radiative lifetimes cf several milliseconds. This behavior can be compared with organic fluorescent molecules where molar extinction coefficients of tens of thousands and radiant lifetime in the nanosecond range are common. The term luminescence is used to refer to radiative emission in general. Fluorescence and phosphorescence are used to refer to the singlet-singlet and triplet-singlet emission of organic molecules, respectively.

Figure 7-1 shows the energy levels of the five central members: Sm^{3+}, Eu^{3+}, Gd^{3+}, Tb^{3+}, and Dy^{3+}. The other members rarely luminesce in solution and, therefore, are not useful as emissive sources in coordination chemistry. This is because there are no large gaps between potential emissive levels and acceptor levels of the ground manifold. These smaller energy level differences are sufficiently bridged by nonradiative processes. For Eu^{3+} and Tb^{3+} ions, the energy gaps between the lowest lying

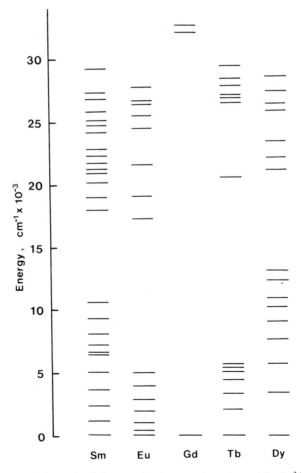

Figure 7-1 ■ Energy level diagrams for the central members of the Ln^{3+} ions.

emissive level and the highest level of the ground manifold are 12,150 and 14,800 cm^{-1}, respectively. In fact, only Eu^{3+} and Tb^{3+} have been used in this area because of their relatively strong emission. The emissive levels of Eu^{3+} ($^5D_0$) and Tb^{3+} ($^5D_4$) lie 17,250 cm^{-1} (580 nm) and 20,500 cm^{-1} (488 nm) above the ground level.

There are two basic spectroscopic procedures available for the study of luminescence. In emission spectroscopy, the sample is excited at a given wavelength λ_{ex}, while the emission monochromater is scanned. Emission spectra give information about the probability and structure of the transitions from the emissive level to the various terminal levels of the ground manifold. In excitation spectroscopy, on the other hand, the emission is monitored at a particular wavelength λ_{em}, while scanning the excitation monochromater. Excitation spectroscopy gives information about transition from the ground state to the emissive state and higher levels; therefore, an excitation spectrum is similar to the absorption spectrum with the addition of nonradiative transition probability effects.

Luminescence techniques have an advantage over other experimental methods; that is, emission from a sample can be detected with high sensitivity, which allows the study to be made at relatively low concentration of the luminescent species ($\sim 10^{-6}$ M).

These techniques have been widely used in the study of biomolecular structure.[10] However, in this chapter, we discuss mainly our investigation of the luminescence techniques to study:

1. Structure of ionomers.
2. Ion binding properties of synthetic polyanions in both solid and solution.
3. The number of coordination water molecules of metal–polyelectrolyte complexes in aqueous solutions.

7-2. Use of Lanthanide Metal Probes to Elucidate the Structure of Ionomers

The properties of synthetic polymers containing metal ions have been extensively investigated in recent years.[11,12] An important class of these polymers consists of the ethylene-carboxylic acid copolymers, which have been either partially or wholly neutralized with group I and II cations. After neutralization of the carboxylic acid, it is found that properties such as optical clarity, tensile strength, impact resistance, and melt viscosity are dramatically enhanced. It has been proposed that these polymers contain microscopic aggregates of ionic groups.[13] Various structures of the ionic aggregates have been reported and these depend on the composition of polymers, nature and concentration of metal ions, temperature, and so forth.

The excitation and emission spectra of film and powder samples for the copolymer of styrene and acrylic acid (PSAA)-Eu^{3+} complexes were found to be identical, as shown in Fig. 7-2.[14] The bright red emission lines of Eu^{3+} at ca. 600 nm were assigned to transition from the $^5D_2$ level to all the multiple levels of ^7F. This emission spectrum was obtained under excitation at 396 nm, corresponding to the $^5L_7$ level of the Eu^{3+} ion. Similar luminescence spectra of the PSAA-Tb^{3+} complex were obtained as shown in Fig. 7-3. The excitation spectrum for the Tb complex is broad compared with that of the Eu complex and the maximum is located at 305 nm. The

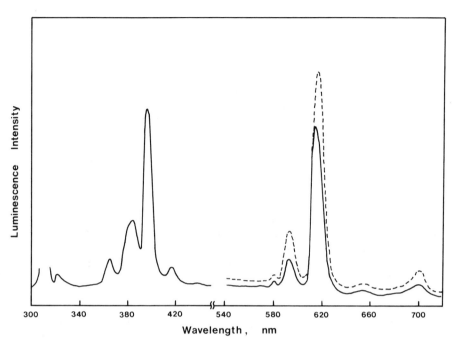

Figure 7-2 ■ Excitation and emission spectra of copolymers of styrene (91 mol%) and acrylic acid (9 mol%) (PSAA)-Eu salt. _____, powder; _ _ _ _ _, film. $\lambda_{em} = 617$ nm; $\lambda_{ex} = 396$ nm.

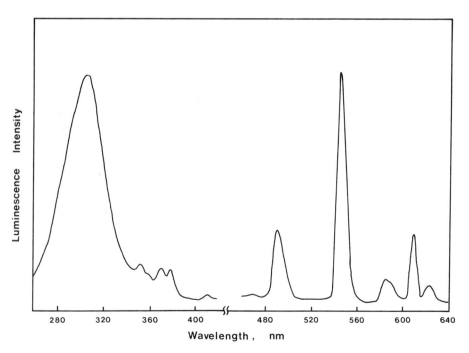

Figure 7-3 ■ Excitation and emission spectra of PSAA-Tb^{3+} salt. $\lambda_{em} = 545$ nm; $\lambda_{ex} = 305$ nm.

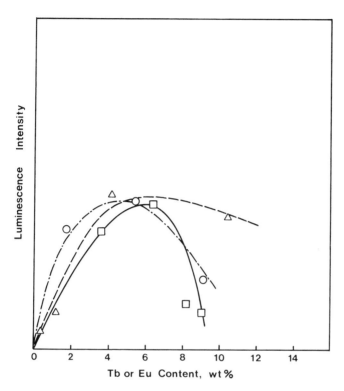

Figure 7-4 ▪ Relationship between luminescence intensity and metal content. ○, PSAA-Eu salt (COOH content 13 mol%); △, PSAA-Eu salt (COOH content 9 mol%); □, PSAA-Tb salt (COOH content 13 mol%).

relationship between luminescence intensity and Eu or Tb content of PSAA complexes is shown in Fig. 7-4. The intensity increases with increasing lanthanide ion content, reaches a maximum at 4–6 wt% of the ions, and then decreases with further increase of the ion content. The same tendency was also found in the copolymer of methyl methacrylate(93.2%)-methacrylic acid(6.8%)-Eu complexes (Fig. 7-5). However, the lanthanide salts of poly(acrylic acid) and poly(styrene-maleic acid) show a linear increase of luminescence intensity with increasing lanthanide ion content up to around 15 wt%. Linear relationships were also observed in samples in which $Eu(OAc)_3$ is uniformly dispersed in polymers such as polystyrene and poly(methylmethacrylate) (Fig. 7-5).

The tendencies in the relationship between luminescence intensity and lanthanide content in PSAA and PMAA/MA complexes are typical of concentration quenching of luminescence as described in the literature.[15] Deviation from linearity and concentration quenching clearly suggest formation of ionic aggregates in which metal ions are close together.

A study of the glass transition temperatures of ionomers suggests that both the charge and the size of the ion may affect clustering.[12] However, it is interesting to

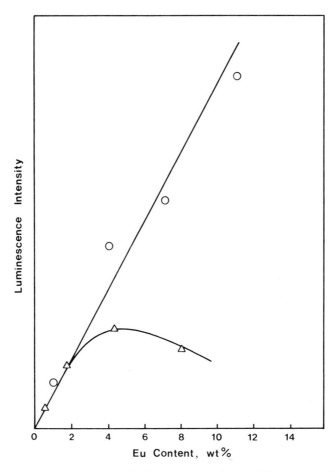

Figure 7-5 ■ Relationship between luminescence intensity and Eu content. ○, PMMA-Eu(OAc)$_3$ system; △, PMM/MA-Eu^{3+} system.

note that the concentration at which the maximum of the luminescence intensity of the PSAA-Eu complex (6 wt%; 4.0×10^{-4} mol/g of salt) occurs is in the same range of ion concentration where the aggregate formation started in poly(styrene-sodium methacrylate) ($\sim 5.7 \times 10^{-4}$ mol/g of salt).[16]

Lundberg and Makowski[17] prepared partially sulfonated and carboxylated polystyrenes. They studied the differences of physical properties of these ionomers and found that the ionic association in the sulfonate is much stronger than in the carboxylate.

Typical luminescence spectra of Eu and Tb salts of the partially sulfonated (SPS) and carboxylated (CPS) polystyrenes are shown in Figs. 7-6 and 7-7. The SPS-Eu complex has the same sharp excitation peak at 376 nm as PSAA-Eu and PMM/Ma-Eu

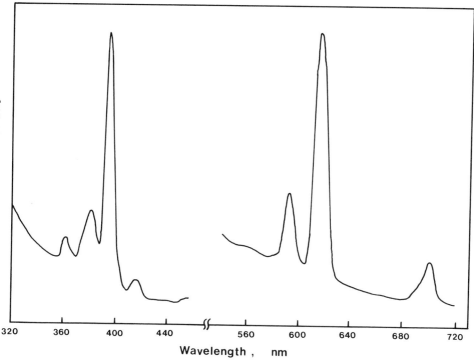

Figure 7-6 ■ Luminescence spectra of sulfonated polystyrene(sulfonic acid content 6.1 mol%)-Eu^{3+} salt. $\lambda_{em} = 615$ nm; $\lambda_{ex} = 396$ nm.

complexes, which corresponds to the excitation of the Eu^{3+} ion itself. However, CPS-Eu shows broad peaks and a different excitation maximum at shorter wavelength ($\sim$ 371 nm). This indicates that an energy transfer from the ligand, that is, the benzoate moiety to Eu^{3+}, occurred, as in Eu^{3+}-o-benzyl benzoate.[18] However, in the case of the SPS-Eu complex no such energy transfer was observed.

Typical relationships between luminescence intensity and Eu or Tb content in SPS and CPS are shown in Figs. 7-8 and 7-9. The luminescence intensity of the SPS-ion complex increased linearly with increasing ion content. In the case of the CPS, the intensity also increased with ion content and did not show the strong concentration quenching observed in PSAA- and PMM/MA-ion complexes. The results suggest that lanthanide ions in the SPS and CPS systems are homogeneously dispersed throughout the polymer matrix. On the basis of these measurements, it is concluded that partially carboxylated and sulfonated polystyrene ionomers do not contain ion aggregates up to 6 mol% ion.[14]

Considerable effort has been devoted to the study of nonradiative energy transfer between luminescent lanthanide donor–acceptor pairs in the solid state, particularly in glasses.[19-23] When the emission spectrum of the donor overlaps the absorption of the acceptor, excitation energy absorbed by the donor can be transferred to the

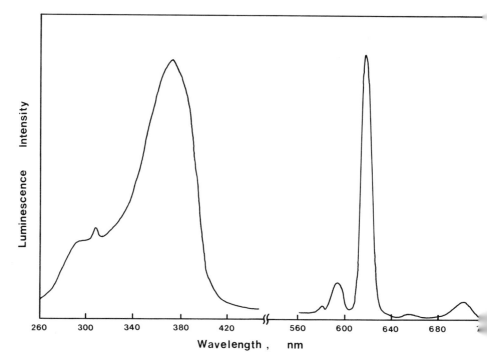

Figure 7-7 ■ Luminescence spectra of carboxylated polystyrene(carboxylic acid content 6.4 mol%)-Eu^{3+} salt. $\lambda_{em} = 616$ nm; $\lambda_{ex} = 371$ nm.

acceptor over considerable distances. Thus, lanthanide ions were applied as luminescence probes in the study on conformations of biological molecules.[24] The metal ions that are native to metalloenzymes, for example, Ca^{2+}, Mg^{2+}, and Zn^{2+}, are spectroscopically inactive. The substitution of trivalent lanthanides for these ions has been utilized in the elucidation of metal binding sites. Moreover, if some ions that act as donor or acceptor against the lanthanide ions are introduced, the distance between the metal binding sites may be estimated by measurements of energy transfer efficiency.

As a "spectroscopic ruler," the donor–acceptor pairs, Tb-Eu, Tb-Co, and UO_2-Eu were used to elucidate the structures of various ionomers. Tb-Eu and Tb-Co metal containing poly(styrene-acrylic acid) and partially carboxylated and sulfonated styrenes were prepared and the energy transfers from Tb^{3+} to Co^{2+} and Eu^{3+} were measured by the Tb^{3+} luminescence quenching behavior.

The relationship between luminescence intensity of Tb^{3+} at 545 nm and Eu^{3+} content in PSAA-Tb/Eu is shown in Fig. 7-10. At an equimolar content of Eu and Tb ions, the luminescence of Tb was found to be quenched more than 90%. When the Tb^{3+} and Co^{2+} ions in the PSAA system were used as a probe, very efficient quenching of the Tb luminescence was also observed as shown in Fig. 7-11. These

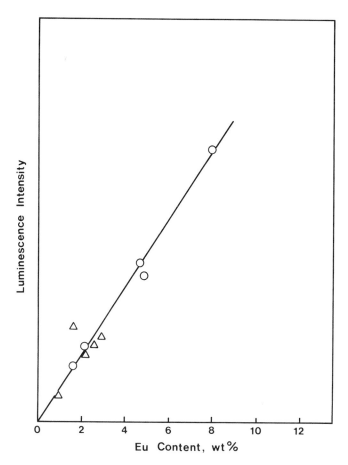

Figure 7-8 ■ Luminescence intensity versus Eu content in sulfonated polystyrene-Eu^{3+} salt: △, sulfonic acid content 2.6 mol%; ○, sulfonic acid content 6.1 mol%.

results indicate that Tb^{3+} and Co^{2+} ions are located very closely to each other and further confirm the existence of ion aggregates in the PSAA system.[14]

The emitting level of the UO$_2$ ion is about 480 nm and the luminescence extends from 480 to about 570 nm. Three transitions of Eu^{3+} ion are observed at room temperature within this energy range. It has been found that UO$_2^{2+}$ transfers energy to Eu^{3+} in borosilicate glass, resulting in a fivefold increase in the Eu^{3+} luminescence.[25] If the ion-containing polymer forms ion aggregates, it is very likely that Eu^{3+} and UO$_2^{2+}$ pairs are located close together within the aggregates and that efficient energy transfer would be obtained. Thus, the Eu^{3+} and UO$_2^{2+}$ mixed salts of PSAA and PMMA/MA were prepared. The luminescence intensity was found to be greatly increased (∼ 30-fold) when equal molar concentrations of UO$_2^{2+}$ and Eu^{3+} were present in these polymer systems and UO$_2^{2+}$ was excited.[26]

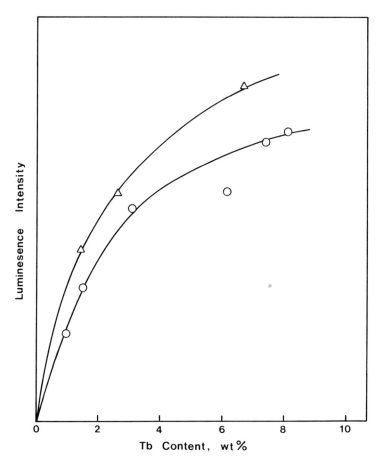

Figure 7-9 ■ Luminescence intensity versus Tb content in carboxylated polystyrene-Tb^{3+} salt: ○, carboxylic acid content 5.0 mol%; △, carboxylic acid content 6.4 mol%.

These results further support the existence of ion aggregate structures in the ionomers, such as PMM/MA and PSAA, in which Eu^{3+} and UO$_2^{2+}$ are located close together, facilitating the energy transfer between these ions.

The luminescence probe technique was also utilized[27] to elucidate the structure and thermal behavior of Nafion, the perfluorinated sulfonate membrane (registered trademark of E. I. du Pont de Nemours and Co.). When the Nafion membrane was soaked in Eu^{3+} solution, exchange between H$^+$ and Eu^{3+} occurred. It was observed that the luminescence of Eu^{3+} was enhanced when Eu^{3+} was bound to the Nafion membrane. The enhancement of the luminescence of the Eu^{3+} in the Nafion may be due to the formation of asymmetric bonding between Eu^{3+} and locally concentrated SO^{3-} groups. Similar results were also obtained when Tb^{3+} was bound to the Nafion membrane.[27]

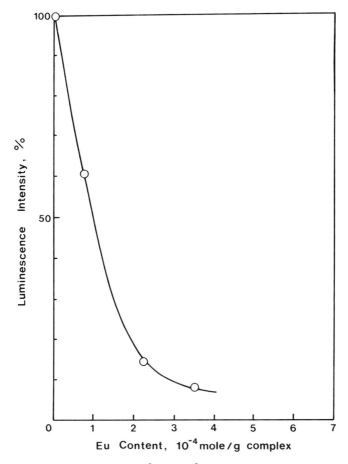

Figure 7-10 ■ Energy transfer from Tb^{3+} to Eu^{3+} in PSAA (COOH content, 9 mol%); $[Tb] = 3.9 \times 10^{-4}$ mol/g salt.

Nafion Eu^{3+} and UO_2^{2+} salts were prepared by immersing the membrane in an aqueous solution that contained both Eu^{3+} and UO_2^{2+} ions. The luminescence intensities of Eu^{3+} at 616 nm were found to be greatly increased when UO_2^{2+} was present, as shown in Fig. 7-12. These results support the existence of ion cluster structure in the Nafion membrane in which Eu^{3+} and UO_2^{2+} are close together, facilitating energy transfer from UO_2^{2+} to Eu^{3+}. When the Tb^{3+} and Co^{2+} ions are present together in the Nafion membrane, very efficient quenching of Tb^{3+} luminescence is observed as described previously in the styrene and acrylic acid copolymer system (Fig. 7-11).

Gierke and Hsu[28] investigated the cluster morphology of Nafion alkali salts (1200 EW) and showed that the cluster does exist in the dry polymer and that the diameter is ~ 19 Å, containing about 26 ion-exchange sites. When the H^+ of Nafion was

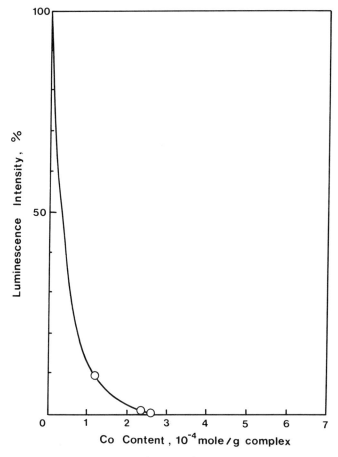

Figure 7-11 ■ Energy transfer from Tb^{3+} to Co^{2+} in PSAA (COOH content, 9 mol%); [Tb] = 3.9×10^{-4} mol/g salt.

exchanged mostly with equimolar concentrations of Tb^{3+} and Co^{2+} ions, the luminescence intensity of Tb^{3+} was found to decrease by 80% (Fig. 7-12). Exact calculation of the distances between these metal ions is a rather complex problem. However, according to Föster,[29] the quenching of luminescence owing to dipole–dipole radiationless energy transfer is given by

$$\frac{F}{F_0} = \frac{1}{1 + (R_0/R)^6} \tag{7-1}$$

where R is the distance between donor and acceptor and R_0 is the critical distance for 50% energy transfer, given by

$$R_0^6 = 8.78 \times 10^{-25} K^2 Q n^{-4} J \text{ cm}^6 \tag{7-2}$$

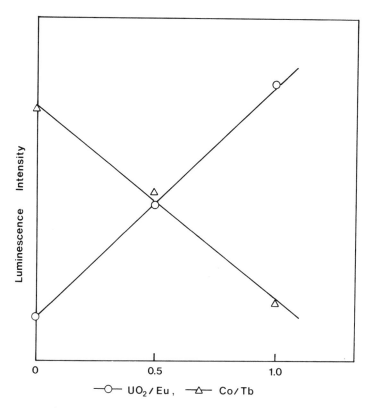

Figure 7-12 ■ O, Energy transfer from UO_2^{2+} to Eu^{3+} in Nafion membrane at room temperature: membrane, 0.19 g; 0.4×10^{-4} mol of SO_3^-; $[Eu^{3+}] = 0.5 \times 10^{-5}$ mol; emission at 616 nm;
excitation at 320 nm. △, Energy transfer from Tb^{3+} to Co^{2+} in Nafion membrane at room
temperature: membrane, 0.27 g; 0.54×10^{-4} mol of SO_3^-; $[Tb^{3+}] = 1.5 \times 10^{-5}$ mol; emission at
545 nm.

In Eq. (7-2), K is the dipole–dipole orientation factor, Q is the donor quantum yield
in the absence of energy transfer, n is the refractive index of the investigating
medium, and J is the spectral overlap integral, given by

$$J = \int_0^\infty F(\nu)\varepsilon(\nu)^{-4}\,d\nu \qquad (7\text{-}3)$$

where $F(\nu)$ is the spectral distribution of the donor emission normalized to unity, $\varepsilon(\nu)$
is the molar extinction coefficient of the acceptor per square centimeter per molar,
and ν is the frequency per centimeter. Using Eq. (7-2), the critical distance R_0 for the
donor–acceptor pairs was calculated[27] as 19.4 Å, assuming the refractive index 1.35
(Teflon) as a first approximation. The luminescence yield in the presence and absence

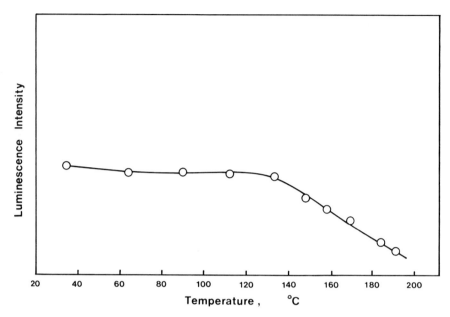

Figure 7-13 ■ Effect of temperature on the luminescence intensity of Nafion Eu^{3+} and UO_2^{2+} salts. $[Eu^{3+}] = 0.5 \times 10^{-4}$ mol; $[UO_2^{2+}] = 0.5 \times 10^{-4}$ mol; emission at 616 nm; excitation at 320 nm.

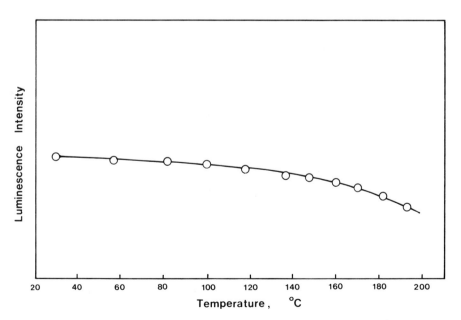

Figure 7-14 ■ Effect of temperature on the luminescence intensity of Nafion Tb^{3+} and Na^+ salts. Tb^{3+} content $= 5.6 \times 10^{-6}$ mol/g of Nafion; emission at 545 nm; excitation at 305 nm.

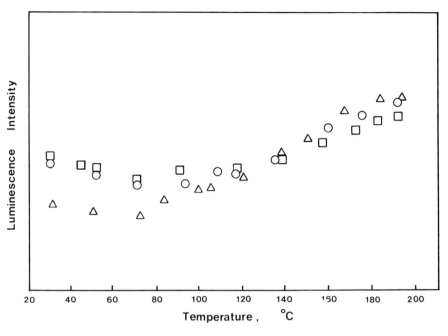

Figure 7-15 ■ Effect of temperature on the luminescence intensity of Nafion Tb $^{3+}$ and Na$^+$ salts. Tb^{3+} content $= 4.0 \times 10^{-5}$ mol/g of Nafion.

of acceptor, F/F_0, was 0.2. Thus, using Eq. (7-1), the average Tb and Co distance in this system was calculated as 15.4 Å. This value is comparable to the cluster diameter obtained by Gierke and Hsu.[28]

When samples of Nafion Eu^{3+} (or Tb^{3+}) and UO$_2^{2+}$ salts were heated, the luminescence intensities of Eu^{3+} (or Tb^{3+}) decreased sharply at around 130°C (Fig. 7-13). This indicated that the distance between Eu^{3+} and UO$_2^{2+}$ was increased, and as a result the energy-transfer efficiency from UO$_2^{2+}$ to Eu^{3+} ions was decreased.

Nafion salt containing a small amount of Tb^{3+} (5.6 × 10^{-6} mol/g of Nafion) was prepared and heated gradually to 200°C. The luminescence intensities of Tb^{3+} (at 545 nm) were found to decrease with increasing temperature (Fig. 7-14). This may be due to thermal quenching of Tb^{3+} luminescence. However, Nafion Na salt containing a higher concentration of Tb^{3+} (4.0 × 10^{-5} mol/g of Nafion) was heated similarly, and the luminescence was found to increase sharply beginning around 130°C. This behavior was also reproducible on thermal cycling of the sample, as shown in Fig. 7-15. As described previously when Eu^{3+} or Tb^{3+} were introduced into ionomers having ion aggregates, the luminescence intensities reached a maximum at certain concentrations of these ions and decreased with further increase in these ion concentrations. These results were accounted for by concentration quenching.[14] The Eu^{3+} or Tb^{3+} ions in the aggregates would create locally high ion concentrations whose interactions lead to concentration quenching.

When Nafion Tb^{3+} salt was heated to 150°C, the data of the increasing luminescence intensity suggested[27] that the distances between ions were increasing by about

~ 1.2 Å. These investigations suggest that the lanthanide luminescence probe technique is simple and useful for studying ionomer structures.

7-3. Use of Lanthanide Metal Probes to Study the Ion Binding Properties of Polyelectrolytes in Aqueous Solutions

In aqueous media the lanthanide ions are subject to hydrolysis and a distinct lowering of pH is noted when salts of these elements are dissolved in water. The formation of aquo complex $[Ln(OH_2)_n]^{3+}$ (where n is larger than 6, perhaps 8 or 9) takes place. The extent of lowering of pH depends on the concentration of the lanthanide; the heavier lanthanide ions with smaller ionic radii have greater tendency to hydrolyze.

The aquo complex has n H_2O molecules (e.g., n-9 for Tb^{3+}) coordinating in the inner coordination sphere of the central ion. Therefore, when $LnCl_3$ salts are dissolved in water there is very little interaction between $[Ln(OH_2)_n]^{3+}$ and the solvated Cl^- ion. Unless the other ions or ligands have a strong coordinating power, the coordinated water molecules protect the lanthanide ions from the influence of other anions or ligands. Frank and Evans[30] and Freed[31] showed that nitrate ions have greater interaction tendency than Cl^- ions. When complexes are formed, the approach of a ligand interferes with the hydration shell and the ordered geometry breaks down. The multidentate ligands presumably have a stronger influence than the unidentate ones.

Thus the nature of the metal ions and the complexing agents can affect the formation of the complexes; moreover the environmental factors, such as solvent, temperature, and pressure are also important. The stability of the complexes may depend on the concentration. Substitutions on the ligands strongly affect the stability of the complexes. Multidentate ligands usually form stable complexes, due to the chelate ring formation.

The lanthanide ion complexes with synthetic polyelectrolytes in solution have not been investigated in full detail, whereas the transition metal ion complexes have been well studied. Several studies on the Ln-polymer complexes employed the Bjerrum titration method to determine the formation constant.[32-34]

This method, however, provides some practical difficulties. It is known both theoretically and experimentally that the addition of a neutral salt, such as NaCl, will give a shift of pH in polyelectrolyte solutions, even when there is little possibility of complex formation. Particularly when the complexing tendency is very small, it is difficult to distinguish between the pH shift due to true complex formation and that resulting from such an electrostatic effect. Usually this error can be neglected if a sufficient amount of neutral salt is used in the experiment to make the ionic strength constant.

The pK_a values strongly depend on the nature of the polymer, on the ionic strength of the solution, and on the charge density of the polyelectrolyte chain. In the presence of metal ions, which to some extent screen the fixed polymer charges, these proton association constants may, in principle, differ from those of the pure acid; therefore the exact value cannot be determined. An incorrect choice of pK_a values may result in a completely misleading evaluation of the extent of binding.

Similar difficulties are found in the demands on the accuracy of pH determinations, not only due to uncertainties connected with the definition of pH in polyelectrolyte solutions, but because the experimental uncertainties, though small, are usually too large for an accurate determination of both the amount of metal ions bound and of the stability constants of the complexes.

Nagata and Okamoto[35] used the luminescence technique to obtain the overall formation constants K (M^{-1}) of the Tb-polyacrylate complexes in different ionic media. They assumed that three carboxylate groups are bound to the central Tb^{3+} ion. This assumption was based on the existence of the $Tb^{3+}(acetate)_3$ complex. However, in the polymeric environment, it is possible for the Tb^{3+} ion to have more than three carboxylates attached, because in the polymer domain the carboxylate ligands are so concentrated compared to the monomeric system, such as acetate, that multichelate formation may be largely enhanced (cf., polymer cooperative effect) (see later discussion).

The extremely low activity coefficients of counterions in polyelectrolyte solutions are attributed to condensation of counterions on the polyion. One extreme type of condensation is referred to as delocalized binding or ion atmosphere binding. In this case the counterions are delocalized in the area of the polyelectrolyte domain, and retain the fully hydrated state while having unrestricted mobility along the length of the polyion chain. When alkali metal and alkaline earth cations are condensed on extended polynucleotides, they are almost fully hydrated and delocalized on the polymer surface. This behavior is expected because singly charged phosphate monomers do not interact strongly with cations. Each phosphate group is monovalent and rigidly held at a great distance.[36] A very small dilatometric effect was observed for the binding of Mg^{2+} to DNA, which indicated that Mg^{2+} was bound to DNA with almost no dehydration. Another extreme type of condensation is "site binding," where the counterions are directly coordinated to several neighboring charged groups and fully dehydrated. Site binding is plausible for a polyelectrolyte with monovalent charged groups only if the chain is sufficiently flexible, or the nearest neighbor charge spacing is sufficiently close so that two or more groups can cooperate in holding the cation.[37] Evidence for site binding has accumulated from studies of dialysis equilibrium,[38] dilatometric measurements,[39,40] UV spectroscopy,[41,42] and potentiometric titrations of synthetic polyelectrolytes.[43] The ion binding can be characterized quantitatively by phenomena that are directly related to the release of water molecules coordinated to the cations. Complexation, resulting in dehydration of the counterion, leads to an increase in volume[39,40] and a characteristic decrease in refractive index.[44]

The Tb^{3+} aquo ion is known to contain nine water molecules in its inner coordination sphere as long as no cation–anion association takes place.[45] It has been established by Kropp and Windsor[46,47] that the emissive $^5D_4$ excited state could experience effective nonradiative quenching via energy transfer to the $-OH$ vibrational manifold of the coordinated water molecules. Because complexation of the Tb^{3+} ion with suitable ligands must result in the expulsion of some or all of the coordinated water molecules, it follows that formation of a stable Tb^{3+} complex must be accompanied by enhanced luminescence intensities and lifetimes.[48] Thus the complexation of Tb^{3+} ions with various polycarboxylates was investigated by monitoring its luminescence intensity and lifetime.[49]

The addition of sodium polyacrylate to an aqueous Tb^{3+} solution induces marked effects on the luminescence spectra of the Tb^{3+} ion (Fig. 7-16). The effect of Tb^{3+} concentration on the magnitude of luminescence intensity and lifetime in the Tb(PAA) complexes is illustrated in Fig. 7-17. The Tb:polymer ratio ranged from 0.006 to 0.1,

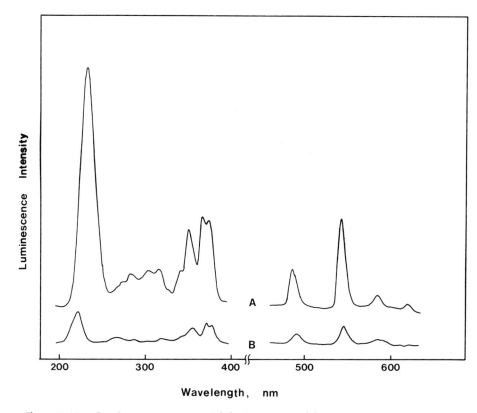

Figure 7-16 ■ Luminescence spectra of (A) Tb-PAA and (B) TbCl$_3$ in aqueous solutions: [Tb^{3+}] = 2 × 10^{-3} M, [PAA] = 2 × 10^{-2} N; pH = 4.0 for Tb^{3+}, 9.5 for Tb-PAA; excitation spectra (left) λ_{em} = 545 nm; emission spectra (right) λ_{ex} = 370 nm. The intensity scale is arbitrary.

and the data were acquired by using an excitation wavelength of 370 nm and an emission wavelength of 545 nm. It was observed that although the emission intensity increased linearly with the Tb^{3+} concentration, the luminescence lifetime remained constant at 780 μs over the entire range. This value may be compared with the lifetime of 420 μs measured for the Tb^{3+} aquo ion. Comparison of theoretically calculated and experimentally determined decay curves for Tb-PAA complexes indicated the existence of single exponential decays, which may signify the presence of essentially one type of Tb^{3+} emitting species. If several complexes exist in the system, and are in rapid equilibrium, the luminescence lifetime will be the average of all the complex species.

These results are strong evidence for the existence of a well defined metal-ion-binding site[48] whose nature is independent of the amount of Tb^{3+} condensed on the polymer. Morawetz et al. obtained similar results for Cu^{2+} binding to PAA.[38,50] Their spectroscopic studies showed that the intensity of Cu^{2+} absorption is a function of the

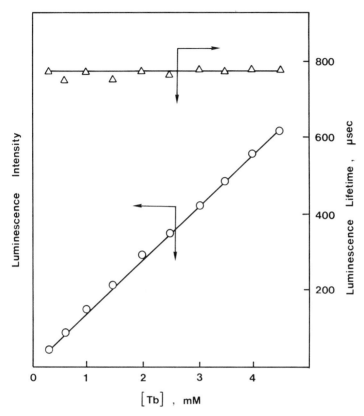

Figure 7-17 ■ ○, Tb^{3+} luminescence intensity ($\lambda_{ex} = 370$ nm; $\lambda_{em} = 545$ nm) and △, luminescence lifetime versus Tb^{3+} ion concentration in the presence of PAA in aqueous solution; [PAA] = 5×10^{-2} N; pH = 9.5. The intensity scale is arbitrary.

degree of neutralization of the polyion, but the location of the absorption maximum is independent of the degree of neutralization. Therefore, it is likely that only a single type of complex is formed. By comparing the location of the absorption maximum with the absorption spectra of cupric acetate complexes, they concluded that Cu(COO$^-$)$_4$ is the complex formed.

The luminescence properties of Tb^{3+} have been extremely useful to the study of metal ion coordination within biological systems[51-53] and to the characterization of metal ion binding polymers.[14] It has been observed that the luminescence intensity of Tb^{3+} is greatly enhanced by complex formation with poly(acrylate) and poly(styrenesulfonate), and that this intensity enhancement is useful in the elucidation of metal ion binding tendencies.[14,54]

To investigate the utility of the Tb^{3+} luminescence probe further, the chain structure of the polycarboxylates has been systematically varied to deduce the relation between polymer structure and Tb^{3+} luminescence properties.[49] The polymer studied

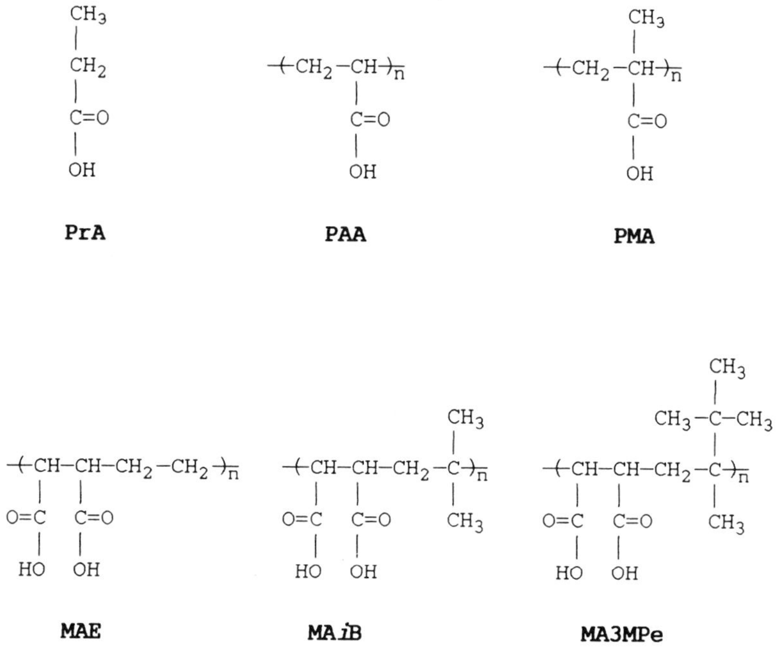

```
      CH3                                              CH3
       |                                                |
      CH2              +(CH2-CH)n-                   +(CH2-CH)n-
       |                   |                              |
      C=O                 C=O                            C=O
       |                   |                              |
      OH                   OH                             OH

      PrA                  PAA                            PMA
```

```
                                            CH3                        CH3
                                             |                          |
                                                                   CH3-C-CH3
                                                                        |
  +(CH-CH-CH2-CH2)n-    +(CH-CH-CH2-C)n-      +(CH-CH-CH2-C)n-
      |   |                |   |    |              |   |    |
     O=C  C=O            O=C  C=O  CH3          O=C  C=O   CH3
      |   |                |   |                    |   |
     HO   OH              HO   OH                  HO   OH

      MAE                  MAiB                      MA3MPe
```

Figure 7-18 ■ Monomeric and polymeric ligands investigated.

had a series of structural variations in which the stiffness of the polymer chain was gradually increased. These were poly(acrylic acid) (PAA), poly(methacrylic acid) (PMA), and the 1:1 alternation copolymers of maleic acid with ethylene (MAE), with isobutene (MAiB), and with 2,4,4-trimethyl-1-pentene (MA3MPe). The structures of the polymers are shown in Fig. 7-18. It is anticipated that an increase in the stiffness of the polymer backbone would affect the nature of the metal-ion-binding site that could be formed through a polymer conformation change.

The luminescence properties of the Tb^{3+} ion greatly change when bonded with polycarboxylates. This behavior is illustrated in Fig. 7-19, where the Tb^{3+} excitation spectra obtained for the Tb-MAE and Tb-PMA complexes are contrasted with that of the aquo ion. Binding of Tb^{3+} by all polycarboxylate ligands resulted in a strong increase in the luminescence intensity, but the most dramatic effects were noted in the excitation spectra. In most of the Tb-polymer complexes, a strong excitation feature located around 300 nm was observed. It has been shown that observation of a broad band in this spectral region is linked to the binding of Tb^{3+} in an oligomeric environment.[55] This band appears to be a hypersensitive Tb^{3+} absorption ($^7F_6 \rightarrow {}^5H_4$), which is only observed when the ion is bound by a polymer whose binding site is asymmetric. The hypersensitive nature of this excitation band is consistent with the accepted view of these transitions.[56]

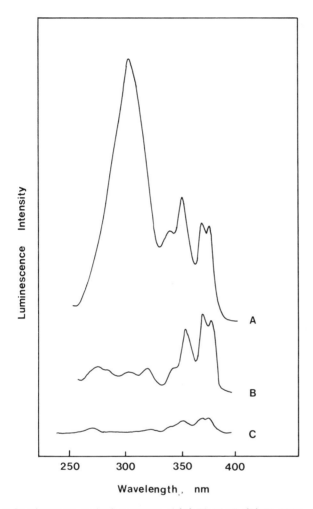

Figure 7-19 ■ Luminescence excitation spectra of (A) Tb-MAE, (B) Tb-PMA, and (C) TbCl$_3$ aqueous solution; $\lambda_{em} = 545$ nm; [— COOH] $= 5 \times 10^{-2}$ N; pH $= 9.5$. The intensity scale is arbitrary.

The relative intensities of Tb^{3+} excitation peaks measured at 305 nm, as observed for the different polycarboxylate complexes, are shown in Table 7-2. With the exceptions of Tb-PAA and Tb-PMA, all of the complex systems exhibited higher luminescence intensities when excited at 305 nm than when excited at 370 nm. The intensity order (for 305 nm excitation) was found to be MA3MPe > MAiB > MAE > PAA > PMA. Because the linear charge density of all polymers was the same, it follows that it was the actual distribution of the carboxylate groups on the chain and their mutual orientation that led to the excitation trend.

Table 7-2 ■ Relative Luminescence Intensity of the Hypersensitive Band
of Tb^{3+}-Polymer Complexes

Polycarboxylate System[a]	Relative Luminescence Intensity[b]
PMA	0.2
PAA	1
MAE	2
MAiB	5
MA3MPe	8

[a]Obtained at $[-COO^-Na^+] = 10$ mN, $[Tb^{3+}] = 1$ mM, and at pH 9.5 ~ 10.
[b]$\lambda_{ex} = 305$ nm and $\lambda_{em} = 545$ nm.

On passing from MAE to MAiB and to MA3MPe, several effects take place. They can be summarized as follows:

1. In the alkaline region, the second acid dissociation constant of the MA copolymers becomes increasingly weaker.[57,58]
2. The chelating ability of MA copolymers for divalent cations increases.[59]
3. The rigidity of the chain increases strongly due to increasing steric hindrance associated with interaction among side chains.
4. The increase in the bulk of the alkyl side chains results in an increase in the hydrophobicity of the chain.

In regard to points 1 and 2, it should be noted that the increase in bulkiness of the alkyl residues on the olefinic comonomer decreases the pK_a of the first dissociation constant, whereas it increases the second pK_a, becoming more basic. Because lanthanide metal ions are known to form stronger complexes when ligand basicity increases, it is reasonable, from these observations, to assume that the order of Tb-polymer complex stability would be MA3MPe > MAiB > MAE > PAA > PMA.[49] It is also reasonable to assume that not only does Tb^{3+} form stronger complexes with the MA copolymers (especially MAiB and MA3MPe), but that the ion binding site of these is more asymmetric than in the other polymers. This conclusion is supported by the fact that the MAiB and MA3MPe polymers would be more rigid than the other polymers, and less likely to adjust their conformational preference to suit the Tb^{3+} ion.

7-4. Determination of Coordinated Water Molecules using Lanthanide Metal Ion Probes

The existence of a deuterium isotope effect in the luminescence properties of lanthanide compounds has been known for some time and has been studied in detail by a number of investigators.[60-62] The OH vibration (3500 cm^{-1}) of coordinated water molecules presents an efficient mechanism whereby the Tb^{3+} excited state may undergo radiationless deactivation, but the O^2H vibration (2800 cm^{-1}) is far less efficient. Horrocks and Sudnick[62] placed these observations on a quantitative scale and developed a method whereby the hydration state of a lanthanide complex could be determined. Luminescence lifetimes measured in H_2O and ^2H_2O are used to obtain the apparent rate constants for radiative decay, and the difference between these two constants gives the number of coordinated water molecules.

Upon addition of poly(sodium acrylate) to an aqueous solution of Tb^{3+}, it was found that the Tb^{3+} luminescence intensity was greatly increased.[35] Repeat of the same experiment in ^2H_2O led an approximate order of magnitude increase in luminescence intensity.[63] The luminescence spectra are shown in Figs. 20 and 21. Addition of simple carboxylic acids leads to the observation of similar effects, although the magnitudes are not as dramatic. These experiments demonstrate that the binding of both simple and polymeric carboxylates significantly perturbs the Tb^{3+} luminescence properties, enabling these quantities to be used in studies of complex formation.

The observed luminescence decay constant k_{obs} is the reciprocal of the measured emission lifetime. In a mixed $H_2O/^2H_2O$ solvent, this quantity may be expressed as[61,62]

$$k_{obs} = k_0 + k_{H_2O} X_{H_2O} \qquad (7\text{-}4)$$

k_0 is the portion of the rate constant consisting of all radiative and nonradiative decay processes not involving solvent deactivation of the excited state. k_{H_2O} is the rate constant for the transfer of energy to the O-H vibrations of coordinated water, and is expected to be proportional to the number of coordinated water molecules. X_{H_2O} is the mole fraction of H_2O in the $H_2O/^2H_2O$ solvent mixture. Horrocks and Sudnick have plotted k_{obs} against X_{H_2O} for a wide variety of Tb^{3+} complexes and have obtained proportionality constants C from the slopes of these plots. The value C obtained for Tb^{3+} by these investigators was 4.2 ms^{-1}.[61] Thus, the number of coordinated water molecules q can be calculated by

$$q = C(k_{H_2O} - k_{^2H_2O}) \qquad (7\text{-}5)$$

The luminescence lifetimes of Tb^{3+} bound to poly(acrylic acid) (PAA), poly[(maleic acid)-co-ethylene] (MAE), poly(methacrylic acid) (PMA), and poly[(acrylic acid)-co-acrylamide] (PAAm) (acrylic acid:acrylamide = 1:2.5) were obtained in both H_2O and ^2H_2O solutions.[63] In addition, analogous data were obtained on several simple carboxylic acids that could be taken as model compounds for the polymers: propionic acid (PrA), glutaric acid (GA), 2,3-dimethylsuccinic acid (DSA), and dimethylmalonic acid (DMA). All pertinent data and computed results are summarized in Tables 7-3 and 7-4.

The number of water molecules bound to Tb^{3+} in its polyacrylate complex was found to be 3.5. The analogous carboxylate monomer to PAA was propionate, and it was observed that the Tb-PrA complex contained 6.0 coordinated water molecules. If it is assumed that the Tb^{3+} aquo complex contains nine water molecules of hydration,[61] then the data of Tables 7-3 and 7-4 imply that about five or six water molecules are expelled upon formation of the Tb-PAA complex, whereas only three are expelled when the Tb-PrA complex is formed. The larger amount of water expelled by the PAA ligand indicates that the Tb^{3+} ion is more strongly bound by the polymer, and this increase in binding ability could be attributed to the polymer cooperative effect.[64]

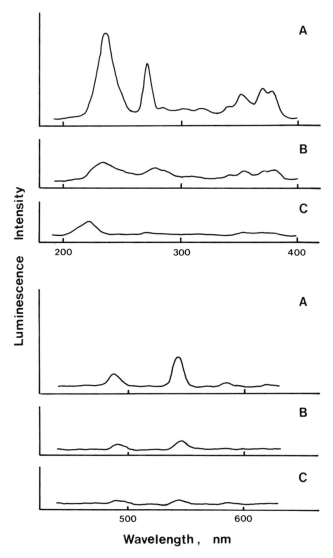

Figure 7-20 ■ (Top) Luminescence excitation spectra of Tb^{3+} complexes in H$_2$O solution; λ_{em} = 545 nm. (Bottom) Luminescence emission spectra of Tb^{3+} complexes in H$_2$O solution; λ_{ex} = 368 nm. (A) Tb-PAA; (B) Tb-PrA; (C) TbCl$_3$. The intensity scale is arbitrary.

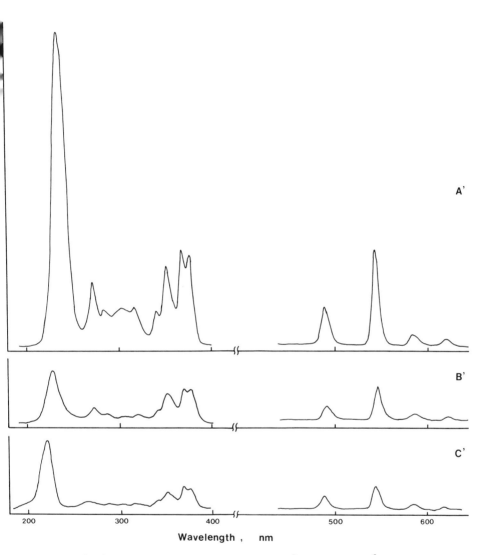

Figure 7-21 ■ (Left) Luminescence excitation spectra of Tb^{3+} complexes in ^2H_2O solution; λ_{em} = 545 nm. (Right) Luminescence emission spectra of Tb^{3+} complexes in ^2H_2O solution; λ_{ex} = 370 nm. (A′) Tb-PAA; (B′) Tb-PrA; (C′) TbCl$_3$. The intensity scale is arbitrary.

Table 7-3 ▪ **Number of Coordinated Water Molecules and Photophysical Parameters for Tb^{3+}-Polymeric Carboxylate Complexes**

Ligand[a]	Solvent	τ (μs)	k_{obs} (m s^{-1})	Δk_{obs} (m s^{-1})	Number of H_2O[b]
PAA[c]	H_2O	780	1.28	0.84	3.5
	^2H_2O	2300	0.44		
MAE[d]	H_2O	740	1.36	0.85	3.5
	^2H_2O	1980	0.51		
PAAm[e]	H_2O	690	1.44	0.91	3.8
	^2H_2O	1900	0.53		
PMA[f]	H_2O	760	1.32	0.93	3.9
	^2H_2O	2560	0.39		

[a] $[-COO^-Na^+] = 50$ mN and $[Tb^{3+}] = 2$ mM.
[b] 4.2 m s^{-1} used as the constant of proportionality.
[c] PAA: poly(acrylic acid).
[d] MAE: poly[(maleic acid)-*alt*-ethylene].
[e] PAAm: poly[(acrylic acid)-*co*-acrylamide] (acrylic acid:acrylamide = 1:2.5).
[f] PMA: poly(methacrylic acid).

The copolymer of maleic acid-ethylene (MAE) is also known to exhibit excellent metal binding properties.[59] A Tb^{3+} hydration number of 3.5 was measured from the luminescence lifetime determinations and it is quite close in magnitude to the value obtained for the PAA complex. The relative magnitudes of Tb^{3+} emission intensities in the PAA and MAE systems were also quite similar, indicating that the metal ion binding properties were comparable in the two polymers. These observations agree with the results of the dilatometric investigation by Begala and Strauss.[40] They found

Table 7-4 ▪ **Number of Coordinated Water Molecules and Photophysical Parameters for Tb^{3+}-Monomeric Carboxylate Complexes**

Ligand[a]	Solvent	τ (μs)	k_{obs} (m s^{-1})	Δk_{obs} (m s^{-1})	Number of H_2O[b]
PrA[c]	H_2O	530	1.87	1.45	6.0
	^2H_2O	2370	0.42		
DMA[d]	H_2O	460	2.17	1.52	6.3
	^2H_2O	1550	0.65		
DSA[e]	H_2O	550	1.83	1.39	5.8
	^2H_2O	2260	0.44		
GA[f]	H_2O	550	1.83	1.35	5.6
	^2H_2O	2100	0.48		

[a] $[COO^-Na^+] = 50$ mN and $[Tb^{3+}] = 2$ mM.
[b] 4.2 m s^{-1} used as the constant of proportionality.
[c] PrA: propionic acid.
[d] DMA: dimethylmalonic acid.
[e] DSA: 2,3-dimethylsuccinic acid.
[f] GA: glutaric acid.

that the volume changes upon the binding of alkaline-earth metal ions such as Mg^{2+} and Ba^{2+} to PAA and MAE were fairly close. They concluded that the chelate formation of the metal ions with two adjacent carboxyl groups is of minor importance because seven- or eight-membered rings would be involved in chelation for these polyelectrolytes.

When Tb^{3+} is bound to poly(methacrylate) (PMA), slightly less enhancement of Tb^{3+} luminescence intensity (relative to the Tb-PAA complex) was observed. At the same time, the number of coordinated water molecules was found to be 3.9. These results indicate that the Tb^{3+} binding property of PMA is slightly different from the analogous binding properties of PAA or MAE. Strauss and Leung also found that the volume change on complexation of alkaline-earth metal ions with PMA was smaller than that with PAA.[39] When a methyl group is attached to the polymer chain, the polymer becomes less flexible.[65] This inflexibility of the PMA chain interferes somewhat with the formation of a multidentate metal ion binding site because the polymer cannot fold in such a manner to present an equivalent site. Interaction with a Tb^{3+} ion becomes more difficult, and hence PMA cannot form a strong Tb^{3+} complex, like those of PAA and MAE.

In the case of the Tb-PAAm complex, the number of coordinated water molecules was found to be 3.8, and the enhancement of Tb^{3+} luminescence intensity was comparable to that found in PAA- and MAE-Tb^{3+} complexes. These results indicate that although the carboxylic groups are randomly distributed in the polymer chain, the Tb^{3+} ion binding site of the copolymer is similar to that of the homopolymer PAA. This is probably because at this monomer composition (AAm/AA = 2.5), spacing between carboxylates is not large enough to prevent strong complex formation. However, the spacer, acrylamide, reduces electric repulsion among the carboxylates and consequently makes the polymer chain of the copolymer more flexible than that of the homopolymer. As a result, the copolymer shows strong Tb^{3+} binding ability, comparable with that of the homopolymer of acrylic acid.

The Tb^{3+} binding properties of several monomeric dicarboxylic acid ligands were also studied, and the results are summarized in Table 7-4. The Tb^{3+} complexes of glutaric, 2,3-dimethylsuccinic, and dimethylmalonic acids all contain approximately six waters of hydration, as did the propionic acid complex. The luminescence intensities of these complexes were comparable, and the results of these investigations imply the formation of similar complexes in each case. The luminescence lifetime results indicate that probably three waters of hydration are expelled upon formation of the complex.

The results show that the number of water molecules bound to Tb^{3+}-monomeric carboxylate complexes is significantly larger than to the polymeric carboxylate. These values provide strong evidence that the flexible polymers wrap around the Tb^{3+} ion, forming a strong and definite metal ion binding site. Thus the large enhancement of the luminescence intensity of Tb^{3+} in the flexible polymer-Tb^{3+} complexes in the aqueous solution can be accounted for by (1) the decrease of coordinated water molecules on Tb^{3+} and (2) the strong and also asymmetrical binding of carboxylate to Tb^{3+}.

These results show that the lanthanide ion luminescence probe technique is also useful for the elucidation of the ion binding properties in aqueous polyelectrolyte solutions.

References

1. F. A. Cotton and G. Willkinson, *Advanced Inorganic Chemistry*, 4th ed., Interscience, New York (1980), Chapter 23.
2. C. K. Jørgensen, *Orbitals in Atoms and Molecules*, Academic, New York (1962), Chapter 11.
3. T. Moeller, *The Chemistry of the Lanthanides*, Reinhold, New York (1963), Chapters 3 and 4.
4. G. Schwarzenbach, *Advances in Inorganic Chemistry and Radiochemistry*, Vol. 3, H. J. Emeleus and A. G. Sharpe, eds., Academic, New York (1961), p. 265.
5. E. Nieboer and W. A. E. McBryde, *Can. J. Chem.*, **51**, 2511 (1973).
6. R. J. P. Williams and J. D. Hale, *Struct. Bonding*, **1**, 249 (1966).
7. R. G. Pearson, *J. Chem. Educ.*, **45**, 581 (1968).
8. G. Klopman, *J. Am. Chem. Soc.*, **90**, 223 (1968).
9. T. Moeller, *Inorganic Chemistry Series One*, K. W. Bagnall, ed., University Park Press, Baltimore (1972), Vol. 7, p. 275.
10. F. S. Richardson, *Chem. Rev.*, **82**, 541 (1982).
11. L. Holliday, ed., *Ionic Polymers*, Applied Science, London, (1975).
12. A. Eisenberg and M. King, *Ion-Containing Polymers*, Academic, New York (1977).
13. A. Eisenberg, *Polym. Prep., Am. Chem. Soc. Div. Polym. Chem.*, **20**(1), 286 (1979).
14. Y. Okamoto, Y. Ueba, N. F. Dzhanibekov, and E. Banks, *Macromolecules*, **14**, 17 (1981).
15. L. G. Van Uitert, *J. Electrochem. Soc.*, **107**, 803 (1960).
16. A. Eisenberg and M. Navratil, *Macromolecules*, **6**, 604 (1973).
17. R. D. Lundberg and H. S. Makowski, *Polym. Prep., Am. Chem. Soc., Div. Polym. Chem.*, **19**, 287 (1978).
18. S. P. Tanner and D. L. Thomas, *J. Am. Chem. Soc.*, **96**, 706 (1974).
19. R. Reisfeld, E. Greenberg, R. Velapoldi, and B. Barnett, *J. Chem. Phys.*, **56**, 1698 (1972).
20. R. W. Dawson, L. J. Kropp, and M. W. Windsor, *J. Chem. Phys.*, **45**, 2410 (1966).
21. E. Nakazawa and S. Shinoya, *J. Chem. Phys.*, **47**, 3211 (1967).
22. R. Reisfeld and L. Boehm, *J. Solid State Chem.*, **4**, 417 (1972).
23. R. Reisfeld, E. Greenberg, and E. Biron, *J. Solid State Chem.*, **9**, 224 (1974).
24. W. D. Horrocks, Jr., B. Holmquist, and B. L. Vallee, *Proc. Nat. Acad. Sci. USA*, **72**, 4764 (1975).
25. D. DeShazer and A. Y. Cabezas, *Proc. IEEE*, **52**, 1355 (1964).
26. Y. Okamoto, Y. Ueba, I. Nagata, and E. Banks, *Macromolecules*, **14**, 807 (1981).
27. I. Nagata, R. Li, E. Banks and Y. Okamoto, *Macromolecules*, **16**, 903 (1983).
28. T. D. Gierke and W. Y. Hsu, *Macromolecules*, **15**, 101 (1982).
29. T. Föster, *Discuss. Faraday Soc.*, **27**, 7 (1959).
30. H. S. Frank and M. W. Evans, *J. Chem. Phys.*, **13**, 507 (1945).
31. S. Freed, *Rev. Mod. Phys.*, **14**, 105 (1942).
32. T. Maekawa, H. Nishide, E. Tsuchida, H. Ohmichi, and J. Okamoto, *J. Appl. Polym. Sci.*, **29**, 3795 (1984).
33. H. Nishide, T. Izushi, N. Yoshioka, and E. Tsuchida, *Polymer Bull.*, **14**, 387 (1985).
34. J. Kido, H. Akiba, H. Nishide, E. Tsuchida, H. Ohmichi, and J. Okamoto, *Kobunshi Ronbunshu*, **43**, 31 (1986).
35. I. Nagata and Y. Okamoto, *Macromolecules*, **10**, 773 (1977).
36. W. K. Olson, *Biopolymers*, **14**, 1775 (1975).
37. U. P. Strauss and A. Siegel, *J. Phys. Chem.*, **67**, 2683 (1963).
38. A. M. Kotliar and H. Morawetz, *J. Am. Chem. Soc.*, **77**, 3692 (1955).
39. U. P. Strauss and Y. P. Leung, *J. Am. Chem. Soc.*, **87**, 1476 (1965).
40. A. J. Begala and U. P. Strauss, *J. Phys. Chem.*, **76**, 254 (1972).
41. F. T. Walls and S. J. Gill, *J. Phys. Chem.*, **58**, 1128 (1954).
42. R. J. Eldridge and F. I. Treloar, *J. Phys. Chem.*, **74**, 1446 (1970).

43. H. Morawetz and R. H. Gobran, *J. Polym. Sci.*, **12**, 133 (1954).
44. A. Ikegami, *J. Polym. Sci.*, *Part A*, **2**, 907 (1964).
45. J. M. Brewer, L. A. Carreira, R. M. Irwin, and J. I. Elliot, *J. Inorg. Biochem.*, **14**, 33 (1981).
46. J. L. Kropp and M. W. Windsor, *J. Chem. Phys.*, **39**, 2769 (1963).
47. J. L. Kropp and M. W. Windsor, *J. Chem. Phys.*, **42**, 1599 (1965).
48. G. Stein and E. Wruzberg, *J. Phys. Chem.*, **62**, 208 (1975).
49. Y. Okamoto, J. Kido, H. G. Brittain, and S. Paoletti, *J. Macromol. Sci. Chem.*, **A25**, 1385 (1988).
50. H. Morawetz, *J. Polymer Sci.*, **17**, 442 (1955).
51. M. S. Kayne and M. Cohn, *Biochemistry*, **13**, 4149 (1974).
52. J. M. Wolfson and D. R. Kearns, *Biochemistry*, **14**, 1436 (1975).
53. J. M. Brewer, L. A. Carreira, R. M. Irwin, and J. I. Elliot, *J. Inorg. Biochem.*, **14**, 33 (1981).
54. V. Crescenzi, H. G. Brittain, N. Yoshino, and Y. Okamoto, *J. Polym. Sci.*, *Polym. Phys. Ed.*, **23**, 437 (1985).
55. A. Rudman, S. Paoletti, and H. G. Brittain, *Inorg. Chem.*, **24**, 1283 (1985).
56. D. E. Henrie, R. L. Fellows, and G. R. Choppin, *Coord. Chem. Rev.*, **18**, 199 (1976).
57. E. Bianchi, A. Ciferri, R. Parodi, R. Rampone, and A. Tealdi, *J. Phys. Chem.*, **74**, 1050 (1970).
58. V. Crescenzi, F. Delben, S. Paoletti, and J. Skerjanc, *J. Phys. Chem.*, **78**, 607 (1974).
59. S. Paoletti, F. Delben, and V. Crescenzi, *J. Phys. Chem.*, **80**, 2564 (1976).
60. W. D. Horrocks, Jr., G. F. Schmidt, D. R. Sudnick, C. Kittrel, and R. A. Bernheim, *J. Am. Chem. Soc.*, **99**, 2378 (1977).
61. W. D. Horrocks, Jr. and D. R. Sudnick, *J. Am. Chem. Soc.*, **101**, 334 (1979).
62. W. D. Horrocks, Jr. and D. R. Sudnick, *Science*, **206**, 1194 (1979).
63. J. Kido, H. G. Brittain, and Y. Okamoto, *Macromolecules*, **21**, 1872 (1988).
64. M. Kaneko and E. Tsuchida, *Macromol. Rev.*, **16**, 467 (1980).
65. M. Shirai, T. Nagatsuka, and M. Tanaka, *Makromol. Chem.*, **178**, 37 (1977).

Local Structure of Macromolecule-Metal Complexes in Solution

Hitoshi Ohtaki and Hideki Masuda

Coordination Chemistry Laboratories, Institute for Molecular Science, Myodaiji-cho, Okazaki, 444

The structure of metal complexes of macromolecular ligands has been widely studied, especially in connection with biochemically important materials. Most of the structures have been determined by the single crystal X-ray diffraction method for solid samples, and some other methods such as Mössbauer and Raman spectroscopies have been used for studies on the structures and structural changes of various complexes. In such cases the structures of the complexes to be referred to are usually known from X-ray crystallographic determinations.

The structure of complexes in solution has so far been discussed by measuring NMR chemical shifts, broadening and splitting of the signals, Raman spectra, thermodynamic properties, and UV and visible spectra of complexes in solution, and so forth, on the basis of the knowledge of the structure of the complexes in the solid state. The solution X-ray diffraction method becomes important to determine the structure of complexes dissolved in solution, although there still remain experimental difficulties requiring a relatively large amount of a concentrated solution of samples. This method, however, has a great advantage in that it is not necessary to isolate complexes from solution as a single crystal for the structure determination.

Recently developed extended X-ray absorption fine structure (EXAFS) and X-ray absorption near edge structure (XANES) methods give us very useful information on the local structure and the electronic state of the central atoms. These methods are applicable to solution, powder, and amorphous samples.

In this chapter we describe the methods of determination of the local structure of metal complexes in solution by means of solution X-ray diffraction, EXAFS, and XANES, and the results obtained by these methods are compared with those determined by the X-ray crystallographic method where possible.

8-1. Methods of Measurement of the Local Structure of Complexes in Solution

The crystallographic X-ray diffraction method is already described in many textbooks, but the solution X-ray diffraction and EXAFS and XANES methods are still not too familiar to most readers. Thus, we briefly describe the principles of the methods in this section.

8-1-1. Solution X-Ray Diffraction Method

There are, in principle, two methods for determining the solution structure. One is the reflection method, in which X-ray beams radiated to a sample solution are reflected at the surface of the solution. The other is the transmission method, in which X-ray beams are transmitted through a sample solution and diffracted.

The Reflection Method. In this method a sample solution with a free surface is placed in a vessel that is situated in a sample chamber in order to prevent evaporation. X-ray beams pass through a window of the chamber that is nearly transparent to X-rays irradiating the surface of the sample, and intensities of the diffracted X-rays are measured with a detector. The scheme of the apparatus is illustrated in Fig. 8-1.

In this method the X-ray tube and the detector move in opposite directions in the same plane vertical to the sample surface by keeping the angular relationship of θ for both incident and diffracted X-rays over the whole angle region examined. X-rays are monochromatized before or after the diffraction. Mo $K\alpha$ radiation ($\lambda = 71.07$ pm) is used for most cases, but Ag $K\alpha$ radiation ($\lambda = 56.08$ pm) is occasionally used. The wavelength of Cu $K\alpha$ radiation ($\lambda = 154.05$ pm) is so long that a reasonably wide range of the scattering vector $s = 4\pi\lambda^{-1}\sin\theta$ can hardly be covered in this method, and thus rather serious errors are often introduced when the Fourier transform is carried out for reduced intensities to obtain radial distribution curves.

The Transmission Method. This method is similar to the powder X-ray diffraction method: X-ray beams are transmitted through a sample solution, which usually has either a cylindrical or a thin plate form in a cell with thin walls.

In Fig. 8-2 a schematic picture of the transmission method is shown, in which a planar monochromatized X-ray beam emitted from an X-ray tube irradiates a cylin-

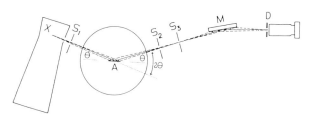

Figure 8-1 ▪ A schematic picture of an X-ray diffractometer of the reflection type. X, X-ray tube; S_1, S_2, S_3, slits; M, monochromator; A, sample solution; D, detector. The θ-θ relationship is held throughout the measurement.

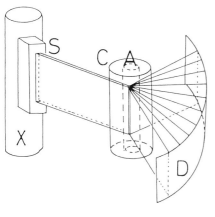

Figure 8-2 ▪ A schematic picture of a cell arrangement of the transmission method.

drical sample in a tube and diffracted X-rays are detected by a semicircular detector. Alternatively, a detector can sweep over a reasonably wide angle range to detect diffracted X-ray beams. The detector should be protected from direct X-ray beams from the source.

The measurement of the intensity of scattered X-rays at various angles is called the *angular dispersive method*. Because the value $s = 4\pi\lambda^{-1}\sin\theta$ is obtainable as a function of both λ and θ, we can use white X-rays of various wavelengths at a given θ value instead of using monochromatized X-rays at various angles. In this case a multichannel analyzer for analyzing different energies of diffracted X-rays must be used. This method is called the *energy dispersive method*. The former method is employed for most cases, but the latter method has an advantage in measurement over a wide range of s values as a single measurement in a relatively short time. Instead, elimination of incoherent X-rays due to the Compton effect is practically impossible. If fluorescent X-rays are produced by irradiation of a sample by white X-ray, the fluorescent X-rays thus produced can hardly be eliminated in the course of the measurement. Thus, the energy dispersive method may not be suitable for solutions containing heavy atoms, which may exist in metal complexes.

Measured intensities of scattered X-rays are corrected for polarization, absorption, multiple scatterings, and incoherent scatterings and are scaled to absolute intensities. The corrected intensities are Fourier transformed to obtain the radial distribution function, from which we can reduce the structure data of complexes in solutions. The structural data thus obtained are usually refined by applying a least-squares method to the corrected intensities.

8-1-2. EXAFS and XANES Measurements

Each atom can absorb X-rays with characteristic wavelengths to emit photoelectrons from their electron shells. The energies of the X-rays absorbed by a given atom are affected by the electronic state of the atom and surrounding atoms that interact with it. The emitted electrons are scattered by electrons of surrounding atoms. Thus, the

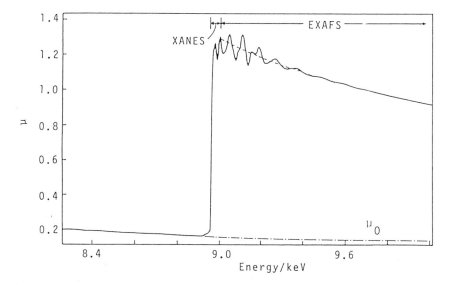

Figure 8-3 ▪ X-ray absorption spectrum of copper metal. The background curve μ_0 is drawn by extrapolating the curve in the low-energy region. The broken line shows the assumed absorption curve without the fine structure (modulation) in the spectrum.

position of the absorption edge of X-rays and the shape of the absorption spectrum of a given atom change depending on the oxidation state, the local structure around the central atom, and the structure of the surrounding ligands.

A typical pattern of the absorption spectrum is shown in Fig. 8-3. The spectrum near the absorption edge is called the X-ray absorption near edge structure (XANES), which corresponds to the Kossel structure in the X-ray absorption spectra of crystals. The rest of the spectrum is called extended X-ray absorption fine structure (EXAFS), which may be compared with the Kronig structure in the X-ray absorption spectra of solids. The absorptivity μ is measured as a ratio of intensities of the incident beam and the beam after absorption by a sample. The background curve μ_0 is drawn by extrapolating the curve before absorption starts. Extraction of the modulation from the EXAFS part of the spectrum is carried out by estimating a smooth background curve with an assumption of no fine structure in the spectrum.

XANES and EXAFS spectra allow us to obtain structural information around the central atom, which absorbs a characteristic X-ray.

Strong X-ray beams containing various wavelengths can be produced by the synchrotron orbital radiation (SOR), and thus they are very suitable to measure XANES and EXAFS spectra with high accuracy. Recently a small scale EXAFS apparatus, which can be used in laboratories, has been developed. An example of the laboratory scale EXAFS apparatus[1] is shown in Fig. 8-4. X-rays emitted from an X-ray generator constructed from a rotatory anode are reflected at a bent crystal, and then the intensity I_0 of the incident beam is measured by an ion chamber before entering the sample. The intensity of the beam after absorption by the sample is

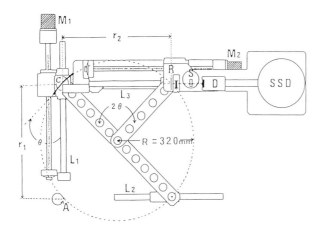

Figure 8-4 ■ A schematic picture of a laboratory scale EXAFS apparatus. A, X-ray source of a rotatory anode; C, a bent crystal; R, receiving slit; I, ion chamber; S, sample; D, SSD detector; L_1, L_2, L_3, moving arms; M_1, M_2, pulse motors. [K. Tohji, Y. Udagawa, T. Kawasaki, and K. Masuda, *Rev. Sci. Instrum.*, **54**, 1482 (1983).]

determined by a detector. Thus the intensity ratio $\mu = I/I_0$, where I and I_0 are the intensities of the incident beam and of the beam after absorption by a sample, respectively, is obtained at various energies of the incident X-rays.

The modulation is extracted from the EXAFS spectrum after the estimation of a reasonable background, and then the data are Fourier transformed to obtain a function that corresponds to the radial distribution function in the X-ray diffraction method. The Fourier transform thus obtained gives preliminary information about the structure of complexes in the system, and the structural data are subsequently subjected to the least-squares refinement.

8-2. Local Structure of Metal Complexes with Relatively Large Ligands in Solution

This section surveys structural data of metal complexes with organic ligands larger than simple inorganic ligands, which may be regarded as moieties of biochemically interesting macromolecules.

8-2-1. Metal Complexes with Amino Acids

Amino acids are the most fundamental ligands that interact with metal ions in biochemical systems. Although a number of structures of metal complexes with amino acids have so far been determined by the X-ray crystallographic method, systematic investigations of structures of a series of complexes with one kind of amino acid and a given metal ion have scarcely been carried out because of the difficulty of preparation

Table 8-1 ▪ The Bond Lengths between Metal Ions and Ligating Atoms in the Amino Acid Complexes of Divalent Metal Ions in Aqueous Solutions at 25°C (in picometers)

Complex	Ni^{2+}	Cu^{2+}	Zn^{2+}
$[M(gly)(H_2O)_4]^+$	Ni—OH$_2$: 208 Ni—O: 209 Ni—N: 209	Cu—OH$_2$(eq): 198 Cu—OH$_2$(ax): 227 Cu—O(eq): 199 Cu—N(eq): 199	Zn—OH$_2$: 212 Zn—O: 212 Zn—N: 212
$[M(gly)_2(H_2O)_2]^0$	Ni—OH$_2$: 206[a] (210[b]) Ni—O: 206[a] (206[b]) Ni—N: 206[a] (208[b])	—	Zn—OH$_2$: 208[a] Zn—O: 208[a] Zn—N: 208[a]
$[M(gly)_3]^-$	Ni—O: 203 (203[a]) Ni—N: 214 (212[a])	Cu—O: 202 Cu—N: 202	Zn—O: 212 (214[a]) Zn—N: 212 (214[a])
$[M(\alpha - ala)(H_2O)_4]^+$	—	—	Zn—N: 202 Zn—O: 214
$[M(\alpha - ala)_2(H_2O)_2]^0$	—	—	Zn—N: 203 Zn—O: 214
$[M(\alpha - ala)_3]^-$	—	—	Zn—O: 202 Zn—N: 213

[a] EXAFS data, ref. 5.
[b] X-ray crystallographic data, ref. 7.

of single crystals of a series of the complexes with varying numbers of ligands. Because there is an advantage in the solution X-ray diffraction method in which one need not prepare single crystals for the structure determination, one can study the structures of a series of complexes of a metal ion with one kind of ligand and various ligand numbers. Nevertheless, very limited structural investigations of amino acid complexes of metal ions in solution have been conducted so far.

The structures of mono- and tris(glycinato) complexes of nickel(II),[2] copper(II),[3] and zinc(II)[4] ions have been studied by the solution X-ray diffraction method. The structures of the bis(glycinato) complexes of nickel(II) and zinc(II) ions in water have been determined by the EXAFS method[5] and the structure of the tris(glycinato) complexes of these ions has also been examined by the EXAFS method[5] in order to compare the structural data with those obtained by the solution X-ray diffraction method.

The structure of α-alaninato complexes of zinc(II) ions[6] has also been determined by the solution X-ray method. Distances between the metal ions and the ligand atoms of these complexes are summarized in Table 8.1. The X-ray crystallographic data of the bis(glycinato) complexes of nickel(II) ions are also shown in the table for comparison.[7]

In all cases, the amino acid complexes of the metal ions are octahedral. The results are different from those found for the ethylenediamine complexes in which the bis- and tris- complexes of zinc(II) ions are tetrahedral,[8] whereas those of the other divalent metal ions are octahedral[9-11] (no structural investigation has been done for the monocomplexes of ethylenediamine of the divalent metal ions).

In crystalline form the bis(glycinato)zinc(II) complex is octahedral, although no bond lengths in the complex have been reported.[12] Monoaspartatozinc(II) complex has two water molecules in the crystalline state, and the structure around the zinc(II) ion is octahedral.[13] The $Zn-N$ bond length is determined to be 207.7 pm, which is close to the $Zn-N$ bond length in the glycinato and α-alaninato zinc(II) complexes in solution and in the solid state (see Table 8-1). However, $Zn-O$ lengths in the monoaspartatozinc(II) complex are found to be 206.0–221.2 pm, and the average $Zn-O$ bond length in the complex is 214.6 pm, which is much longer than the $Zn-O$ bond length found in the amino acid complexes in solution.

Zinc(II) complexes of glutamic acid,[14] L-serine,[15] and triglycine[16] have a penta-coordinate structure where the average bond lengths of the Zn-ligand atom are 206.9, 207.0, and 205.8 pm, respectively, which are shorter than the bond lengths in the complexes of the hexa-coordinate structure. These complexes have one or two nitrogen atoms in the coordination sphere.

Carboxypeptidase A has two nitrogen atoms and three oxygen atoms in the coordination sphere, and the average bond length of all metal-ligand pairs is 214 pm.[17, 18] The long Zn-ligand bond length in carboxypeptidase A suggests that the interaction between the central metal ion and the coordinated ligands is weaker than that in the usual amino acid complexes that have no enzymatic properties, and thus activities of the enzyme may partly arise from a higher net charge of the central zinc(II) ion than that in the complexes with simple amino acids.

Thermolysin has two nitrogen and two oxygen atoms in the coordination sphere of the central zinc(II) ion[19] and the average metal-ligand bond length is 203.5 pm. In the complex, the $Zn-OH_2$ length is very short (188 pm) compared with the other three bonds (208–210 pm). Among zinc(II) amino acid complexes with two nitrogen atoms as ligands, no tetra-coordinated complex has been observed so far in either solution or crystal. No structural information has been reported for diamminezinc(II) and monoethylenediaminezinc(II) complexes. Although a definite conclusion cannot be drawn from such insufficient information, one can suppose that thermolysin may have unsaturated character in the coordination structure of the zinc(II) ion, and the enzymatic activity of thermolysin may be partly due to a high reactivity of the zinc(II) ion with an unsaturated coordination shell.

8-2-2. Metal Complexes with Macrocyclic Ligands

Metal complexes with macrocyclic ligands are found in many biological systems, and thus macrocyclic ligands of relatively small molecular weight can be good model compounds for biochemical investigations of naturally occurring macromolecular macrocyclic ligand complexes. For example, macrocyclic polyethers act as model compounds of the ionophores nonactin and valinomycin and they have a high selectivity for alkali metal ions. Nitrogen-containing macrocyclic ligands, such as cyclam (1, 4, 8, 11-tetraazacyclotetradecane) and its relatives, can also be regarded as models of porphyrin and its derivatives. In the complex formation reactions of macrocyclic ligands with metal ions, the ligands sometimes undergo ring conforma-

Table 8-2 ■ The Interatomic Distances in the 18-Crown-6 Complexes
with Alkali Metal Ions

r	Na^+		K^+		Cs^+	
(pm)	Solution	Crystal	Solution	Crystal	Solution	Crystal
M—O	245	255^a	277	280^a	$\begin{cases} 312 \\ 344 \end{cases}$	306^a
M...C	340	339^a	364	362^a	$\begin{cases} 384 \\ 410 \end{cases}$	$\begin{cases} 377^a \\ 398^a \end{cases}$

[a]Average of the interatomic distances in crystal.

tional changes. The flexibility of the ring may play an important role in the complex formation reaction and in anlogous biochemical reactions. However, the solution X-ray diffraction data and EXAFS spectra are rather insensitie to changes in the ring conformation compared with the methods by Raman and NMR spectroscopy, and thus only the local structure around the central metal ions in the complexes can usually be determined by the solution X-ray diffraction and EXAFS methods.

In this section we will discuss the local structure of 18-Crown-6 complexes with alkali metal ions and 1, 4, 8, 11-tetraazacyclotetradecane with nickel(II) and copper(II) ions in aqueous solution.

Alkali Metal Complexes with 18-Crown-6. From the analysis of solution X-ray radial distribution curves[20] we obtained the metal–oxygen and nonbonding metal–carbon distances in the alkali metal complexes of 18-Crown-6 in water. The interatomic lengths are summarized in Table 8-2.

In the crystal structure, the metal–oxygen and metal–carbon lengths vary, and thus, they are averaged in order to compare the lengths obtained in the complexes in solution with those in the crystal. In the sodium[21] and potassium[22] complexes no significant difference has been observed between the metal–oxygen and metal–carbon distances in solid and in solution. The cesium complex of 18-Crown-6 was crystallized with thiocyanate as the counterion. It has a dimeric structure in the solid state,[23] but in solution the complex has a monomeric structure. The structures estimated from the solution X-ray diffraction data[20] and other supporting data from Raman spectroscopic measurements[24] and molecular mechanical calculations[25] are depicted in Fig. 8-5.

The cone angles shown in the figure indicate possible regions of coordination of water molecules to the central metal ions, which are estimated from the van der Waals radii of oxygen, carbon, and hydrogen atoms in the ligand. In the sodium complex, which has the C_1 conformation of the ring, one water molecule may interact with the sodium ion in the complex [Fig. 8-5(a)]. In the potassium complex the crown ring has the D_{3d} conformation, and thus the complex has enough space to be coordinated with water molecules from above and below the crown ring as shown in the Fig. 5(b). Because the cesium ion has a large ionic radius, it cannot situate at hte center of the crown ring and is located apart from the mean plane of the ring. The

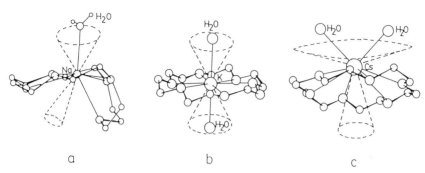

a　　　　　　　　b　　　　　　　　c

Figure 8-5 ▪ A proposed structure of hydrated 18-Crown-6 alkali metal complexes. Broken lines indicate cone angles above and below the metal ion in the complexes. (a) Sodium complex with C_1 conformation of the 18-Crown-6 ring. (b) Potassium complex with D_{3d} conformation. (c) One of the possible structures of the cesium complex that has the D_{3d} conformation. [K. Ozutsumi, M. Natsuhara, and H. Ohtaki, *Bull. Chem. Soc. Jpn.*, **62**, 2807 (1989).]

conformation of the ring structure was not conclusive enough to be either C_1 or D_{3d}; both models fit the experimental results reasonably well. Because the ring may be more flexible than that in the sodium and potassium complexes, both conformations may coexist in solution. The frequency factor for the Cs—O interactions obtained from the intensity data indicates that the cesium ion may be solvated with one to two water molecules in solution. The cone angle shown in Fig. 8-5(c) suggests a possibility of the coordination of two water molecules to the central cesium ion in the complex from one side of the crown ring.

Nickel(II) and Copper(II) Complexes with Cyclam. Cyclam $(1, 4, 8, 11$-tetra-azacyclotetradecane) forms stable complexes with various divalent metal ions, and reactivities of the complexes have been widely studied. The crystal structures of cyclam complexes have also been determined for many systems; however, the structures of the complexes in solution have not been thoroughly investigated compared with those in the solid state. Although these complexes are expected to be solvated in solution, no structural data have been reported for solvated complexes in solution. It is obvious that knowledge of the hydration structure of such complexes is very important for elucidating reaction mechanisms of cyclam complexes.

According to the result[26] of a solution X-ray diffraction measurement, the square-planar nickel(II) cyclam complex, which is in equilibrium with the paramagnetic octahedral complex though the equilibrium is shifted very much toward the square-planar complex, has no water molecules, as expected from the magnetic and spectroscopic properties. The Ni—N bond length was determined to be 197.6 pm, which can be compared with the X-ray crystallographic data as shown in Table 8-3.

Upon addition of ammonia to a nickel(II)-cyclam solution, the color changes, and the cyclam-ammine mixed ligand complex of nickel(II) forms. This complex has a regular octahedral structure having an Ni—N bond length of 209.0 pm.[26] The bond

Table 8-3 ▪ The Ni—N and Cu—N Bond Lengths in the Cyclam Complexes in Solution and Crystalline States

Metal	Ligand	Counterion	$r_{M—N}$ (pm)	Form[a]	Ref.
Ni^{2+}	cyclam	Cl^-	197.6	sq, sol	25
	cyclam, NH_3	Cl^-	209.0	oh, sol	25
	cyclam	I^-	195	sq, cry	26
	cyclam	ClO_4^-	197	sq, cry	27
	cyclam	$ZnCl_4^{2-}$	190, 194	sq, cry	28
	cyclam, H_2O	Cl^-	209–211	oh, cry	28
	cyclam, Cl^-	Cl^-	207	oh, cry	29
	cyclam, NCS^-	SCN^-	$\begin{cases} 206\text{–}208 \text{ (plane)} \\ 208\text{–}213 \text{ (axis)} \end{cases}$	oh, cry	29
	cyclam, NO_3^-	NO_3^-	205–206	oh, cry	30
Cu^{2+}	cyclam, H_2O	Cl^-	201.3	dis-oh, sol	25
	cyclam, ClO_4^-	ClO_4^-	202	dis-oh, cry	31
	cyclam, $SC_6F_5^-$	$C_6F_5S^-$	201	dis-oh, cry	32

[a] sq = square-planar, sol = solution, oh = octahedral, cry = crystal, dis-oh = distorted octahedral.

length is compared with those of other octahedral complexes of the nickel(II)-cyclam system[27-31] in Table 8-3.

The copper(II)-cyclam complex is hydrated with water molecules to form a distorted octahedral structure. However, the axial Cu—OH$_2$ bond is so long, 277 pm, that the complex can be regarded as essentially a square-planar one.[26] The Cu—N bond length of the complex found in water is essentially the same as those in the solid state.[32, 33]

From the structure determination of the cyclam complexes in water and the comparison of the structural data of the complex in solution with those in crystals, it can be concluded that the metal–nitrogen bond lengths in these systems are not significantly affected by the counterions and the states (to be either in solution or in the solid). The invariable metal–ligand bond length may be due to strong metal–nitrogen interactions in the complex and the ions can be accommodated at the center of the ring. However, it should be noted that the Ni—N bond length in the square-planar complex is shorter than that of the octahedral complex. The results indicate that the metal–nitrogen bond length is still affected by the coordination structure and, probably, by the electron donation from the ligands axially coordinated to the central metal ions.

The copper(II) cyclam complex has a distorted octahedral structure, but the structure can also be regarded as practically square-planar because of the long Cu—OH$_2$ bond at the axial position. Nevertheless, the Cu—N bond is still longer than that in the bis(ethylenediamine) complexes,[10] which can also be regarded as a square-planar structure. The result suggests that the cyclam ring can expand slightly but is difficult to shrink due to steric congestion. The expansion of the ring may be accompanied by the deformation of the whole ring structure.

8-2-3. Metal Complexes with Sugar-Type Ligands

Interactions between nucleic acid and metal ions are essential to living organisms. Many investigations have been carried out for the structure determination of metal complexes of the purine and pyrimidine bases of nucleic acids, nucleosides, and nucleotides by the X-ray crystallographic method. The structures of the complexes may be too complicated to be analyzed by the solution X-ray diffraction method, but the EXAFS method is applicable to solution samples of some complexes. Metal complexes with sugar components of nucleic acids have also been studied by the X-ray diffraction method for crystals, but no attempts have been made for the structure determination of the complexes in solution except by the EXAFS method. Other methods, such as ESR, Mössbauer, and visible spectroscopy, are very useful in combination with the EXAFS method to elucidate the structure of metal complexes with sugar-type ligands.

In this section we show results for the local structure of some metal complexes with sugar-type ligands in solution that have been done by the EXAFS and XANES methods.

Iron(III) Complexes with Mannitol, Fructose, and Gluconate. XANES spectra and Fourier transforms of EXAFS data of the iron(III) complexes with mannitol, fructose, and gluconate ions are shown, respectively, in Figs. 8-6 and 8-7. In the figures, the Fe K-edge XANES and Fourier transformed EXAFS spectra of the complexes in the solid state (powder) and those of the hexaaquairon(III) complex in crystal and iron

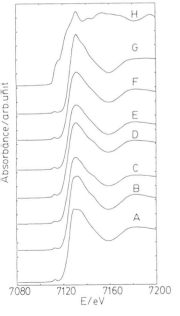

Figure 8-6 ■ XANES spectra of Fe(III)-mannitol [A, aqueous solution (aq) and B, powder (pw)], -fructose (C, aq and D, pw), -gluconato complexes (E, aq and F, pw), $FeNO_3 \cdot 9H_2O$ (G), and Fe foil (H). [L. Nagy, H. Ohtaki, T. Yamaguchi, and M. Nomura, *Inorg. Chim. Acta*, **159**, 201 (1989).]

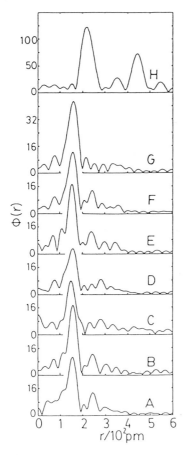

Figure 8-7 ■ Fourier transforms $\phi(r)$ of EXAFS data of Fe(III)-mannitol (A, aq and B, pw), -fructose (C, aq and D, pw), -gluconato complexes (E, aq and F, pw), $FeNO_3 \cdot 9H_2O$ (G), and Fe foil (H) (uncorrected for the phase shifts). [L. Nagy, H. Ohtaki, T. Yamaguchi, and M. Nomura, *Inorg. Chim. Acta*, **159**, 201 (1989).]

foil as reference are also shown.[34] From the analysis of the peak position and the shape of the XANES spectra and the main peak of the Fourier transforms of the EXAFS spectra, the following conclusions may be drawn:

1. The iron(III) ion in the complexes may have a slightly distorted octahedral structure according to the shape analysis of the XANES spectra, although the distortion was not observable from the EXAFS data. The Fe—O length is about 195 pm irrespective of the sugar ligands. The local structure of the complexes in solution seems to be very similar to that of the solid.

2. In the fructose complexes the Fe...Fe length was observed at about 310 pm, indicating dimerization of the complex probably through a double hydroxide bridge. Alkoxide oxygen atoms (in the case of the gluconato complex one alkoxide and one carboxylate oxygen atom) coordinate to the iron(III) ion. The remaining two coordination sites of the iron(II) ion may be occupied by two oxygen atoms either of OH^- ions or of sugar OH groups to form an elongated octahedral arrangement (Fig. 8-8).

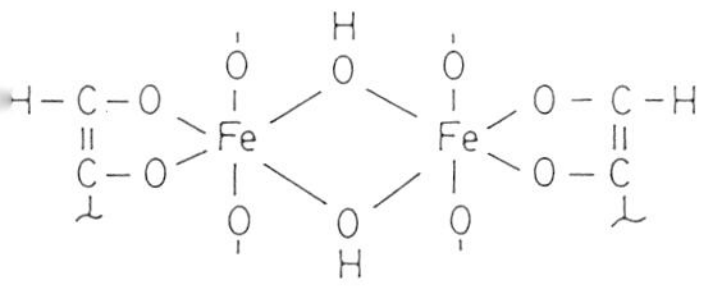

Figure 8-8 ▪ The most likely structure of the binuclear Fe(III)-sugar complex.

3. In the case of the mannitol and gluconato complexes the Fe...Fe length was not determinable from the analysis of the EXAFS spectra. In these cases the dimer species may not be the predominant one and, moreover, the Fe...Fe length may become close to the Fe...C length in the course of dimerization of the complexes. The results are summarized in Table 8-4.

Copper(II) Complexes with D-Ribose and D-Glucosamine. When sugars are added to a copper(II) nitrate aqueous solution, they interact to form a 1:1 complex at high pH. ESR spectroscopic measurements indicated that the complex is dimeric, because the ESR signal disappeared in a strongly alkaline solution. It is concluded from ESR measurements that D-glucosamine molecules form chelate rings around the copper(II) ion via NH_2 and O^- donor sites of the ligand.[35] The remaining sites of the copper(II) ion may be occupied by hydroxide ions in solutions of high pH as expected from potentiometric studies of the copper(II)-D-glucosamine and -L-α-alaninehydroxamato systems.[35] Visible spectra of the solutions showed the copper(II) ions in distorted octahedral (or practically planar) structures, and they do not form tetra-coordinated tetrahedral structures even at the highest pH.[36]

The XANES spectra of the solutions showed that the copper(II) ions are not reduced to +1 or 0 oxidation states, but are kept as +2 in the complex. The shape of the spectra[36] (Fig. 8-9) is similar to that of the (creatinium)$_2$CuCl$_4$ complex.[37]

The EXAFS spectra shown in Fig. 8-10 were well explained in terms of a dimerized local structure for the complexes (Fig. 8-11). Due to their long length, the

Table 8-4 ▪ Results of the Curve-fitting Analysis for the Iron(III) Sugar Complexes in Aqueous Solution (aq) and in the Solid State (pw)[a]

	Fe—O			Fe...C			Fe...Fe		
Ligand	r (pm)	σ (pm)	n	r (pm)	σ (pm)	n	r (pm)	σ (pm)	n
A Mannitol aq	194	9.4	6.5	287	5.5	2.6			
B Mannitol pw	198	7.3	6.1	285	5.0	2.2			
C Fructose aq	192	9.8	6.0	277	6.9	2[b]	310	6.9	0.5
D Fructose pw	194	10.1	5.7	285	9.3	2[b]	304	9.3	0.7
E Gluconate aq	195	8.3	6.3	285	5.2	2.4			
F Gluconate pw	196	8.7	6.3	283	5.0	2.3			
G Water pw	196	8.7	6[b]						
H Fe foil							249	6.88	8[b]

[a] r, σ, and n represent the interatomic distance, the root mean squares displacement, and the coordination number, respectively.
[b] Fixed.

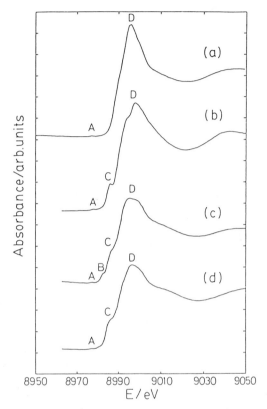

Figure 8-9 ■ XANES spectra of (a) copper(II) nitrate aqueous solution, (b) copper(II) acetate monohydrate crystal, (c) Cu(II)-D-ribose aqueous solution at pH 12, and Cu(II)-D-glucosamine aqueous solution at pH 10. (L. Nagy, T. Yamaguchi, M. Nomura, T. Pali, and H. Ohtaki, unpublished.)

Cu...Cu interactions are not observable in the EXAFS spectra and their Fourier transforms, but the dimerized structure was verified from the ESR spectra of the complexes. The structural parameters of the complexes are summarized in Table 8-5.

8-3. The Structure of Macromolecule-Metal Complexes by X-Ray and EXAFS Measurements

X-ray crystallographic studies on the structures of macromolecule-metal complexes have been extensively carried out for various systems, especially for biochemically important substances. The X-ray diffraction studies for hemoglobin and myoglobin are examples of the outstanding work in this field.

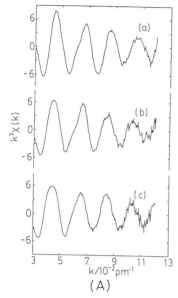
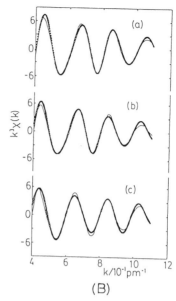

Figure 8-10 ■ (A) k^3-weighted EXAFS pattern $\chi(k)$ (k denotes the photoelectron wave vector) and (B) Fourier-filtered k^3 weighted $\chi(k)$ values (dots) and the fitted curves after the least-squares refinements of (a) Cu(II)-D-ribose, (b) Cu-D-glucosamine, and (c) copper(II) nitrate aqueous solutions. (L. Nagy, T. Yamaguchi, M. Nomura, T. Pali, and H. Ohtaki, unpublished.)

Figure 8-11 ■ Proposed local structures of binuclear Cu-sugar complexes. (a) Cu-D-glucosamine complex and (b) Cu-D-ribose complex.

Table 8-5 ▪ The Structural Parameters of the Cu(II)-D-Ribose
and Cu(II)-D-Glucosamine Complexes in Aqueous Solution[a]

Ligand (L)	L:Cu	pH	Interaction	N	r (pm)	σ (pm)
D-ribose	1:1	12	$Cu-O_{eq}$	4^b	191	7.5
			$Cu-O_{ax}$	2^b	230	13
			$Cu\cdots C$	2^b	271	10^b
D-glucosamine	1:1	10	$Cu-O_{eq}$	3^b	193	7.7
			$Cu-N_{eq}$	1^b	193	7.7
			$Cu-O_{ax}$	2^b	234	13
			$Cu\cdots C$	2^b	274	10^b

[a] r, N, and σ represent the distance, coordination number, and root-mean-square deviation of the distance of the atom pair, respectively.
[b] Fixed.

The structural information given by X-ray crystallographic analysis provides us with the most fundamental and reliable knowledge of the structure of substances, and changes in the structure can thus be discussed from spectroscopic, magnetic, and other data. The structure of complexes thus determined are often discussed in connection with their biochemical functions and reactivities. However, structural information of macromolecule complexes of metal ions in solution has been offered very rarely. In recent years some attempts have been made to determine the structures of biochemically interesting macromolecule complexes in frozen and glassy samples by using the EXAFS method. In this section the local and molecular structures of some biochemically attractive metal complexes will be shown and the structures will be discussed in connection with reactivities of the complexes. When EXAFS data are available, they are quoted to compare them with X-ray results. Unfortunately no solution X-ray data have been reported for the structure of macromolecule complexes of metal ions in solution.

8-3-1. Iron-Porphyrin Complexes

Iron porphyrins serve as prosthetic groups of some metalloproteins and enzymes known as hemoproteins. Hemoglobin attracts special attention as an oxygen carrier, which is a tetramer made up of two α and two β chains (subunits) containing 141 and 146 amino acid residues, respectively. Each chain contains one heme. The α and β chains contain seven and eight helical segments, respectively, interrupted by nonhelical ones.

Myoglobin is a similar protein consisting of a single chain constructed with 153 amino acid residues and one heme. Cytochrome c as an electron carrier has also been extensively studied. Cytochrome P-450 and cytochrome c oxidase are also important from the viewpoint of enzymes of oxygen utilization. These hemoproteins contain iron-porphyrin moieties with a very similar structure (heme), but physicochemical properties, such as the redox potentials of the iron atom in the complexes, are strongly influenced by the structure of polypeptides and coordinating ligand molecules along the axis vertical to the porphyrin ring. In hemoglobin and myoglobin a histidine moiety coordinates to the central iron atom as the fifth ligand, and the sixth site of the iron atom, which also weakly interacts with another histidyl imidazole, serves as the oxygen binding site. In cytochrome c the fifth and sixth coordination sites of the iron atom are coordinated with methionine and histidine, respectively, and these coordinations are essential for the electron transport from one protein to another in the respiratory chain.

Such manipulations by the heme groups give rise to various spin states of the iron atom. In other words, investigations of the spin states of the iron atom in the hemoproteins can elucidate the structure of the hemoproteins. Systematic X-ray diffraction studies on several synthetic iron porphyrins have revealed the relationship between the structure and the spin state.

In general synthetic and natural iron-porphyrin complexes are classified into two categories: ferrous [Fe(II)] and ferric [Fe(III)]. Furthermore, the electronic states of the iron atoms are classified into the high- and low-spin states. Besides the two states an intermediate-spin state is possible for each of Fe(II) and Fe(III). The oxidation state of Fe(IV) (ferryl) sometimes appears in certain iron proteins. In Table 8-6 the spin states, the theoretical values for their effective magnetic moments, and bond lengths between an iron and ligating atoms at different coordination numbers in various iron(II) and iron(III) hemes are summarized.

It is seen from Table 8-6 that the $Fe-N_p$ length (N_p denotes porphinatonitrogen atom) is independent of the coordination number but significantly affected by the spin state of the central iron atom. This bond length variation is different from the general rule that the bond length is shorter in a complex with a smaller coordination number. On the other hand, the bond length between the central iron atom and the ligand coordinated at the axial position tends to be shorter in the penta-coordinated complexes than in the hexa-coordinated complexes with the same oxidation and spin states of the iron atom. In the hexa-coordinated complexes the iron atom is situated at almost the center of the plane constructed by the four nitrogen atoms of the porphyrin ring; in the penta-coordinated complexes the iron atom is displaced from the plane by about 20–50 pm. In a complex with low-spin state, the displacement of the iron atom from the center of the plane is smaller than that in a complex with the high-spin state. Therefore, one may obtain approximate information about the local structure of complexes containing a heme-iron atom by measuring the effective magnetic moment and the stretching frequency of the Fe-imidazole bond by using infrared and/or Raman spectroscopy under favorable conditions. However, in most cases frequency measurements of marker bands, which arise from distortion of the porphyrin ring in the penta-coordination, give more reliable and easily identifiable information about the coordination structure of the complexes.

In fact, occupation of antibonding $d_{x^2-y^2}$ and d_{z^2} orbitals by electrons in the high-spin heme give rise to an expansion of the porphinato core in the hexa-coordinated complexes and extrude the iron out of the porphinato plane in the penta-coordinated complexes. Therefore, the $Fe-N_p$ bond length increases markedly from that of the low-spin complexes. When an electron is introduced into the d_{z^2} orbital while the $d_{x^2-y^2}$ orbital is kept vacant resulting in the intermediate-spin state, the axial Fe-imidazole bond length increases by about 30 pm, whereas the $Fe-N_p$ distance does not change remarkably compared to the bond length in the complexes with the low-spin state.

Table 8-6 ■ The Electronic, Magnetic, and Bonding States of Heme Irons in Porphyrins

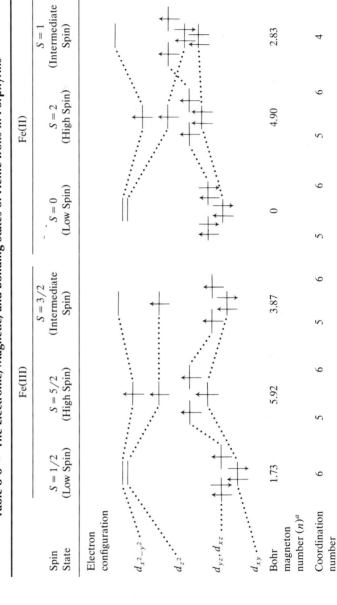

	Fe(III)			Fe(II)		
Spin State	$S = 1/2$ (Low Spin)	$S = 5/2$ (High Spin)	$S = 3/2$ (Intermediate Spin)	$S = 0$ (Low Spin)	$S = 2$ (High Spin)	$S = 1$ (Intermediate Spin)
Electron configuration						
Bohr magneton number $(n)^a$	1.73	5.92	3.87	0	4.90	2.83
Coordination number	6	5	5	6	5	4

Bond
length (pm)

Fe—N$_p^b$	197–200	206–209	205	199–200	199.5	198–200	198–201	207–210	205.7	197.2
Fe—C$_t^c$	0–10	39–54	0	26–28	0	21–23	0–11	40–52	0	0
Fe—L^d	196–201	210–240	208–232	203	232	172	201	210–216	235	—

[a] Bohr magneton number: $n = 2\sqrt{S(S+1)}$.
[b] N$_p$: porphinatonitrogen atom.
[c] C$_t$: center of the best plane constructed by the four N$_p$s.
[d] L: the axial ligand (imidazole or another ligand) coordinating to the Fe atom.

(X): the sixth ligand when the complex forms the six-coordination structure.

Because the spin state of the complexes is determined by the strength of the axial ligand field, it is possible to prepare a spin crossover ($S = 1/2 \leftrightarrow 5/2$) complex and a quantum admixed state ($S = 3/2, 5/2$) complex by selecting suitable ligands coordinated at the axial position.

Some examples of complexes containing heme-iron atoms follow:

Fe(III)-porphinato complexes
  Low-spin state ($S = 1/2$).
    n (coordination number) = 6: metmyoglobin, methemoglobin, cytochrome c, horseradish peroxidase (HRP), and cytochrome P-450 having CN^-, NH_3, imidazole, pyridine, or OH^- as the sixth ligand, cytochrome c without external ligand, octaethylporphinatobis(imidazole)iron(III) [$OEPFe(Im)_2^+$], tetraphenylporphinatobis(imidazole)iron(III) [$TPPFe(Im)_2^+$].
  High-spin state ($S = 5/2$).
    $n = 5$: OEPFeCl, TPPFeCl.
    $n = 6$: metmyoglobin, methemoglobin, HRP, and cytochrome P-450 with H_2O, CH_3SH, SCN^-, or $HCOO^-$ as the sixth ligand, $TPPFe(DMSO)_2^+$ (DMSO = dimethylsulfoxide), $TPPFe(H_2O)_2^+$.
  Intermediate-spin state ($S = 3/2$).
    $n = 5$: cytochrome c', $OEPFeClO_4$, and $TPPFeClO_4$.
    $n = 6$: $OEPFe(THF)_2ClO_4$ (THF = tetrahydrofuran), $OEPFe(EtOH)_2ClO_4$.
Fe(II)-porphinato complexes.
  Low-spin state ($S = 0$).
    $n = 5$: TPPFe(NO).
    $n = 6$: myoglobin and hemoglobin with O_2, CO, or NO as the sixth ligand.
  High-spin state ($S = 2$).
    $n = 5$: deoxymyoglobin and deoxyhemoglobin.
    $n = 6$: $TPPFe(THF)_2$.
  Intermediate-spin state ($S = 1$).
    $n = 4$: TPPFe.

8-3-2. Hemoglobin

Hemoglobin is one of the most typical iron porphyrin containing materials and is the major component of red blood cells for oxygen transport from lungs to all parts of the organism. The iron atom in the heme is coordinated by four pyrrole nitrogen atoms of protoporphyrin IX and an imidazole nitrogen of a histidine residue from the globin chain (proximal histidine). The sixth coordination site of the iron atom is occupied by another histidine, but the Fe—N(histidine) length is so long that the interaction is very weak (distal histidine). The iron is ferrous in the physiological form of hemoglobin, as well as in myoglobin, which is similar to hemoglobin, and can bind ligands such as a small molecule of oxygen or carbon monoxide at the sixth site. The Fe—N(porphyrin) bond length changes with the change in the spin state of the iron from high to low spin, which allows efficient oxygen transport.

The most interesting physiological property of hemoglobin is its ability to combine reversibly with oxygen molecules and the cooperativity of dioxygen binding to hemoglobin. The equilibrium curve for the oxygen absorption for hemoglobin is

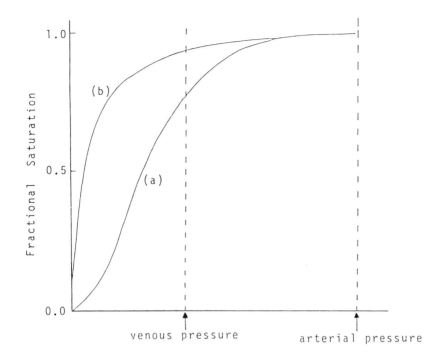

Partial Pressure of Oxygen

Figure 8-12 ∎ Typical oxygen equilibrium curves for (a) hemoglobin and (b) myoglobin.

shown in Fig. 8-12, together with the curve for myoglobin for comparison, where the fraction of the heme groups bearing oxygen (fractional saturation) is measured as a function of the partial pressure of oxygen.

The curve for myoglobin is hyperbolic, as expected from the combination of one myoglobin molecule with one oxygen molecule. However, the curve for hemoglobin is a more complicated sigmoidal shape, reflecting more effective oxygen transport to the tissues by hemoglobin than myoglobin.

The X-ray data for deoxy-[38] and oxyhemoglobins[39] are summarized in Table 8-7, together with those of deoxy-[40] and oxymyoglobins[41] and some synthetic model hemes.[42] In Fig. 8-13 the structural changes between the deoxy- and oxyhemoglobins are illustrated. In deoxyhemoglobin the porphyrins are domed, and the iron atoms are displaced by about 40 pm from the mean plane of the porphyrin nitrogen atoms toward the proximal histidines [T (tense) state]. Upon oxygenation the iron atoms move toward the heme planes, the proximal histidyl imidazoles move toward the heme groups by ca. 60 and 50 pm in the α and β subunits, respectively [R (relaxed) state], the $Fe-N_p$ bond lengths contract, and the porphyrin rings change from domed to planar. Such a conformational change of the hemes occurs also in myoglobin and synthetic iron porphyrins.

Table 8-7 ■ Structural Data of Hemoglobin (Hb), Myoglobin (Mb), and Synthetic Hemes[a]

	Bond Length (pm)			Angle (deg)		$Fe-O_1$ (pm)	$Fe-O_1-O_2$ (deg)
	$Fe-N_p$	$Fe-N_{im}$	$Fe-C_t$	θ	ϕ		
Human deoxy-Hb (T state)							
α	208(3)	216(6)	40(5)	18	78		
β	205(3)	209(6)	36(5)	23	79		
Human oxy-Hb (R state)							
α	199(5)	194(9)	12(8)	11	84	166(8)	153.0(7)
β	196(6)	207(9)	−11(8)	27	91	187(13)	159.0(12)
Sperm whale deoxy-Mb	203(10)	222	42	12	87		
Sperm whale oxy-Mb	195(6)	207(6)	18	1	90	183(6)	115.0(5)
Fe(PP)(2-MeIm) (deoxy T-state)	207.2(5)	209.5(5)	40				

| Fe(PP)(2-MeIm)-
(O₂)
(oxy R-state) | 199.6(4) | 201.7(4) | 8.6 | 189.8(7) | 129.0 |

[a] Fe—C_t: displacement of Fe from the mean plane of porphyrin N; N_p: porphyrin nitrogen atom; N_{Im}: imidazole nitrogen atom; θ: angle between the plane of imidazole projected onto the porphyrin and line N_1-N_3; ϕ: angle between the line C_2-C_4 of imidazole and porphyrin normal; Fe(PP)(2-Meim): (2-methylimidazole) (meso-tetrakis- (1, 1, 1-o-pivalamidophenyl) (porphinato)-iron(II); PP = picketfence porphyrin.

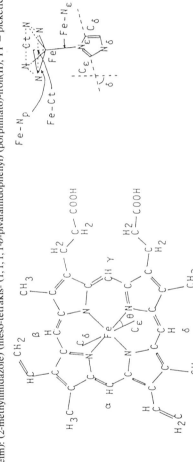

Heme with imidazole of His-F8 projected onto its plane (top). Interatomic parameters around Fe in heme. The diagram illustrates the meanings of angles θ and ϕ in the table (bottom) [M. F. Perutz, G. Fermi, and B. Luisi, *Acc. Chem. Res.*, **20**, 309 (1987)].

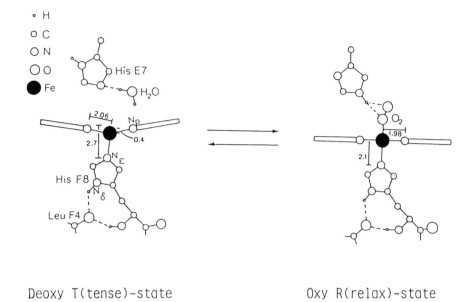

Deoxy T(tense)-state   Oxy R(relax)-state

Figure 8-13 ▪ A schematic diagram of changes in heme stereochemistry on binding of O_2. [M. F. Perutz, G. Fermi, and B. Luisi, *Acc. Chem. Res.*, **20**, 309 (1987).]

Thus, O_2 binding to a high-spin heme induces electronic changes in the coordination groups to give a low-spin heme, and the change in spin state provides the stereochemical trigger for the cooperativity between the hemes of the T and the R states.

8-3-3. Cytochrome c'

Ferrihemoproteins and heme enzymes can be classified into high- and low-spin complexes. It is noted that ferricytochrome c' isolated from photosynthetic and denitrified bacteria has an intermediate-spin state $(S = 3/2)$.[43] The electronic structure of cytochrome c' is rather abnormal: The oxidized form in alkaline solution is in a low-spin state and below pH 11 it is converted to the high-spin state.[44] The magnetic moment shows an intermediate property between the low- and high-spin states.[45] Mössbauer spectra show a larger quadrupole splitting than those observed for low- and high-spin complexes.[46] In order to elucidate the anomalous properties of cytochrome c' some investigations have been attempted to simulate spectroscopic properties by using synthetic hemes, and the unusual magnetic moments and Mössbauer spectra of cytochrome c' and synthetic hemes have been interpreted in terms of an intermediate-spin state.[47]

Detailed examinations of X-ray structure analysis for the synthetic hemes $(\text{OEPFeClO}_4,[48]$ $\text{OEPFe(THF)}_2\text{ClO}_4,[49]$ and $\text{TPPFeClO}_4[50])$ suggested that the spin state requires a weak axial ligand field, which compensates for a strong ligand field

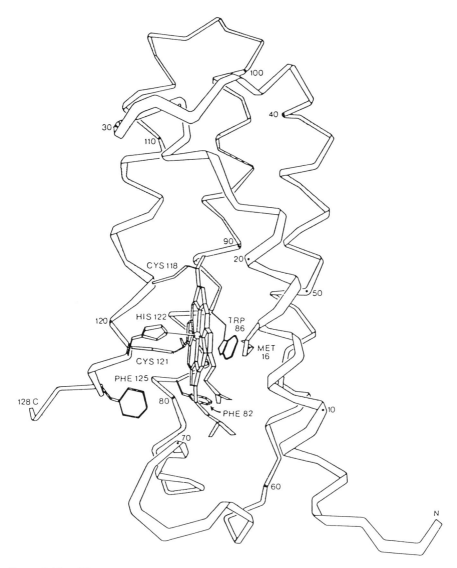

Figure 8-14 ▪ The structure of the heme environment of cytochrome c'. [P. C. Weber, F. R. Salemme, F. S. Mathews, and P. H. Bethge, *J. Biol. Chem.*, **256**, 7702 (1981).]

due to the equatorial porphyrin in the complex.[51] This suggests that the local structure of the heme in cytochrome c' is a five-coordinate and the fifth histidine imidazole ligand is weakly coordinated to the central iron atom. The X-ray structural determination of cytochrome c' at 2.5×10^2 pm resolution clearly show the five coordination structure[52] as shown in Fig. 8-14. The sixth site of the iron atom is vacant. The distal heme surface is surrounded by hydrophobic aromatic amino acid residues.

8-3-4. Hemocyanin

Hemocyanin is a large copper-containing nonheme protein that easily dissolves in the hemolymph of many invertebrate animals and serves to transport molecular oxygen to living organisms. Two types of hemocyanins exist, molluscan and arthropodan hemocyanins, which have completely different molecular architectures. Arthropodan hemocyanin consists of hexamers or multihexamers with subunits of molecular weight of ca. 75,000, whereas molluscan hemocyanins are cylindrical oligomers with subunits of molecular weight of about 400,000. The oxygen-binding centers of hemocyaninins contain a pair of copper ions (binuclear copper sites) and are considered to be quite

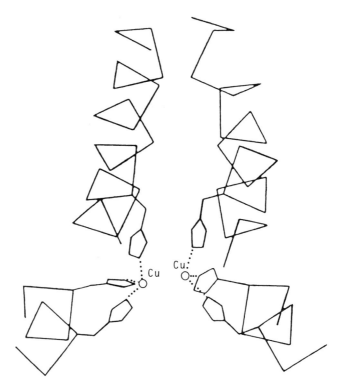

Figure 8-15 ■ The local structure of the copper binuclear active site of arthropodan hemocyanin. [A. Valbeda and W. G. J. Hol, *J. Mol. Biol.*, **206**, 531 (1989).]

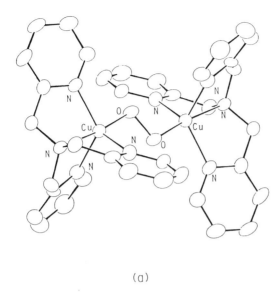

(a)

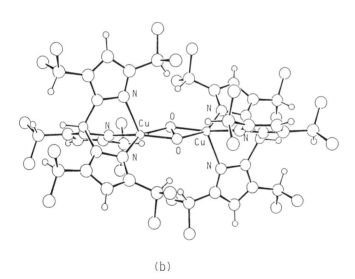

(b)

Figure 8-16 ■ Molecular structures proposed as a model complex for the active site of hemocyanin. (a) [{tris(2-pyridylmethylamine)Cu}$_2$(O$_2$)] [R. R. Jacobson, Z. Tyeklar, A. Farooq, K. D. Karlin, S. Liu, and J. Zubieta, *J. Am. Chem. Soc.*, **110**, 3690 (1988)] and (b) [{hydrotris-(3,5-diisopropyl-1-pyrazolylborato)Cu}$_2$(O$_2$)] [N. Kitajima, K. Fujisawa, Y. Moro-oka, and K. Toriumi, *J. Am. Chem. Soc.*, **111**, 8975 (1989).]

similar in both molluscan and arthropodan. Furthermore, the binuclear copper sites in oxytyrosinase, ascorbate oxidase, laccase, and ceruloplasmin are also believed to be similar to those in the oxyhemocyanins.

Recently the structure of deoxyhemocyanin of the spiny lobster *Panulirus interrupts*, an arthropodan, has been determined by the single crystal X-ray diffraction method at the 3.2×10^2 pm resolution level.[53] According to the result, the binuclear copper site is situated at the center of the second domain and is in a hydrophobic environment that is rich in tryptophan and phenylalanine protein residues. The distance between the two copper atoms is $3.5-3.6 \times 10^2$ pm. This agrees well with the value obtained by an EXAFS measurement, 339 pm, for deoxyhemocyanin.[54]

Each copper(I) ion in the binuclear center is coordinated with imidazole moieties of three histidine groups and the binding site is rather distorted. The two copper-imidazole nitrogen distances are about 195–277 pm[53] (Fig. 8-15). The electron density map shows no protein bridging ligand to facilitate the strong diamagnetic coupling between the unpaired Cu(II) electron spins in oxyhemocyanin.

Oxyhemocyanin can be prepared by the addition of O_2 to a solution containing deoxyhemocyanin and has been known to give a peroxo(O_2^{2-})dicopper(II) complex with a Cu...Cu distance of 355–367 pm according to an EXAFS measurement.[54] In order to account for this geometry, two types of copper-dioxygen complexes have been synthesized. One is [{tris(2-pyridylmethylamine)Cu}$_2$(O$_2$)][55] and the other is [{hydrotris(3,5-diisopropyl-1-pyrazolylborato)Cu}$_2$(O$_2$)].[56] The local structures of the complexes are shown in Fig. 8-16. The copper atom in the former complex is penta-coordinated with a distorted trigonal bipyramidal geometry, and the peroxo-oxygen ions occupy the axial sites [Fig. 8-16(a)]. The Cu...Cu separation is 435.9(1) pm[55] and is longer than that in oxyhemocyanin, although the complex has no bridging ligand. The copper atom in the latter complex has a tetrahedrally coordinated structure, and two copper atoms are bridged with the peroxo-oxygen ions with the μ-η^2:η^2 coordination mode [Fig. 8-16(b)]. The Cu...Cu separation of 356.0(3) pm[56] is easily compatible with the proposed structure of the active sites in oxyhemocyanin. The isopropyl groups attached to pyrazolyl groups may contribute to the preparation of the stable dioxygen-copper complex through the formation of the hydrophobic field.

8-3-5. Superoxide Dismutase

The superoxide dismutases (SOD) are metalloproteins that catalyze the very rapid two-step dismutation of peroxide anion radical ($O_2^{\cdot-}$) to molecular oxygen and hydrogen peroxide through alternating reduction and oxidation of the copper atom at the active site. The dismutation reaction can be written as

$$2O_2^{\cdot-} + 2H^+ \rightleftarrows H_2O_2 + O_2$$

Superoxide dismutases are found in microbes, plants, and animals, and have been assumed to play an important role in the protection of cells against the destructive effects of superoxide ion radicals. Three different SODs have been isolated and characterized by their metal content and the source: Cu—Zn SOD, Mn SOD, and Fe SOD. The Cu—Zn SOD enzyme has been most extensively studied. In solution it is a dimer of two identical subunits, each of them having a molecular weight of about

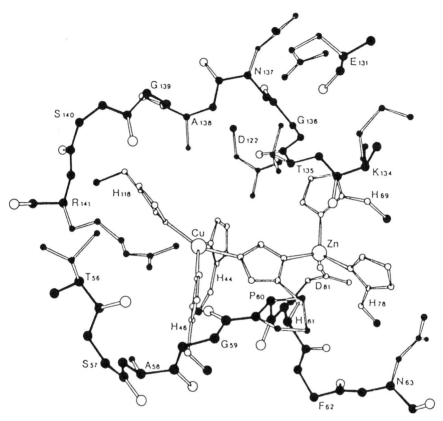

Figure 8-17 ■ The local structure of the active site in the Cu—Zn SOD enzyme. [J. A. Tainer, E. D. Getzoff, K. M. Beem, J. S. Richardson, and D. C. Richardson, *J. Mol. Biol.*, **160**, 181 (1982).]

16,000. Each subunit contains one copper and one zinc atom. The structure of the active site of Cu—Zn SOD[57] is shown in Fig. 8-17. Each enzyme subunit is composed primarily of eight antiparallel β strands and three external loops. The Cu(II) and Zn(II) atoms at the active site lie 6.3×10^2 pm apart at the bottom of a long channel between two large loops (loop 6,5 and loop 7,8) on the external surface of the β barrel.[58]

The copper(II) ion is coordinated with four histidine imidazoles (His-44, -46, -61, and -118) and the binding sites around the copper(II) ion are strongly distorted from the square-planar structure to have different Cu—N lengths.[58] The distortion is caused by the orientation of the histidine rings around the copper(II) ion, and thus the distorted square-plane (or distorted tetrahedron) of the copper coordination opens the axial position toward solvent molecules. One histidine imidazole His-61 attached to a copper(II) ion forms a bridge to a zinc(II) ion, and two histidine imidazole moieties, His-69 and -78, and an oxygen atom of aspartic acid, Asp-81, combine with the zinc(II) ion to construct a largely distorted tetrahedral or distorted trigonal pyramidal structure having the Asp-81 oxygen at the apex of the pyramid. The bridging histidine imidazole ring may function as a proton carrier to facilitate protonation of $O_2^{\cdot-}$.

The Cu—N and Zn—N bond lengths are 2.1×10^2 pm on average, whereas the Zn—O bond length is shorter than 2.0×10^2 pm.[58] The zinc(II) atom is buried in the protein structure but the copper(II) atom is solvent accessible. The position of the copper(II) ion is appropriate to bind $O_2^{\cdot-}$ because the Cu—H_2O bonding within the complex may be weak.

Neutralization of Arg-141, which is separated from the copper(II) ion by about 6×10^2 pm by using phenylglyoxal, caused a 90% reduction in activity of the superoxide dismutase.[59] The result suggests that $O_2^{\cdot-}$ binds to the copper(II) and a nitrogen atom in the guanidinium moiety of Arg-141.[59]

A recent study of the local structure around copper and zinc atoms in the oxidized and reduced forms of bovine superoxide dismutase by the EXAFS method using Cu K and Zn K edges shows that the copper atom is most probably coordinated with three imidazole groups at the distance Cu—N = 194 pm in the reduced form, whereas the local structure of the zinc site remains unchanged.[60] The bond between His-61 and the Cu atom is apparently broken down during enzyme catalysis. Based on such facts and SOD biochemistry, a proposed enzymatic mechanism is shown in Fig.

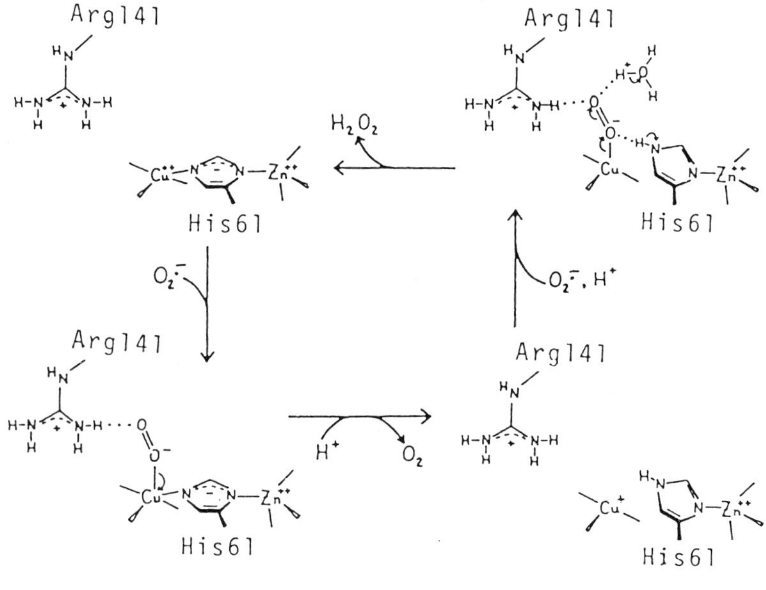

Figure 8-18 ■ A proposed mechanism for two-step dismutation deformed: molecular oxygen and hydrogen peroxide by Cu—Zn SOD. [J. A. Tainer, E. D. Getzoff, J. S. Richardson, and D. C. Richardson, *Nature* (*London*), **306**, 284 (1983).]

8-18.[57] The proposed mechanism is supported by other biophysical and biochemical studies.

8-3-6. Rubredoxin

Rubredoxin, which has one or two Fe(III) ions surrounded by four cysteine thiolate ligands to form a tetrahedral structure, is widely distributed in anaerobic bacteria. It is thought to function as an electron carrier in the redox cycle of Fe(III)/Fe(II), although the physiological role in anaerobic organisms is still unknown. The X-ray crystal structure analysis of rubredoxin isolated from *Clostridium pasteurianum* has been studied at a 1.5×10^2 pm resolution level.[61] The analysis revealed that the Fe ion is located near C and N termini, is coordinated with four cysteine thiolate ligands, and that the iron to sulfur distances are 224, 232, 234, and 205 pm. On the other hand, using the X-ray crystal structure results as a guide, the EXAFS data of a solution of *Peptococcus aerogenes* rubredoxin (a protein similar to *Clostridium pasteurianum* rubredoxin) were fitted to a model in which three lengths of the iron-sulfur bonds had the same value R_3 and the fourth had a different value R_1.[62] The mean square error for various sets of the R_3-R_1 values had a broad minimum between $R_3 = 221.7$, $R_1 = 238.9$ and $R_3 = 226.8$, $R_1 = 210.8$ pm. The average distance from EXAFS was in excellent agreement with the average distance from the X-ray crystal structure determination,[61] although the agreement of the curve obtained with $R_3 = 224.6$ and $R_1 = 221.7$ pm with the calculated curve using all four crystal structure values was very poor. This difference may be due to the different proteins being compared or different structures in the solution and in the crystals.

8-3-7. Xanthine Oxidase

Xanthine oxidase is a large molybdenum containing enzyme (molecular weight = 283,000) that oxidizes xanthines under molecular oxygen as an electron acceptor according to the reaction

The enzyme is a mixture of two molybdenum containing forms. However, it is isolated as a mixture of an active form and a catalytically inactive form (desulfo form). The local structure of the molybdenum coordination in xanthine oxidase has been investigated[62] in the molybdenum K-edge region of the EXAFS spectra for complete desulfo and mixed (desulfo + active) xanthine oxidases; that for the pure active form has been obtained from the difference between the two EXAFS spectra. EXAFS spectra were also measured for freeze-dried samples of milk xanthine oxidase solutions. Satisfactory simulations of the spectrum of the desulfo enzyme have been carried out by assuming that the molybdenum is coordinated with two terminal oxygen atoms ($Mo=O$), two sulfur atoms (presumably from cysteine residues, $Mo-S$), and one sulfur atom (presumably from a methionine, $Mo-S$), as shown[63] in Fig. 8-19(a). The EXAFS spectrum of the active enzyme differs appreciably from

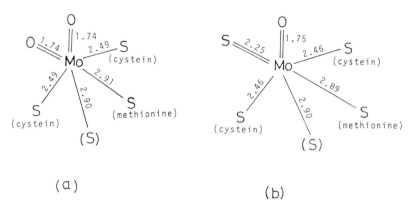

Figure 8-19 ■ A pictorial representation of EXAFS results for the (a) desulfo and (b) active xanthine oxidase.

that of the desulfo enzyme, and the main difference between active and desulfo enzymes is due to the replacement of the terminal sulfur in the active enzyme by a terminal oxygen[63] [Fig. 8-19(b)]. In both desulfo and active enzymes the number of sulfur atoms located at about 290 pm would be two rather than only one.

8-3-8. Cytochrome c Oxidase

Cytochrome c oxidase is a terminal enzyme in the mitochondrial respiratory chain. It contains four redox centers per functional unit: two hemes and two copper atoms. The two hemes are those of type a, but they are structurally and functionally different. They are referred to as heme a and heme a_3. The two copper atoms are also different from each other and they are referred to as Cu_A and Cu_B. The heme a and Cu_A are functionally associated, and the heme a_3 and Cu_B are magnetically coupled in the oxidized state. In the respiratory chain, four electrons are transferred from cytochrome c to cytochrome c oxidase, which in turn reduces molecular oxygen to H_2O, and energy is conserved. The heme a–Cu_A site is thought to act as an initial electron accepting site and an electron pool, and the heme a_3–Cu_B binuclear center is thought to bind O_2 and reduce it to H_2O. A very brief outline of the reaction scheme is shown in Fig. 8-20.

Until recent EXAFS studies, very little was known about the detailed local structure of the metal active sites of cytochrome c oxidase, because the magnetically interacting binuclear active site cannot be observed by other spectroscopic techniques, and the crystal structure is too large (molecular weight = ~ 200,000) to be analyzed by the X-ray diffraction method.

The proximity of the metal atoms of heme a and Cu_A and heme a_3–Cu_B binuclear active sites is uniquely identified[64] at a low temperature from the copper and iron EXAFS studies of the oxidized resting native oxidase, the iron EXAFS study of the oxidized resting copper-depleted oxidase, and their comparison with those of model compounds. These approaches in the fully oxidized resting site gave the following results (Fig. 8-21): the Cu_A has three (or two) sulfur atoms at a normal Cu—S bond length 227(3) pm and one (or two) nitrogen atoms at 197(3) pm. The structure of

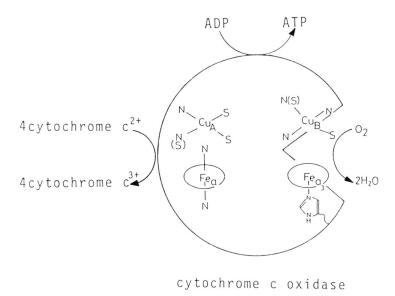

cytochrome c oxidase

Figure 8-20 ■ A schematic representation of cytochrome c oxidase. Each of the metal centers is expected to undergo single electron transfers, that is, between Fe^{2+} and Fe^{3+} and between Cu^+ and Cu^{2+}.

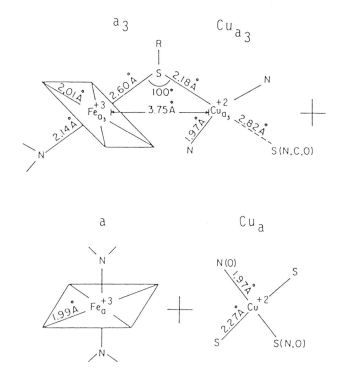

Figure 8-21 ■ EXAFS results for the oxidized resting state. [L. Power, B. Chance, Y. Ching, B. Muhoberac, S. Weintraub, and D. Wharton, *FEBS Lett.*, **138**, 245 (1982).]

heme a showed a hexa-coordinated form having $Fe-N = 199(3)$ pm, which is in agreement with the bond length found in $TPPFe(Im)_2^+$. The Cu_B has been shown to be similar to that of stellacyanin, a type I copper protein, which has a short $Cu-S$ length of 218(3) pm, two nitrogen ligands, and a very long $Cu-S(C,N,O)$ length of 282(3) pm. The heme a_3 is similar to methemoglobin with a long $Fe-S$ bond [260(3) pm], which has a sulfur atom at an apex instead of an oxygen atom. The separation between the iron and copper atoms in the binuclear active site has been found to be 375(5) pm. The result suggests that the two metal atoms are bridged by a sulfur atom having an Fe_a-S-Cu_B bond angle of $\sim 103°$, a value typical of the sp^3 bonding required for a cysteine sulfur.

8-4. Conclusion

In spite of the importance of knowledge of the structure of metal complexes with biologically interesting ligands under living conditions, very little reliable information has been obtained in the life sciences. Only the knowledge of the structure in crystals of such substances is reported, and it is usually assumed that the structure is retained in the reacting system of biofluids. However, it is obvious that the structure found in crystals is not always the same as that in solution. Recently developed solution X-ray diffraction techniques and EXAFS measurements may provide useful information for the structure of such substances in solution, but only limited cases of relatively simple systems are yet possible to study by these methods.

It is hoped that the problem of elucidating the structure of biochemically important complexes will be better solved by these methods in the future in combination with other suitable techniques. More important knowledge may be the structure of intermediate species in biochemical reactions, but the goal of this problem is still very far from the present situation of elucidating reaction mechanisms of biochemical reactions.

Acknowledgment. The authors wish to express their thanks to Professor C. J. Burrows for her helpful comments in preparing this manuscript.

References

1. K. Tohji, Y. Udagawa, T. Kawasaki, and K. Masuda, *Rev. Sci. Instrum.*, **58**, 1482 (1983).
2. K. Ozutsumi and H. Ohtaki, *Bull. Chem. Soc. Jpn.*, **56**, 3635 (1983).
3. K. Ozutsumi and H. Ohtaki, *Bull. Chem. Soc. Jpn.*, **57**, 2605 (1984).
4. K. Ozutsumi and H. Ohtaki, *Bull. Chem. Soc. Jpn.*, **58**, 1651 (1985).
5. K. Ozutsumi, T. Yamaguchi, H. Ohtaki, K. Tohji, and Y. Udagawa, *Bull Chem. Soc. Jpn.*, **58**, 2786 (1985).
6. T. Radnai, K. Inoue, and H. Ohtaki, *Bull. Chem. Soc. Jpn.*, in press.
7. H. C. Freeman and J. M. Guss, *Acta Crystallogr.*, Sect. B, **24**, 1133 (1968).
8. T. Fujita, T. Yamaguchi, and H. Ohtaki, *Bull. Chem. Soc. Jpn.*, **52**, 3539 (1979).
9. T. Fujita and H. Ohtaki, *Bull. Chem. Soc. Jpn.*, **55**, 455 (1982).
10. T. Fujita and H. Ohtaki, *Bull. Chem. Soc. Jpn.*, **56**, 3276 (1983).
11. T. Fujita and H. Ohtaki, *Bull. Chem. Soc. Jpn.*, **53**, 930 (1980).

12. B. W. Low, F. L. Hirshfield, and F. M. Richards, *J. Am. Chem. Soc.*, **81**, 4412 (1959).

13. L. Kryger and S. E. Rasmussen, *Acta Chem. Scand.*, **27**, 2674 (1973).

14. C. M. Gramaccioli, *Acta Crystallogr.*, **21**, 600 (1966).

15. D. van der Helm, A. F. Nicholas, and C. D. Fisher, *Acta Crystallogr., Sect. B*, **26**, 1172 (1970).

16. D. van der Helm and H. B. Nicholas, Jr., *Acta Crystallogr., Sect. B*, **26**, 1858 (1970).

17. D. C. Rees, M. Lewis, R. B. Honzatko, W. N. Lipscomb, and K. D. Hardman, *Proc. Nat. Acad. Sci., USA*, **78**, 3408 (1981).

18. D. C. Rees, M. Lewis, and W. N. Lipscomb, *J. Mol. Biol.*, **168**, 367 (1983).

19. M. A. Holms and B. W. Mattews, *J. Mol. Biol.*, **160**, 623 (1982).

20. K. Ozutsumi, M. Natsuhara, and H. Ohtaki, *Bull. Chem. Soc. Jpn.*, **62**, 2807 (1989).

21. M. Dobler, J. D. Dunitz, and P. Seiler, *Acta Crystallogr., Sect. B*, **30**, 2741 (1974).

22. P. Seiler, M. Dobler, and J. D. Dunitz, *Acta Crystallogr., Sect. B*, **30**, 2744 (1974).

23. M. Dobler and R. P. Phizacherley, *Acta Crystallogr., Sect. B*, **30**, 2748 (1974).

24. H. Takeuchi, T. Arai, and I. Harada, *J. Mol. Struct.*, **146**, 197 (1986).

25. G. Whipff, P. Weiner, and P. A. Kollman, *J. Am. Chem. Soc.*, **104**, 3249 (1982); G. Wipff, P. A. Kollman, and J. M. Lehn, *J. Mol. Struct.*, **93**, 153 (1983).

26. H. Ohtaki and H. Seki, *J. Macromol. Sci. Chem.*, in press.

27. L. Prasad and A. McAulay, *Acta Crystallogr., Sect. C*, **39**, 1175 (1983).

28. L. Prasad and S. C. Nyburg, *Acta Crystallogr., Sect. C*, **43**, 1038 (1987).

29. E. K. Barefield, A. Bianchi, E. J. Billo, P. J. Connolly, P. Paoletti, J. S. Summers, and D. G. van Derveer, *Inorg. Chem.*, **25**, 4197 (1986).

30. T. Ito, M. Kato, and H. Ito, *Bull. Chem. Soc. Jpn.*, **57**, 2641 (1984).

31. V. J. Thom, C. C. Fox, J. C. A. Boeyens, and R. D. Hancock, *J. Am. Chem. Soc.*, **106**, 5947 (1984).

32. P. A. Tasker and L. Sklar, *J. Cryst. Mol. Struct.*, **5**, 329 (1975).

33. A. W. Addison and A. Sinn, *Inorg. Chem.*, **22**, 1225 (1983).

34. L. Nagy, H. Ohtaki, T. Yamaguchi, and M. Nomura, *Inorg. Chim. Acta*, **159**, 201 (1989).

35. B. Kurzak, K. Kurzak, and J. Jezierska, *Inorg. Chim. Acta*, **125**, 77 (1986).

36. L. Nagy, T. Yamaguchi, M. Nomura, T. Pali, and H. Ohtaki, unpublished.

37. N. Kosugi, T. Yokoyama, and H. Kuroda, in *EXAFS and Near Edge Structure*, III, K. O. Hodgson, B. Hedman, and J. E. Penner-Hahn, eds., Springer, Berlin (1984), p. 55.

38. G. Fermi, M. F. Perutz, B. Shaanan, and R. Fourme, *J. Mol. Biol.*, **175**, 159 (1984).

39. B. Shaanan, *J. Mol. Biol.*, **171**, 31 (1983).

40. T. Takano, *J. Mol. Biol.*, **110**, 569 (1977).

41. S. E. V. Phillips, *J. Mol. Biol.*, **142**, 531 (1980).

42. G. B. Jameson, F. S. Molinaro, J. A. Ibers, J. P. Collman, J. I. Brauman, E. Rose, and K. S. Suslick, *J. Am. Chem. Soc.*, **102**, 3224 (1980).

43. H. Iwasaki and T. Mori, *J. Biochem.*, **42**, 375 (1955); V. P. Vernon and M. D. Kamen, *J. Biol. Chem.*, **211**, 643 (1954); K. Das, H. D. Klerk, R. G. Bartsch, T. Horio, and M. D. Kamen, *Proc. Nat. Acad. Sci. USA*, **57**, 367 (1967).

44. T. Horio and M. D. Kamen, *Biochim. Biophys. Acta*, **48**, 266 (1961).

45. A. Ehrenberg and M. D. Kamen, *Biochim. Biophys. Acta*, **102**, 333 (1965); A. Tasaki, J. Ohtsuka, and M. Kotani, *Biochim. Biophys. Acta*, **140**, 284 (1967).

46. T. H. Moss, A. J. Bearden, R. G. Bartsch, and M. A. Cusanovich, *Biochemistry*, **7**, 1583 (1968).

47. D. H. Dolphin, J. R. Sams, and T. B. Tsin, *Inorg. Chem.*, **16**, 711 (1977); M. M. Maltempo, T. H. Moss, and M. A. Cusanovich, *Biochim. Biophys. Acta*, **342**, 290 (1974); M. Maltempo, *Biochim. Biophys. Acta*, **379**, 95 (1975).

48. H. Masuda, T. Taga, K. Osaki, H. Sugimoto, Z. Yoshida, and H. Ogoshi, *Inorg. Chem.*, **19**, 950 (1980).

49. H. Masuda, T. Taga, K. Osaki, H. Sugimoto, Z. Yoshida, and H. Ogoshi, *Bull. Chem. Soc. Jpn.*, **55**, 3891 (1982).

50. C. A. Reed, T. Mashiko, S. P. Bentley, M. E. Kastner, W. R. Scheidt, K. Spartalian, and G. Lang, *J. Am. Chem. Soc.*, **101**, 2948 (1979).

51. H. Masuda and T. Taga, *J. Cryst. Soc. Jpn.*, **25**, 289 (1983).

52. P. C. Weber, R. G. Bartsch, M. A. Cusanovich, R. C. Hamilin, A. Howard, S. R. Jordan, M. D. Kamen, T. E. Meyer, D. W. Weatherford, Ng H. Xuong, and F. R. Salemme, *Nature*, **286**, 302 (1980); P. C. Weber, F. R. Salemme, F. S. Mathews, and P. H. Bethge, *J. Biol. Chem.*, **256**, 7702 (1981).

53. W. P. J. Gaykema, W. G. J. Hol, J. M. Vereijken, N. M. Soeter, H. J. Bak, and J. J. Beintema, *Nature* (*London*), **309**, 23 (1984); W. P. J. Gaykema, A. Volbeda, and W. G. J. Hol, *J. Mol. Biol.*, **187**, 255 (1986); **209**, 249 (1989).

54. G. L. Woolery, L. Powers, M. Winkler, E. I. Solomon, and T. G. Spiro, *J. Am. Chem. Soc.*, **106**, 86 (1984).

55. R. R. Jacobson, Z. Tyeklàr, A. Farooq, K. D. Karlin, S. Liu, and J. Zubieta, *J. Am. Chem. Soc.*, **110**, 3690 (1988).

56. N. Kitajima, K. Fujisawa, Y. Moro-oka, and K. Toriumi, *J. Am. Chem. Soc.*, **111**, 8975 (1989).

57. J. A. Tainer, E. D. Getzoff, J. S. Richardson, and D. C. Richardson, *Nature* (*London*), **306**, 284 (1983).

58. J. A. Tainer, E. D. Getzoff, K. M. Beem, J. S. Richardson, and D. C. Richardson, *J. Mol. Biol.*, **160**, 181 (1982).

59. D. P. Malinowski and I. Fridovich, *Biochemistry*, **18**, 5909 (1979).

60. N. J. Blackburn, S. S. Hasnain, N. Binsted, G. P. Diakun, C. D. Garner, and P. F. Knowles, *Biochem. J.*, **219**, 985 (1984).

61. K. D. Watenpaugh, L. C. Sieker, J. R. Herriott, and L. H. Jensen, *Acta Crystallogr.*, *Sect. B*, **29**, 943 (1973).

62. R. G. Shulman, P. Eisenberger, W. E. Blumberg, and N. A. Stombaugh, *Proc. Nat. Acad. Sci. USA*, **72**, 4004 (1975).

63. J. Bordas, R. C. Bray, C. D. Garner, S. Gutteridge, and S. S. Hasnain, *Biochem. J.*, **191**, 499 (1980).

64. L. Power, B. Chance, Y. Ching, and P. Angiolillo, *Biophys. J.*, **34**, 465 (1981); L. Power, B. Chance, Y. Ching, B. Muhoberac, S. Weintraub, and D. Wharton, *FEBS Lett.*, **138**, 245 (1982).

Part IV   Multielectron Transfer Processes

Intracomplex and Intramolecular Electron Transfer in Macromolecules

Michael A. Cusanovich

Department of Biochemistry, University of Arizona, Tucson, Arizona 85721

9-1. Introduction

Biological electron transfer can occur by two general mechanisms: collision dependent processes and intramolecular electron transfer within preexisting complexes. In fact, collision dependent processes consist of two steps: formation of a transient complex and, subsequently, intracomplex electron transfer. The main focus of the discussion presented here will be on intracomplex and intramolecular electron transfer. But first, the basic features of the collisions leading to transient complex formation are outlined because these collisions determine, to a large extent, the principal features of subsequent events.

In biological systems, collision dependent physiologically relevant electron transfer can occur between two soluble proteins, between a soluble protein and a membrane-bound protein (or protein complex), or between two membrane-bound proteins (soluble in the membrane). To date, extensive structural information is available only for soluble electron transfer proteins, hence limiting our detailed understanding of these processes to only those systems. It should be noted that very recently the three-dimensional structure of a true membrane redox protein complex has been determined. Thus, bacterial photosynthetic reaction centers containing bacterio-chlorophyll and quinones—with or without bound cytochromes—should provide a means to begin to understand, in molecular terms, biological electron transfer in membrane-bound redox complexes. However, much effort is needed to bring these systems to the level of understanding of the soluble systems.

Soluble redox proteins have been extensively studied in regard to both structure and mechanism. The principal focus to date has been on the c-type cytochromes, particularly mitochondrial cytochrome c, but also including the cytochromes c' and

c_3. A number of recent reviews exist that provide an excellent summary of our understanding at this time.[1-4]

Moreover, a variety of other types of redox proteins, including heme proteins, iron-sulfur proteins, copper proteins, and flavoproteins, have been well characterized (for example, see refs. 5–10). Nevertheless, the discussion to follow will focus principally on the cytochromes c and their interactions with a variety of electron donors and acceptors. Moreover, no effort has been made to provide a comprehensive review of the literature. Rather, a variety of examples are used to illustrate concepts and features and to examine prospects for future studies. Inherent in the approach used here is an effort to point out areas where our understanding is incomplete.

Perusal of existing data and an understanding of redox protein structures suggests that a number of factors can, to varying degrees, control the direction and kinetics of intracomplex and intramolecular electron transfer. These factors include the distance between redox centers, the relative orientation of the redox centers, the driving force (that is, the difference in the midpoint oxidation reduction potentials of the redox centers), the nature of the intervening media between the redox centers, and the dynamic properties within complexes.

A complete understanding of the factors outlined in the preceding text should permit predicting the kinetics of electron transfer in systems for which structural information is available and, at least in some cases, predicting structure in systems in which the kinetics are well understood. Clearly, such an understanding would permit the design of new molecules or complexes of molecules with specific properties, which would be of use in developing new materials for and in the design of bioelectronic devices.

9-2. Collision Dependent Processes

The formation of transient complexes between interacting redox proteins is controlled by two principal factors that have been well characterized: electrostatics and sterics. It is now clear from a variety of studies that interacting pairs of redox proteins have characteristic interaction domains that represent small parts of their molecular surfaces, that have well defined topologies, and that have (for many cases) complementary electrostatic fields.[1,2] Although only limited structural information is available for protein–protein complexes, there are several such complexes for which the structure of the individual reactants is known and that have been characterized by both kinetic analysis and computer modeling. These include cytochrome c-flavodoxin,[11,12] cytochrome c-cytochrome b_5,[13,14] cytochrome c-cytochrome c peroxidase,[15,16] and, to a lesser extent, cytochrome c-photosynthetic reaction centers.[17,18] In addition, a number of systems where the structure of only one of the reactants is known have been characterized by kinetic analysis. These include cytochrome c_2-ferredoxin:NADP$^+$ reductase, cytochrome c-cytochrome oxidase, and ferredoxin:NADP$^+$ reductase with a variety of iron-sulfur proteins (see ref. 19 for a recent summary).

In all of the preceding examples, the measured second-order rate constants for complex formation approach the expected diffusion controlled limit when corrected for the fraction of the surface defined by the interaction domain at low ionic strength.

Moreover, almost all have second-order rate constants that are ionic strength dependent, consistent with a complementary charge interaction playing an important role in complex formation. This conclusion is strongly supported by computer modeling studies that allow the positioning (docking) of two redox proteins (for example, cytochrome c-cytochrome b_5) and examining variations in complementary charge interactions to optimize complex formation. By analysis of the redox kinetics of flavodoxin semiquinone with a series of structurally homologous cytochromes c with different charged groups in the vicinity of the interaction domain, it has been found that the measured second-order rate constants can vary up to 1000-fold[1,2] due to electrostatic interactions alone.

Data from the studies just described can be corrected for electrostatic interactions; that is, the kinetics of electron transfer at infinite ionic strength can be determined. Under these conditions, variations in the measured second-order rate constants can, within a family of structurally homologous redox proteins, be related to the topology in the region of the interaction domains, correlating topological changes with up to 1000-fold variation in reaction rate constants, as in the case electrostatics. In the case of topology, it is presumed that steric restrictions affect the distance of closest approach of the reactants.

In sum, both electrostatics and sterics can significantly affect the kinetics of complex formation. Moreover, nature can readily modulate these kinetics by varying the amino acid side chains in the interaction domain. Presumably, this is done in physiological systems to control biological specificity; that is, to modulate electron-transfer rates so as to optimize physiologically relevant pathways and to minimize nonproductive side paths.

9-3. Theory

It is well established that biological electron transfer is an outer sphere process that, to the extent we can define parameters, obeys Marcus theory.[1-4] Thus nonadiabatic electron transfer can be described by

$$k_{et} = \nu_{et} \exp(-\Delta G^*/RT) \tag{9-1}$$

where ΔG^* is the free energy of activation and ν_{et} is a frequency. The free energy of activation is related to the driving force of the reaction (ΔG^0) and the reorganizational energy λ as

$$\Delta G^* = (\Delta G^0 + \lambda)^2/4\lambda \tag{9-2}$$

The frequency factor is given by

$$\nu_{et} = \nu \exp(-\beta(d - d_0)) \tag{9-3}$$

where ν is a nuclear frequency factor that can be taken as 10^{13} s^{-1} for a short range electron transfer, β is the rate of decrease of electronic coupling with increasing distance, which is dependent on the nature of the media, d is the electron-transfer distance, and d_0 is the van der Waals contact distance (taken as 3 Å). Clearly, at short distance ($d = d_0$) $\nu_{et} = \nu$. In the case of long-distance electron transfer, the intervening media between redox centers is a tunneling barrier to the electron and ν

can be written as[20]

$$\nu = 2\pi\hbar^{-1}[H_{ab}(d_0)]^2(4\pi\lambda kT)^{1/2} \tag{9-4}$$

where $H_{ab}(d_0)$ describes the donor and acceptor localized electronic states and the tunneling medium.

The literature contains a number of different formulations, all similar in form to Eq. (9-1), but varying in the details [Eqs. (9-2)–(9-4)]. This results from the fact that, as new information becomes available, new models are derived to attempt to describe experimental results. Thus, as pointed out recently,[20] experimental parameters will be dependent on the details of the model used to analyze the data, and much more experimental work is needed to develop fully the degree of sophistication needed to understand biological systems. Nevertheless, to a first approximation, Eqs. (9-1)–(9-4) describe the dependence of electron transfer on driving force (ΔG^0), distance (d), orientation [$H_{ab}(d_0)$ and λ], intervening media [$H_{ab}(d_0)$], and dynamics [$H_{ab}(d_0)$ and λ]. The goal of studies on biological systems is to fix all but one of the parameters and thus isolate individual features of the reaction mechanism for study. As will be shown, this can be done to some extent, but a great deal of work remains before we can understand fully all the factors involved.

9-4. DRIVING FORCE

In a number of systems involving collision dependent processes, it has been shown that Eqs. (9-1) and (9-2) are applicable. That is, a relationship between the driving force and the observed second-order rate constant yields ν_{et} and λ. This would not be expected a priori, because, as described in the section on diffusion-controlled processes, electrostatic and topology both play major roles in defining interaction domains and, thus, second-order rate constants. However, if families of homologous proteins are studied using small uncharged reactants (for example, lumiflavin semiquinone) or second-order rate constants are electrostatically corrected by extrapolating to infinite ionic strength, the contribution of electrostatics, and to a lesser extent topology, can be factored out. This apparently results because, in complex formation followed by rapid intracomplex electron transfer, the apparent second-order rate constant is Kk_{et}, with K and k_{et} as defined in

$$A_{ox} + B_{red} \xrightarrow{K} (A_{ox} \cdots B_{red}) \xrightarrow{k_{et}} (A_{red} \cdots B_{ox}) \tag{9-5}$$

Thus, in a family of homologous proteins where topology is similar and where electrostatics are either absent or corrected for, Kk_{et} is proportional to k_{et} in Eq. (9-1). Put another way, K is approximately the same for all members of the family. Along this vein, we have investigated the interaction of lumiflavin semiquinone (LF·) with the cytochromes c, cytochromes c', high potential iron-sulfur proteins (HIPIPs) and copper proteins and these data are summarized in Table 9-1. It is important to note, however, that direct comparisons between the different families of proteins are not possible, because the topologies are distinct [that is, K in Eq. (9-5) is different]. In fact it is assumed that the differences in ν_{et} shown in Table 9-1 result in part from different distances of electron transfer resulting from topology [see Eq. (9-3)]. It also should be noted that the flavodoxin semiquinone data are much less accurate because

Table 9-1 ▪ Intrinsic Parameters for the Reduction of Redox Proteins by Flavin Semiquinones

Oxidant	Ref.	Reductant	ν_{et} (M^{-1} s^{-1})	λ (kcal/mol)
Cytochromes c	11, 24	LF$^{\cdot}$	1.0×10^8	11.0
		Flavodoxin$^{\cdot}$	5.0×10^9	42.8
HIPIPs	25	LF$^{\cdot}$	9.6×10^7	12.0
		Flavodoxin$^{\cdot}$	2.8×10^9	42.8
Cytochromes c'	21	LF$^{\cdot}$	2.8×10^8	12.0
		Flavodoxin$^{\cdot}$	10^5–10^6	NM^a
Copper proteins	8	LF$^{\cdot}$	4.4×10^7	14.0
		Flavodoxin$^{\cdot}$	2.2×10^9	42.8

^{a}NM = not measured.

this system is much more sensitive to topology and requires electrostatic correction, whereas lumiflavin semiquinone is small and uncharged.

The data summarized in Table 9-1 suggest a number of conclusions. First, the ν_{et} values for lumiflavin semiquinone are generally similar for the four families of redox proteins and are consistent with structural data, which suggest that, in the cases of the cytochromes c, cytochromes c', and copper proteins, the reductant can approach to within van der Waals distances. Moreover, the iron-sulfur center of HIPIP is, at the point of closest approach to the surface, only 4.5 Å from the solvent.[7] Interestingly, λ for the lumiflavin semiquinone reactions is generally similar with the different protein families, suggesting that the solvent rearrangements and movements that must occur prior to electron transfer are similar. This is not surprising because the reductant is the same. What is particularly interesting is that flavodoxin semiquinone studies result in a much larger value of λ. This is consistent with the increased complexity resulting from the association of the protein with flavin in flavodoxin and implies that more solvent rearrangement and nuclear movements are required to reach the transition state. Finally, the large increase in ν_{et} observed when going from the lumiflavin semiquinone to flavodoxin semiquinone system must be related to the specific nature of the flavin-protein interactions that occur at the flavodoxin coenzyme binding site and may reflect the importance of orientation in obtaining optimum interaction. One exception is the flavodoxin-cytochrome c' system where ν_{et} is small. This apparently results from the fact that the cytochrome c' heme is at the bottom of a deep crevice, which is accessible to small molecules, but sterically hinders access to flavodoxin, hence yielding the relatively small value of ν_{et} [see Eq. (9-3)].[21]

Turning to direct measurement of k_{et} as described by Eq. (9-2), as the driving force increases, that is, as ΔG^0 becomes more negative, the rate constant should increase. However, a maximum should be reached (when $\Delta G^0 + \lambda = 0$), and beyond this point, the rate constant should actually decrease. This predicted decrease in rate constant, the so-called inverted region, has been observed in several systems, for example, cytochrome c-cytochrome b_5 where metal substitutions in the cytochrome b_5 heme were used to vary the driving force[22] and in Zn-Fe hybrid hemoglobins.[23] In these systems, the inverted region was observed because the metal substitutions

Table 9-2 ▪ Intracomplex Electron Transfer, Ferrocytochrome c to H_2O_2 Oxidized Cytochrome c Peroxidase[15]

Reductant	$\Delta E_{m,7}$ (mV)	k_{et} $(s^{-1})^a$
Tuna cyto. c	~ 750	950
Yeast iso-1 cyto. c	~ 750	260 (1460)
Yeast iso-2 cyto. c	~ 750	150 (1480)
Yeast iso-1 cyto. c (Ile-13)^a	~ 750	1000 (2240)

^{a}Ionic strength 8 mM. Values in parentheses are ionic strength 200 mM.

allowed a very large range of driving forces. In most, but not all, biological systems, only a small driving force is present (100–400 mV) and the inverted region is not routinely observed. Nevertheless, it is well established that the effect of driving force is well behaved in terms of Eq. (9-2), and thus deviations from the expected relationship must result from the participation of other factors (orientation, distance, intervening media, etc.). An example of this is given in Table 9-2 where the reaction of several different cytochromes c with H_2O_2-oxidized cytochrome c peroxidase [CcP(IV, R·$^+$)] has been studied.[15]

Tuna and the two yeast isocytochromes are structurally homologous and have essentially equivalent oxidation-reduction potentials (260 mV). In addition, the oxidation of a mutation of yeast iso-1-cytochrome c, where arginine-13 has been replaced by isoleucine and is isopotential with the wild-type cytochrome, by H_2O_2 oxidized cytochrome c peroxidase was studied at two ionic strengths. The redox potentials of these cytochromes c are independent of ionic strength over the ionic strength range of the study. Interestingly, the rate constant for electron transfer varies by up to 15-fold (Table 9-2), while the driving force remains constant. Thus, although driving force is an important variable, other factors must dominate in these examples (see the following text).

9-5. Distance

Equation 9-3 predicts a relationship between distance and rate constant. As described in the previous section for the cytochrome c' system, distance can have a very large effect on rate constants.[21] However, the ability to isolate distance as a parameter and to fully quantitate its contribution has not been developed in any depth. Table 9-3 summarizes some biological systems where the distance between redox centers is known or can be estimated. Two types of systems are included in Table 9-3: those where the distance is known from the crystal structure (cytochrome b_2, R. viridis photosynthetic reaction centers, and trimethylamine dehydrogenase) and those where the distance between redox centers has been obtained from computer modeling of the complex, using known structures of the individual reactants. Clearly, those complexes that have been modeled are subject to revision when the structures of the complexes have been determined. However, the estimates appear reasonable, based on complementary charge interactions and the structural constraints. The systems given in Table 9-3 can be further divided into short range electron transfer (van der Waals distances)

Table 9-3 ▪ Intracomplex and Intramolecular Electron Transfer

System	Ref.	d^a (Å)	Angle	$\Delta E_0'$ (mV)	k_{et} (s^{-1})
Flavodoxin → cyto. c	12, 26	3.5	Coplanar	~ 300	85
Cyto. b_5 → cyto. c	14, 22	3.5	Coplanar	~ 250	1400
Trimethylamine dehydrogenase (flavin → Fe-S)	27, 28	3.5	NA	~ 60	~ 60
Cytochrome b_2 (flavin → heme)	29	10	Coplanar	~ 60	> 600
Cyto. c_2 → R. sphaeroides reaction center	17, 18	11	Coplanar	~ 100	500–13,000^b
Cyto. c → cyto. c peroxidase	15, 16	17	Coplanar	~ 750	150–2300^b
R. viridis reaction center	30, 31^c				
heme 1 (c-559) → BChl$^+$		~ 15	Perpendicular	~ 70	3.7×10^6
heme 2 (c-553) → BChl$^+$		~ 25	Coplanar	~ 530	6.0×10^6
heme 3 (c-556) → heme 1(c-559)		~ 20	Perpendicular	~ 70	2.8×10^5

^{a}Distance edge to edge where planar chromophores are involved.
$^b k_{et}$ is dependent on ionic strength (see text).
^{c}Hemes are numbered according to their distance from the special pair chlorophyll.

and long range electron transfer (> 3.5 Å). In the case of short range electron transfer, the range of k_{et} values is 60–1400 s^{-1}, which are generally small values in view of the short distances involved. Clearly, no obvious correlation exists between driving force and k_{et} suggesting that other parameters, such as orientation, electronic coupling, or dynamics—play an important role. The assumption from modeling is that, in both the flavodoxin-cytochrome c and cytochrome b_5-cytochrome c systems, the chromophores are coplanar in the theoretically preferred orientation and yet the rate constants can differ by more than a factor of 10. Thus, orientation does not appear to be a significant factor in these examples. The trimethylamine dehydrogenase system is interesting because the iron-sulfur center can be viewed as symmetrical, and thus orientation does not appear to be a factor to the first approximation. Nevertheless, intracomplex electron transfer is slow even though the redox centers are within van der Waals distance.

In systems where long range electron transfer is observed, the rate constants vary by up to a factor of 10^4 with no obvious correlation with distance. In addition, no obvious correlation with driving force is apparent. The system with the largest driving force, cytochrome c-cytochrome c peroxidase, has an apparent k_{et} value that is much smaller than that of the R. viridis photosynthetic reaction center complex where the driving force is much smaller. However, these long range electron-transfer systems are likely to be strongly influenced by the nature of the intervening media, and hence, distance, orientation, or driving force may be insufficient to explain the kinetics.

Given the limitations of the biological system, Gray and co-workers have pioneered studies in which (NH$_3$)$_5$Ru(III) has been covalently attached to specific

Table 9-4 ▪ Effect of Distance on Intramolecular Electron Transfer in
Myoglobin-Ruthenium Complexes[20,32]

Myoglobin Modified Histidine	E_0' (mV)	Distance (Å)	$k_{et}(5')^a$	
			MgMbdiacid	MgMbdiester
$(NH_3)_5Ru(His-48)$	85	12.7	57×10^3	32×10^3
$(NH_3)_5Ru(His-81)$	61	19.3	134	74
$(NH_3)_5Ru(His-116)$	70	20.1	94	67
$(NH_3)_5Ru(His-12)$	75	22.0	84	48

[a]The rate constants given were extrapolated to correct for the small differences in E_0' among the different modified myoglobins.

histidines on heme containing proteins, such as cytochrome c and myoglobin.[32,33] As shown in Table 9-4, myoglobin can be modified with $(NH_3)_5Ru(III)$ at four different positions with the distance from the X-ray structure given in Table 9-4. By using Mg-porphyrin (either the diacid or the diester), electron transfer to Ru(III) can be initiated photochemically and k_{et} measured. From the data given, it can be seen that, although a rough correlation between distance and k_{et} is observed, the correlation is not strong. Related studies using Zn-porphyrin yield very similar results.[34] One of the points causing the most difficulty is the His-12 substitution in both the Mg and Zn cases where k_{et} is substantially larger than a $\ln k_{et}$ versus distance relation would predict. This is of interest because His-12 is near a tryptophan that, due to its aromatic character, could serve as a path to the heme, resulting in a larger than expected k_{et}. These results suggest that, although a correlation between distance and k_{et} may exist, other factors must also be involved. Clearly, the differences among the various derivatives are in the path from the ruthenium to the heme because the histidines are at different positions on the molecular surface. That is, the intervening media between the redox centers are different with each of the derivatives.

In sum, although distance should and probably does play an important role in controlling the electron-transfer kinetics, its contribution is difficult to quantitate because of the participation of other effects. Thus, to obtain clear correlations between distance and kinetics, it will be necessary to understand the contributions of other factors, such as intervening media, orientation, and dynamics, in order to correct for these parameters.

9-6. Orientation

Because electron coupling between metalloprotein electron donors and acceptors will be affected by their relative orientations, it has been assumed for some time that orientation is an important factor in electron-transfer kinetics. Moreover, at least in heme-heme and heme-flavin systems, it has been assumed that a coplanar orientation would be optimum in order to maximize electronic coupling and obtain the largest rate constant. This, of course, assumes that the largest rate constant is the most biologically relevant. Thus, in the modeled complexes (cytochrome c-cytochrome b_5, cytochrome c-flavodoxin, cytochrome c_2-photosynthetic reaction centers, and cytochrome c-cytochrome c peroxidase), the best coplanar or nearly coplanar fit has been chosen. This approach is supported by the fact that in the cytochrome b_2 system, where the structure has been determined by X-ray crystallography, the heme and flavin moieties are nearly coplanar (angle $\sim 13°$). However, in this example it is notable that k_{et} is not particularly large (~ 600 s^{-1}). More recently, the results with the *R. viridis* tetraheme reaction center cytochrome suggest that a coplanar complex may not be required.[30] Electron transfer from the heme nearest to oxidized bacteriochlorophyll (heme 1) and from the heme third most distant (heme 3) to heme 1 involve chromophores whose orientation is nearly perpendicular yet electron transfer is extremely fast ($> 10^5$ s^{-1}). Although this observation does not prove that electron transfer between orthogonal chromophores is optimum, it does establish that very rapid electron transfer can occur with this orientation. Clearly, much more structural information is required to quantify the effect of orientation.

Studies with both cytochrome c-cytochrome c peroxidase and cytochrome c_2 reaction center systems do suggest that orientation can have an important effect on electron-transfer kinetics. The cytochrome c_2-*R. sphaeroides* reaction center complex has been modeled using complementary charge interactions to guide the fitting process.[18] To a first approximation, this is reasonable because the cytochromes c have a cluster of positive charges in the region of their exposed heme edge (the presumed site of electron transfer) and, in general, they behave as cations in their reaction with physiological reactants.[2,17]

The *R. sphaeroides* reaction center in turn has complementary negative charges near the special pair bacteriochlorophyll. Thus, it follows that at low ionic strength where electrostatics are optimum, complementary charge interactions would lead to an optimum orientation for electron transfer. Moreover, it has been assumed that increasing ionic strength would have one of two effects. It could result in a decrease in k_{et} by masking the most important charge interactions and thus inhibit achieving the optimum orientation, or it could have no effect on k_{et} but greatly decrease the rate constant for complex formation; that is, decrease the probability of finding the most favorable complex but not affecting it once it has formed.

Table 9-5 summarizes some data on k_{et} obtained with *R. rubrum* reaction centers (which are homologous to *R. sphaeroides* reaction centers) and three different c-type cytochromes. In the case of horse cytochrome c and *R. rubrum* cytochrome c_2 k_{et} is independent of ionic strength, an expected situation that suggest distance and orientation are unchanged (see preceding text). In the case of *R. capsulatus* cytochrome c_2, the results were quite different but not unexpected, with k_{et} decreasing 20-fold with increasing ionic strength. This result suggests that the low ionic strength complex in this case is much more favorable for rapid intracomplex electron transfer than is the high ionic strength complex. That is, strong complementary charge interactions appear to result in an orientation and distance that is optimal for k_{et} and that increasing ionic strength masks these interactions, allowing the formation of a complex which is less favorable for rapid intracomplex electron transfer. More recently, a very different and unexpected result has been found with the cytochrome c-cytochrome c peroxidase system.[15] As shown in Table 9-2, wild-type yeast cytochrome c (iso-1 and iso-2) yield k_{et} values that increase with increasing ionic strength, the opposite from the *R. capsulatus* cytochrome c_2 reaction center system.

Table 9-5 ▪ **Kinetics of Reduction of Oxidized R. rubrum Reaction Centers**[a]
by Reduced Cytochrome c_2[17]

Reductant	Driving Force (mV)	k_{et} (s^{-1})
Horse cyto. c	~ 190	6,250[b]
R. rubrum cyto. c_2	~ 125	7,100[c]
R. capsulatus cyto. c_2	~ 81	12,500[d]
	~ 81	556[e]

[a] R. rubrum reaction centers are homologous to those of R. sphaeroides (they do not have a bound cytochrome as in R. viridis).
[b] Ionic strength, 48–168 mM.
[c] Ionic strength, 30–150 mM.
[d] Ionic strength, 30 mM.
[e] Ionic strength, 150 mM.

Interestingly, conversion of Arg-13 on cytochrome c, a side chain thought to be involved in one of the critical complementary charge interactions, results in an increase in k_{et} at both low and high ionic strength. This results suggests that by deleting a critical electrostatic interaction the cytochrome c can be oriented for more favorable electron transfer. This is consistent with the ionic strength effects on the wild-type cytochrome. These results provide support for the argument that orientation is important in electron transfer and that strong electrostatic interactions can result in an unfavorable orientation. It should be noted that the assumption is that ionic strength effects result from reorientation of the reactants within the interaction domain. An alternative interpretation is that increasing ionic strength in the R. capsulatus cytochrome c_2-reaction center and cytochrome c-cytochrome c peroxidase system results in interaction occurring at a different site, possibly increasing the distance between redox centers. However, the effects of increasing ionic strength occur without obvious discontinuities, inconsistent with the notion of two interaction domains (high ionic strength, low ionic strength).

   In summary, the results discussed suggest that orientation can have an affect on k_{et}. However, it is clear that at the present time we have too little information to predict the contribution of orientation to the electron-transfer rate constant in any detail. As more structural information becomes available for complexes containing multiple redox centers, we should be able to address the role of the relative orientation of redox centers.

9-7. Dynamics

The role of λ, the reorganizational energy in intracomplex and intramolecular electron transfer, is defined in Eq. (9-2). The reorganizational energy reflects the movements and rearrangements required to reach the transition state and, as such, reflects the dynamics within protein-protein complexes. In small inorganic or organic redox complexes, λ is generally viewed as movements of water molecules and changes in bond lengths and bond angles; that is, small rearrangements. In protein-protein complexes, it is unclear how large the movements required for electron transfer are. Proteins are quite flexible and capable of undergoing rather large movements, on the order of several angstroms, due to thermal motion. Thus, relatively large motions, which could be quite slow on the time scale for electron transfer, are possible, and in some cases these first-order processes could be rate limiting.

Studies on the interaction of the iron hexacyanides with cytochrome c suggest that a five step mechanism for reversible electron transfer is required.[35] This mechanism includes two second-order collision dependent processes, as shown in

$$\text{Fe}(\text{CN})_6^{-4} + \text{cyto.(III)} \xrightarrow{K_1} C_1 \xrightarrow{K_2} C_2 \xrightarrow{K_3} C_2' \xrightarrow{K_4} C_1' \xrightarrow{K_5} \text{Fe}(\text{CN})_6^{-3} + \text{cyto.(II)}$$
$$(9\text{-}6)$$

Both C_1 and C_1' undergo rapid reversible conformational changes $(1 \times 10^4 \text{ s}^{-1}$ for reduction, $2 \times 10^6 \text{ s}^{-1}$ for oxidation) that lead to the formation of the redox active species (C_2 and C_2'). Both the electron-transfer rate constants for the interconversion of C_2 and C_2' are $> 2500 \text{ s}^{-1}$. The use of a five-step mechanism including two conformational changes is the minimum mechanism required to explain all available data (a more complex mechanism cannot be excluded). In this example, k_{et}, and not the conformational changes, is apparently rate limiting, but it still serves to illustrate the participation of rapid structural changes. Recently, McLendon and coworkers have reported an example where the structural change, not electron transfer, is rate limiting.[36] In this study, k_{et} for electron transfer from the heme moiety of cytochrome b_2 to cytochrome c was measured (the two cytochromes form a stable complex at low ionic strength). For the experiment, derivatives of cytochrome c(Zn-porphyrin and porphyrin) as well as the wild-type (protoheme IX) were used. The three forms of cytochrome c apparently have the same or very similar tertiary structures but provide approximately 800 mV variation in driving force. Thus, it was expected that the measured rate constant would vary as predicted by Eqs. (9-1) and (9-2). However, it was found that the measured first-order process ($\sim 600 \text{ s}^{-1}$) was independent of driving force. Thus, it was concluded that conformational gating[37] is taking place, with formation of the redox active species acting as the rate limiting step. It follows that electron transfer must be greater than or equal to 600 s^{-1}. This gating concept substantially complicates the interpretation of k_{et} (for example, see Table 9-3) because there is no easy way to discriminate between first-order structural changes and first-order electron transfer in kinetic experiments.

An example that illustrates the magnitude of dynamic processes is the flavodoxin-cytochrome c system.[38] From the modeled structure of the flavodoxin-cytochrome c complex, it is apparent that access to the cytochrome c heme would be severely restricted to small molecule reductants due to steric hindrance. Indeed, this is the case, because at low ionic strength where the complex is stable, the second-order rate constants for the reduction of cytochrome c by lumiflavin and riboflavin semiquinone are greatly reduced relative to those of the uncomplexed cytochrome. Moreover, if a somewhat larger flavin, FMN semiquinone, is used, a change in rate limiting step takes place at high complex concentrations. This result is consistent with a rate limiting breathing or gating, which occurs with a first-order rate constant of 2600 s^{-1}. Thus, dynamic motions (2600 s^{-1}) within the flavodoxin-cytochrome c complex are sufficiently large (3–7 Å) to allow entry of the FMN semiquinone into the interface region. Although not a measurement of intracomplex electron transfer, the dynamics of the flavodoxin-cytochrome c complex establishes that relatively large and rapid motions can occur.

It is evident from the discussion presented that dynamics can play an important role in determining the rate constant for electron transfer between redox proteins. These influences range from solvent reorganization and molecular rearrangements, as reflected by λ, to distinct steps in the reaction mechanism. As discussed here, this complicates the interpretation of apparent k_{et} values. Clearly, substantial study is required to identify and define in molecular terms the nature of dynamic properties of protein-protein complexes.

9-8. Intervening Media

The most difficult to understand and yet in many ways the most interesting parameter mediating intracomplex and intramolecular electron transfer is intervening media. In long range electron transfer (> 3.5 Å), the electron must tunnel through a medium made up of water (in some cases), amino acid side chains, and peptide backbone. It is reasonable to assume that nature, in refining the details of physiologically relevant electron transport through evolution, has chosen characteristics for the intervening media that optimize the kinetics of electron transfer. An example of this is the tetraheme reaction center cytochrome from $R.$ $viridis$ (Table 9-3), in which both heme $1 \rightarrow Bchl^+$ and heme $3 \rightarrow$ heme 1 electron transfer occur over relatively long distances (15–20 Å) with a small driving force (~ 70 mV), yet k_{et} is very large. This results suggest a well defined tunneling pathway not observed in most of the other systems studied to date (Table 9-3). This is likely to result because of the need for rapid reduction of the photooxidized bacteriochlorophyll to prevent back reaction. In other systems (for example, cytochrome b_2 and cytochrome c-cytochrome c peroxidase), the need for very fast electron transfer may not be required for biological function. As an alternative, the cytochrome b_2 and cytochrome c-cytochrome c peroxidase systems may be gated to maintain a rate of electron transport that couples them to the overall metabolic pathway in which they function. It is interesting that the other long range electron-transfer system that has relatively large rate constants is the cytochrome c_2-$R.$ $sphaeroides$ reaction center complex (Table 9-3), which is also involved in reducing photooxidized bacteriochlorophyll.

It is also interesting to note that most $(NH_3)_5$-Ru(III)-substituted myoglobins have a low electron-transfer rate constant ($\sim 10^2$ s^{-1}) (Table 9-4). This likely reflects, in part, the fact that this is an artificial electron-transfer system and hence the pathway has not been optimized through evolution. Whatever the case, it is of substantial utility to deduce the factors that control the optimum tunneling pathways in proteins, as such information will lead to insights useful in the design of new biomaterials.

Although a detailed understanding of electron transfer through a protein medium is not currently available, some recent work has addressed this issue. Cowan et al.,[20] have described a model for estimating the decay of the tunneling matrix [$H_{ab}(d_0)$, Eq. (9-4)] per bond traveled by the electron or resulting from a through-matrix jump. The basic idea is that for each bond traveled or for each jump through the matrix, the intrinsic rate of electron transfer is decreased; thus the overall rate constant for electron transfer will depend on the number of bonds or jumps in the path as well as the type of bond. To analyze a path, a line is drawn between the donor and acceptor, which is then overlayed with a cylinder of increasing radius. The pathway is then traced, attempting to minimize the number of covalent and hydrogen bonds that have to be traversed within the cylinder. By assigning a value for the $H_{ab}(d_0)$ decay for each bond or matrix jump, which depends on both the electron tunneling energy and the orbital energies of the bonds along the tunneling path, estimates of the relative rate constants can be obtained for different paths. Because of the energetics, it is proposed that a 4 Å through-matrix gap causes a decay equivalent to 8–9 through-backbone bonds. Thus, tunneling gaps can be used to follow a reasonably straight path but will cost in terms of decay and hence will lead to decreased rate constants. Application of this approach to $(NH_3)_5Ru(III)$ modified myoglobins yields reasonable pathways. For example, the path from ruthenated His-48 to heme can be written as

$$\text{His-48 imidazole} \xrightarrow{8C} C_\alpha\text{Arg-45} \xrightarrow{9C} N\text{Arg-45} \xrightarrow{2H}$$

$$O(\text{propionate of the heme}) \xrightarrow{4C} \text{heme } \pi\text{cloud},$$

where C denotes a covalent bond and H a hydrogen bond. Thus a total of 23 bonds are traversed with no matrix jump in this example. Pathways can be calculated for the other derivatives although they are more ambiguous (His-81, 29 bonds; His-116, 19 bonds, 4 Å through-matrix distance; His-12, 19 bonds, 6.9 Å through-matrix distance). Although much work remains to be done to develop the necessary understanding to apply the approach described, it provides a reasonable model that is experimentally testable and a real opportunity to understand the role of intervening media in biological electron transfer. However, one caveat should be note. It is assumed in the ruthenium-myoglobin system that gating is not occurring. If, in fact, conformation gating is rate limiting in one or more of the derivatives, analysis in terms of intervening media will be incorrect.

In summary, it is clear that electron tunneling through intervening media is required for long range electron transfer in biological systems and that this process has been optimized through evolution. Moreover, models are beginning to appear that should allow us to define in molecular terms the details of this process.

9-9. Summary and Prospects

In recent years, very substantial progress has been made in defining the parameters controlling biological electron transfer. First and foremost, it is now well established that driving force, as predicted by Marcus theory, plays a fundamental role. This is particularly important because it means that the contribution of driving force can be predicted and thus other factors contributing to the electron-transfer kinetics can be identified through deviations from the expected behaviour due to driving force. Distance, although a factor, appears to be dominated by other parameters, such as dynamics and intervening media, and is difficult to isolate and quantitate. The issue of orientation and thus orbital interactions and electronic coupling is undoubtedly important. Unfortunately, at this time, we do not have sufficient structural informa-tion to really clarify and quantify the role of orientation. However, in the next few years, additional structures of proteins and protein complexes containing two or more redox centers should become available and permit a more detailed analysis of the role of orientation.

Dynamics, and thus the reorganizational energy and conformational gating, may turn out to be one of the most important factors in controlling electron-transfer kinetics in biological systems. Until we understand in much more detail the dynamic properties of proteins and protein complexes, it is going to be difficult to understand fully many biological systems. Fortunately, the recent advances in determining protein three-dimensional structures by NMR and the corresponding ability to measure protein dynamics should result in new insights that provide a basis for understanding the role of dynamics in biological electron transfer.

Available data establish that intervening media play a significant role in long range electron transfer. However, much effort will be required to develop the necessary level of understanding. Thus, further work with model systems, the development of the appropriate theory, and more structural data are all required. Nevertheless, there are very real opportunities here to dissect biological systems and, from this information, develop new understandings and likely new technologies.

In addition to the application of NMR for both structural and dynamic data and new X-ray structures, another technique can be expected to contribute to studies on biological electron transfer. This technique is site-directed mutagenesis. A number of redox proteins have been cloned and a variety of mutants are now being analyzed. Clearly, as these studies evolve, we will be able to examine the role of specific amino acid side chains, alter protein dynamics and the properties of the intervening media, expand our understanding of the factors controlling biological oxidation-reduction potentials, and affect distances between redox centers. Thus, the use of site-directed mutagenesis will provide the principal tool to understand the details of biological electron transfer and should allow us to design new biomaterials with properties useful in a wide range of applications, including bioelectronic devices.

Acknowledgment. This work was supported by NIH grant GM2127. The author thanks Dr. T. E. Meyer and Dr. M. J. Halonen for helpful discussions.

References

1. G. Tollin, T. E. Meyer, and M. A. Cusanovich, *Biochim. Biophys. Acta*, **853**, 29–41 (1986).
2. M. A. Cusanovich, T. E. Meyer, and G. Tollin, in *Advances in Inorganic Biochemistry*, Vol. 7, *Heme Proteins*, G. L. Eichorn and L. G. Marzilli, eds., Elsevier, NY (1988), pp. 37–92.
3. H. B. Gray and B. G. Malmström, *Biochem.*, **28**, 7499–7505 (1989).
4. G. McLendon, *Acc. Chem. Res.*, **21**, 160–167 (1988).
5. M. A. Cusanovich, T. E. Meyer, and G. Tollin, *Biochem.*, **24**, 1281–1287 (1985).
6. A. K. Bhattacharyya, T. E. Meyer, and G. Tollin, *Biochem.*, **25**, 4655–4661 (1986).
7. A. K. Bhattacharyya, T. E. Meyer, M. A. Cusanovich, and G. Tollin, *Biochem.*, **26**, 758–764 (1987).
8. G. Tollin, T. E. Meyer, G. Cheddar, E. D. Getzoff, and M. A. Cusanovich, *Biochem.*, **25**, 3363–3370 (1986).
9. G. Cheddar, T. E. Meyer, M. A. Cusanovich, C. D. Stout, and G. Tollin, *Biochem.*, **25**, 6502–6507 (1986).

10. C. T. Przysiecki, A. K. Bhattacharyya, G. Tollin, and M. A. Cusanovich, *J. Biol. Chem.*, **260**, 1452–1458 (1985); A. K. Bhattacharyya, T. E. Meyer, M. A. Cusanovich, and G. Tollin, *Biochem.*, **26**, 758–764 (1987).

11. G. Tollin, G. Cheddar, J. A. Watkins, T. E. Meyer, and M. A. Cusanovich, *Biochem.*, **23**, 6345–6349 (1984).

12. P. C. Weber and G. Tollin, *J. Biol. Chem.*, **260**, 5568–5573 (1985).

13. L. Eltis, A. G. Mauk, J. T. Hazzard, M. A. Cusanovich, and G. Tollin *Biochem.*, **27**, 5455–5460 (1988).

14. F. R. Salemme, *J. Mol. Biol.*, **102**, 563–568 (1976).

15. J. T. Hazzard, G. McLendon, M. A. Cusanovich, G. Das, F. Sherman, and G. Tollin, *Biochem.*, **27**, 4445–4451 (1988).

16. T. L. Poulos and B. C. Finzel, *Pept. Protein Rev.*, **4**, 115–171 (1984).

17. G. K. Rickle and M. A. Cusanovich, *Arch. Biochem. Biophys.*, **197**, 589–598.

18. J. P. Allen, G. Feher, T. O. Yeates, H. Komiya, and D. C. Rees, *Proc. Nat. Acad. Sci. USA*, **84**, 6162–6166 (1987).

19. M. A. Cusanovich, J. T. Hazzard, T. E. Meyer, and G. Tollin, in *Progress in Clinical and Biological Research*, Volume 24: *Oxidases and Related Redox Systems*, T. E. King, H. S. Mason, and M. Morrison, eds., A. R. Liss, Inc., NY (1988), pp. 401–418.

20. J. A. Cowan, R. K. Upmacis, P. N. Beraton, J. N. Onuchic, and H. B. Gray, *Ann. Acad. Sci. N.Y.*, **550**, 68–84 (1988).

21. T. E. Meyer, G. Cheddar, R. G. Bartsch, E. D. Getzoff, M. A. Cusanovich, and G. Tollin, *Biochem.*, **25**, 1383–1390 (1986).

22. G. McLendon and J. R. Miller, *J. Am. Chem. Soc.*, **107**, 7811–7816 (1985).

23. J. McGourty, N. Blough, and B. Hoffman, *J. Am. Chem. Soc.*, **105**, 4470–4472 (1983).

24. T. E. Meyer, J. A. Watkins, C. T. Przysiecki, G. Tollin, and M. A. Cusanovich, *Biochem.*, **23**, 4761–4767 (1984).

25. C. T. Przysiecki, G. Cheddar, T. E. Meyer, G. Tollin, and M. A. Cusanovich, *Biochem.*, **24**, 5647–5652 (1985).

26. R. P. Simondsen, P. C. Weber, F. R. Salemme, and G. Tollin, *Biochem.*, **21**, 6366–6375 (1982).

27. L. W. Lim, N. Shamola, F. S. Mathews, B. J. Steenkamp, R. Hamlin, and N. H. Xuong, *J. Biol. Chem.*, **261**, 15140–15146.

28. J. T. Hazzard, W. McIntire, and G. Tollin, personal communication 1989.

29. Z. X. Xia, N. Shamola, P. H. Bethge, L. W. Lim, H. D. Bellamy, N. H. Xuong, F. Lederer, and F. S. Mathews, *Proc. Nat. Acad. Sci. USA*, **84**, 2629–2633 (1987).

30. R. J. Shopes, L. M. A. Levine, D. Holten, and C. A. Wraight, *Photosyn. Res.*, **12**, 165–180 (1987).

31. J. Deisenhofer, O. Epp, K. Miki, R. Huber, and H. Michel, *Nature*, **318**, 618–624 (1989).

32. J. L. Karas, C. M. Leber, and H. B. Gray, *J. Am. Chem. Soc.*, **110**, 599–600 (1988).

33. T. J. Meade, H. B. Gray, and J. R. Winkler, *J. Am. Chem. Soc.*, **111**, 4353–4356 (1989).

34. A. W. Axup, M. Albin, S. L. Mayo, R. J. Crutchley, and H. B. Gray, *J. Am. Chem. Soc.*, **110**, 435–439.

35. N. Ohno and M. A. Cusanovich, *Biophys. J.*, **36**, 589–605 (1981).

36. G. McLendon, K. Pardue, and P. Bak, *J. Am. Chem. Soc.*, **109**, 7540–7541 (1987).

37. B. M. Hoffman and M. A. Ratner, *J. Am. Chem. Soc.*, **109**, 6237–6243 (1987).

38. J. T. Hazzard, M. A. Cusanovich, J. A. Tainer, E. D. Getzoff, and G. Tollin, *Biochem.*, **25**, 3318–3328 (1986).

Bioelectrocatalysis at Enzyme-Modified Electrodes

Mitsugi Senda and Tokuji Ikeda

Department of Agricultural Chemistry, Kyoto University, Kyoto 606, Japan

Biological oxidation and reduction are characterized by multielectron-transfer processes assisted by various biomacromolecular complexes—the enzymes. Biocatalyst electrodes are enzyme- (that is, redox enzyme-) modified electrodes in which the electrode functions as an electron acceptor or donor for the redox reaction of the enzyme that is immobilized on the electrode surface. Enzyme-modified electrodes of this type are distinguished by their bioelectrocatalytic nature; that is, the electrochemical control or acceleration of the biological processes that are catalyzed by the enzymes. Hence they are called *enzyme electrodes* or, in more general terms, *biocatalyst electrodes*, and they appear promising for such novel applications as sensors, reactors, and fuel cells,[1-5] even in biomolecular electronic devices.[6] In this chapter we review recent studies on biocatalyst electrodes, particularly with reference to the problem of the electron transfer between the electrode and the redox center of the enzyme, and their capabilities for applications to sensors and reactors, with some emphasis on certain subjects concerning our recent research interests.

10-1. Electrode Process at Biocatalyst Electrodes

Figure 1(A) shows a scheme of electrocatalysis at the redox-enzyme-modified electrode where the electrode functions as an electron acceptor for the enzyme immobilized on the electrode surface, E_{ox}/E_{red}, in the oxidation of a substrate S to a product P. In such enzyme-modified electrodes, however, it has been shown experimentally that the rapid (or reversible) direct electron transfer between the electrode and the redox center of the enzyme is generally difficult to achieve. Then the presence of an electron-transfer mediator M_{ox}/M_{red}, which acts as an electron shuttle to provide redox coupling between the electrode and the immobilized enzyme, as shown schematically in Fig. 1(B), is useful to accelerate the electron transfer at the enzyme-modified electrode.

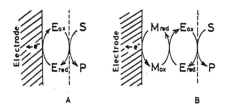

Figure 10-1 ▪ Biocatalyst electrodes. (A) Direct electron transfer and (B) mediated electron transfer between the electrode and the enzyme.

10-2. Biocatalyst Electrode with Mediator

The mediator is an organic or inorganic redox couple, usually of low molecular weight (for review, see refs. 7 and 8). A variety of compounds are used as the mediator for a variety of redox enzymes. The specificity of enzymatic reaction to the mediator is generally not as strict as to the substrate. Here we take glucose oxidase as the representative redox enzyme. For this enzyme, benzoquinones, naphthoquinones, ferricenium ions, phenylenediamines, indophenols, hexacyanoferrate ion, and ruthenium complex ions,[9-12] as well as oxygen[13] (see following text) have been shown to be effective as the mediator. The reaction scheme of the electrooxidation of D-glucose catalyzed by glucose oxidase with a mediator at the electrode can be represented by the ping-pong mechanism:

$$S + E_{ox} \underset{k_{1-}}{\overset{k_1}{\rightleftharpoons}} [ES] \xrightarrow{k_2} P + E_{red} \tag{10-1}$$

$$M_{ox} + E_{red} \underset{k_{3-}}{\overset{k_3}{\rightleftharpoons}} [EM] \xrightarrow{k_4} M_{red} + E_{ox} \tag{10-2}$$

$$M_{red} - ne^- \underset{k_b}{\overset{k_f}{\rightleftharpoons}} M_{ox} \tag{10-3}$$

where S and P are the substrate and product (here D-glucose and D-gluconolactone, respectively), E_{ox} and E_{red} are the oxidized and reduced forms of enzyme, M_{ox} and M_{red} are the oxidized and reduced forms of the mediator, and $[ES]$ and $[EM]$ are the enzyme-substrate and enzyme-mediator complexes, respectively. $k_1, k_{1-}, k_2, \ldots$ are the rate constants of the indicted steps. M_{red} is oxidized to M_{ox} by releasing n electrons (e^-) to the electrode. For simplification the participation of the hydrogen ion is not represented in the scheme.

It has been shown[10] with a glucose-oxidase–immobilized graphite electrode (by irreversible adsorption) that this enzyme-modified electrode can catalytically oxidize D-glucose in the presence of a mediator, such as p-benzoquinone or p-toluquinone, and that the catalytic current can be predicted by the Michaelis–Menten equation with three reaction parameters, k_{cat}, K_1, and K_2 [see Eqs. (10-4) and (10-5)]. The values of the parameters determined by the analysis of the catalytic currents agree well with those obtained from the analysis of the enzyme kinetics in solution (Table 10-1).[10]

Usually the enzyme electrode is constructed by immobilizing or entrapping the enzyme by chemical or physical bonds or by physical entrapment on the electrode surface or in a layer attached to the electrode surface. Also, the immobilized enzyme

Table 10-1 ■ **Enzymatic Reaction Parameters of Glucose Oxidase with *p*-Benzoquinone as a Mediator for the Enzyme-Immobilized Electrode and as an Electron Acceptor for Solubilized Enzyme (25°C, pH 5.0)**

	K_1 (10 mM)	K_2 (0.1 mM)	I_{max} (μA)	V_{max} (10^{-3} mM s^{-1})
Immobilized (on graphite electrode)	6.2	1.1	5.3 ($k_{cat} = 2.3 \times 10^2$ s^{-1})	
Solubilized (in solution)	7.3	4.2		1.8 ($k_{cat} = 3.6 \times 10^2$ s^{-1})

layer is sometimes coated by a film or membrane that is permeable to the substrate (and product) but not to the enzyme. These immobilized enzyme layers are referred to briefly as the enzyme layer (EL) hereafter. The thickness of the EL is usually on the order of micrometers or less (or of a dimension of the mono- to oligomolecular adsorption layer of enzyme in the case of immobilization by adsorption). The species S, P, M_{ox}, and M_{red} are transferred in the EL by diffusion accompanied with enzyme reaction (see Fig. 10-2). M_{red} is converted to M_{ox} at the electrode surface, giving the current as a function of the potential applied to the electrode, whereas S is supplied to and P is removed from the EL by diffusion through the coating film, referred to as the semipermeable membrane (SM) hereafter.

At the steady state the rate of the enzyme reaction [Eqs. (10-1) and (10-2)] per unit volume v_E is given by

$$v_E = k_{cat}[E]/(1 + (K_1/C_s) + (K_2/C_o)) \tag{10-4}$$

with

$$k_{cat} = k_2 k_4/(k_2 + k_4) \tag{10-5a}$$

$$K_1 = k_4(k_{1-} + k_2)/k_1(k_2 + k_4) \tag{10-5b}$$

$$K_2 = k_2(k_{3-} + k_4)/k_3(k_2 + k_4) \tag{10-5c}$$

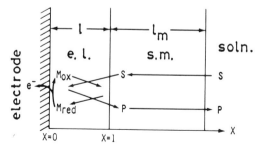

Figure 10-2 ■ Scheme of the electrode. Diffusion and enzymatic reaction of substrate (S), product (P), oxidized mediator (M_{ox}), and reduced mediator (M_{red}) in the immobilized enzyme layer (EL) coated by the semipermeable membrane (SM).

where $[E]$ is the total (that is, $E_{ox} + E_{red}$) concentration of the enzyme, assumed to be uniformly dispersed, in the EL, C_s and C_o are the concentrations of the substrate and the oxidized form of the mediator, respectively. k_{cat}, K_1, and K_2, defined by Eqs. (10-5), are the (maximum) enzyme reaction rate constant, the Michaelis constants for the substrate, and the mediator, respectively. Differential equations for the diffusion accompanied with enzyme reaction for the species involved were solved under appropriate conditions for several cases of reasonable approximation.[13–15] The results can be expressed by the most generalized form of the steady state current I as given by

$$I = I'_{max} / \left(1 + \left(K'_1/{}^lC_s\right) + \left(K'_2/{}^0C_M\right)\right) \tag{10-6}$$

with

$$I'_{max} = nFAk'_{cat}(E)l \tag{10-7}$$

where K'_1 and K'_2 are the apparent Michaelis constants for the substrate S and the mediator M, respectively, k'_{cat} is the apparent enzyme reaction rate constant (here "apparent" means including the correction for the concentration polarization of the species in the EL), l is the thickness of the EL, ${}^lC_s$ is the concentration of the substrate at the surface of the EL toward the SM $(x = l)$, ${}^0C_M$ is the concentration of the oxidized mediator at the electrode surface $(x = 0)$, F is the Faraday constant, and A is the electrode surface area. Equation (10-6) is identical in form with the previous equation [Eq. (1) in ref. 16] of the catalytic current at the biocatalyst electrode with a mediator based on the concept of the enzyme reaction layer.[16] The mathematical analysis[15] has shown that the deviation of the apparent values of the constants from their true values depends on the values of the σ parameter (the Thiele modulus) of S and M_{ox}, respectively defined by

$$\sigma_s^2 = k_{cat}[E]l^2/K_1D_s \tag{10-8a}$$

and

$$\sigma_M^2 = k_{cat}[E]l^2/K_2D_o \tag{10-8b}$$

In these equations D_s and D_o are the diffusion coefficients of S and M_{ox} in the EL, respectively. It is assumed that D_o is equal to D_r, the diffusion coefficient of M_{red}. For sufficiently small values of the σ parameters the apparent values of the constants are reduced to their true values, which means that the concentration polarization of the species in the El becomes negligibly small. Table 10-2 shows how the apparent k'_{cat} and K'_1 values deviate from the true values with increasing σ_s values, as

Table 10-2 ▪ **Dependence of the Ratios of k'_{cat}/k_{cat} and K'_1/K_1 on the σ_s values (modified from Table 1 in ref. 15)**

σ_s	k'_{cat}/k_{cat}	K'_1/K_1
0.17	1.001	1.006
0.5	1.005	1.06
1.0	1.02	1.24
2.0	1.08	1.9
3.0	1.14	3.2
5.0	1.32	7.8

calculated for the case when $K_2/C_M \ll 1$ [that is, for Eq. (10-10)]. Also note that the experimental results[10] in Table 10-1 correspond to the case when both $\sigma_s \ll 1$ and $\sigma_M \ll 1$.

In the preceding discussion (and in the following) it is assumed that the mediator is completely entrapped in the EL, that is, the diffusion of the mediator out of the EL through the SM is negligibly small. When the leakage of mediator out of the EL is not negligible, this effect can be discussed by the collection factor f_c, which represents the ratio of the amount of reduced mediator collected by the electrode to the total amount of reduced mediator produced by enzymatic reaction in the EL. It can be shown[13, 14] that the f_c value is determined mainly by the ratio of the permeability of the SM to the mediator to that of the EL, and that it approaches unity as this ratio becomes negligibly small compared with unity. The last condition is thought of as satisfied with most of the enzyme-modified electrodes, for example, with those that were designed on a mediator-mixed carbon paste electrode with a coating film, as will be described.

10-2-1. Limiting Current

In Eq. (10-6) ^0C_M increases with increasing electrode potential and approaches the total (that is, $M_{ox} + M_{red}$) concentration of entrapped mediator in the EL, $^*C_M$, when the electrode potential becomes sufficiently large [see Eq. (10-12) or (10-15)], which gives the limiting current as expressed by

$$I_l = I'_{max}/\left(1 + \left(K'_1/{}^lC_s\right) + \left(K'_2/{}^*C_M\right)\right) \tag{10-9}$$

and at sufficiently large values of the mediator concentration in the EL, that is, $(K'_2/{}^*C_M) \ll 1$, we have

$$I_l = I'_{max}/\left(1 + \left(K'_1/{}^lC_s\right)\right) \tag{10-10}$$

Usually the EL is covered by the SM, so that

$$I_l = nFAP_m\left({}^mC_s - \left({}^lC_s/\beta\right)\right) \tag{10-11}$$

where P_m is the permeability of the SM to S, ^mC_s is the concentration of S in the solution just outside the SM, and β is the distribution coefficient of S between the EL and the solution. When the solution is stirred well, the concentration polarization of (that is, the diffusion resistance to) the substrate in the solution layer just outside the SM can be made negligibly small compared with that within the SM. Then ^mC_s can approximately be equated to the bulk concentration of S, $^*C_s$.

10-2-2. Current-Potential Curves

When the electrode reaction of the mediator [Eq. (10-3)] is reversible (or nernstian), the surface concentration of the mediator is given by Nernst equation

$$E = E_M^0 + (RT/nF)\ln\left({}^0C_o/\left({}^*C_M - {}^0C_o\right)\right) \tag{10-12}$$

where E is the electrode potential, E_M^0 is the standard electrode potential of the mediator, and $^*C_M - {}^0C_o = {}^0C_r$ is the surface concentration of the reduced mediator because $*C_M = C_o + C_r$ throughout in the EL when the mediator is completely entrapped in the EL, that is, $f_c = 1$). Substituting Eq. (10-12) into Eq. (10-6) and rearranging, we get the equation of the reversible current versus potential curve at the biocatalyst electrode with an entrapped mediator,

$$E = {}_rE_{1/2} + (RT/nF)\ln(I/(I_l - I))$$ (10-13)

where the reversible half-wave potential ${}_rE_{1/2}$ is defined by

$${}_rE_{1/2} = E_M^0 + (RT/nF)\ln\left((K_2'/*C_M)/\left(1 + (K_1'/{}^lC_s) + (K_2'/*C_M)\right)\right)$$ (10-14)

The current versus potential curves of more general cases where the rate of the electrode reaction of the mediator is taken into account have also been discussed.[13, 15, 17] Let the electrode reaction rate be expressed by

$$I = nFA\left(k_f{}^0C_r - k_b{}^0C_o\right)$$ (10-15)

where k_f and k_b are the electrochemical oxidation and reduction rate constants of the mediator, respectively, at the electrode surface and are related to the electrode potential, for instance, by Butler–Volmer-type equations:

$$k_f = k^0 \exp\left[(\alpha_a nF/RT)(E - E_M^0)\right]$$ (10-16a)

$$k_b = k^0 \exp\left[-(\alpha_c nF/RT)(E - E_M^0)\right]$$ (10-16b)

Here k^0 is the standard rate constant (the electrochemical rate constant at $E = E_M^0$), and α_a and α_c are the anodic and cathodic transfer coefficients, respectively. Combination of Eq. (10-15) with Eq. (10-16) (note that $*C_M = {}^0C_o + {}^0C_r$) results in the expression of the current as a function of k_f, k_b, and other kinetic parameters of the enzymatic reaction and this expression can conveniently be formulated by an approximate equation that is given by

$$I = \left\{I_l/\left(1 + \exp\left[-(nF/RT)(E - {}_rE_{1/2})\right]\right)\right\}\{\lambda/(1 + \lambda)\}$$ (10-17)

where I_l and ${}_rE_{1/2}$ are defined by Eqs. (10-9) and (10-14), respectively. The parameter λ is defined by

$$\lambda = \left(1 + (1 - \theta)^{1.13}\right)\{(k_f/k''\theta) + (k_b/k'')\}$$ (10-18)

with

$$k'' = k_{cat}'(E)l/K_2'$$ (10-19)

and

$$\theta = (K_2'/*C_M)/\{1 + (K_1'/{}^lC_s) + (K_2'/*C_M)\}$$ (10-20)

When the parameter $\lambda \gg 1$, Eq. (10-17) is reduced to Eq. (10-13), that is, the equation of the reversible current-potential curve. This condition is satisfied in most cases by the condition that $\Lambda = k^0/k'' \gg 1$, that is, when the electrochemical rate constant (at the standard electrode potential) of the mediator is much larger than the rate constant (with respect to the concentration of the mediator) of the enzymatic reaction in (or more exactly stating, through) the EL. Conversely, when $\Lambda \ll 1$, appreciable current will not be observed at about $E = E_M^0$ (because there $\lambda \ll 1$), but will be observed only at E much more positive than E_M^0 where $k_f \gg k^0$ and also $k_f \gg k_b$ [see Eq. (10-16)]. Thus, Eq. (10-17) is reduced to the equation of the irreversible current-potential curve,

$$I = I_l / \left(1 + \exp\left[-(\alpha_a nF/RT)(E -_{ir} E_{1/2}) \right] \right) \qquad (10\text{-}21)$$

where the irreversible half-wave potential $_{ir}E_{1/2}$ is given by

$$_{ir}E_{1/2} = E_M^0 + (RT/\alpha_a nF)\ln\left[\left(\theta/\left(1 + (1 - \theta)^{1.13} \right) \right)(k''/k^0) \right] \qquad (10\text{-}22)$$

When the EL is covered by the SM, we have also

$$I = nFAP_m\left({}^m C_s - ({}^l C_s/\beta) \right) \qquad (10\text{-}23)$$

Eliminating ${}^l C_s$ from the preceding theoretical equations of the current-potential curve [Eq. (10-17), or Eqs. (10-13) or (10-21)] by means of Eq. (10-23), we can derive the equations of the current-potential curve for the film-coated enzyme-immobilized electrode with entrapped mediator. Usually, ${}^m C_s$ may approximately be equated to ${}^* C_s$ [see the text following Eq. (10-11)].

10-3. Glucose Oxidase Electrode with Entrapped Mediator

The mediator may be added into the solution to achieve the mediated electrocatalysis at the enzyme-modified electrode. However, it is desirable to immobilize or entrap not only the enzyme but also the mediator within the EL attached on the electrode surface, so that we can use the biocatalyst electrode without external addition of mediator in solution. For this purpose two types of the enzyme-immobilized electrode were designed.

10-3-1. Film-Coated Glucose-Oxidase–Immobilized Mediator-Mixed Carbon Paste Electrodes (Film-Coated GOD-M-CPE)

p-benzoquinone (BQ), a mediator, was mixed with graphite powder and paraffin liquid to prepare a carbon paste electrode containing p-benzoquinone (BQ-CPE), and glucose oxidase (GOD) was immobilized on the electrode surface by a casting method (that is, the enzyme solution was syringed onto the electrode surface and the solvent was evaporated), followed by covering the EL with a nitrocellulose film or a dialysis membrane.[16, 18] This film-coated GOD-BQ-CPE worked satisfactorily as a biocatalyst (GOD) electrode to oxidize catalytically D-glucose without addition of external mediator. This indicated that BQ molecules in the BQ-CPE were dissolved in the EL, participating there as a p-benzoquinone/p-hydroquinone (BQ/BQH$_2$) couple in mediating the electron transfer between the electrode and the enzyme. BQ is supplied to the EL from the essentially infinite BQ reservoir of the BQ-CPE to make up the loss due to the small leakage of BQ through the SM. The concentration of BQ in the EL reaches the steady-state value ${}^* C_{BQ}$, which should be determined by the concentration of BQ in the carbon paste electrode m_{BQ} and the permeability of the SM to BQ.[16]

10-3-2. Film-Coated Glucose-Oxidase–Immobilized Metal Gauze Electrodes with Mediator Reservoir (Film-Coated GOD-MGE-M)

Glucose oxidase was immobilized on one face of a gold or platinum gauze electrode, while the paste containing p-benzoquinone (BQ), a reservoir of mediator, was attached to the other face of the gauze electrode, and the surface of the EL toward the solution was covered with a SM (GOD-AuGE-BQ).[13] Also, oxygen [or oxygen/hydrogen peroxide (O_2/H_2O_2) couple] can work as a mediator for glucose oxidase at a platinum electrode. Thus, the paste reservoir of mediator was replaced by air (O_2) with a hydrophobic, oxygen-permeable membrane attached to the platinum gauze electrode [GOD-PtGE-air (O_2)].[13] These film-coated GOD-AuGE-BQ and GOD-PtGE-air(O_2) were proved to work satisfactorily as biocatalyst electrodes to oxidize D-glucose catalytically without external addition of a mediator or in the absence of oxygen in the solution, indicating that the mediator was dissolved into the EL through the pores of the gauze electrode.[13]

10-3-3. Effect of Mediators

Dependence of the catalytic current on the concentrations of the substrate and the mediator was studied with film-coated GOD-BQ-CPEs. The steady-state catalytic current measured at $+0.5$ V versus SCE (that is, in the limiting current range) increased with increasing m_{BQ}, indicating that $^*C_{BQ}$ increased with increasing m_{BQ}, but approached a saturation value for m_{BQ} larger than, for example, 6% (w/w) for a particular film-coated GOD-BQ-CPE. These results are in agreement with the prediction of Eqs. (10-9) and (10-10). Dependence of the limiting current on the concentration of D-glucose C_{glc} in solution was also proved[16] to follow Eqs. (10-10) and (10-11) (see the following text).

Dependence of the catalytic current on the mediator was also studied with a glucose-oxidase–immobilized carbon paste electrode in the presence of the mediator. Twelve quinones:

1. 2-isopropyl-5-methyl-p-BQ
2. 2,5-dimethyl-p-BQ
3. 2-methyl-p-BQ
4. tetrabromo-p-BQ
5. tetrachloro-p-BQ
6. p-BQ
7. 2,5-dichloro-p-BQ
8. 2,6-dichloro-p-BQ
9. tetrabromo-o-BQ
10. 1,4-naphthoquinone (NQ)
11. 1,2-NQ
12. 1,2-NQ-4-sulfonic acid

Table 10-3 ▪ Electrochemical Properties and Mediating Activities of
p-Benzoquinones at Glucose-Oxidase–Immobilized Carbon Paste Electrodes

Mediator	E_{mid}[a] (V)	ΔE_p[b] (V)	$E_{1/2}$[c] (V)	I'_{max}[d] (μA)
2,6-di-Cl-1,4-BQ	0.07	0.06	0.07_3	15
1,4-BQ	0.05	0.31	0.26_5	9.3
2-Me-1,4-BQ	-0.03	0.23	0.06_5	3.8
1,2-NQ	-0.10	0.06	-0.09_5	2.9

[a] Versus SCE, cyclic voltammetry.
[b] Cyclic voltammetry.
[c] Versus SCE, catalytic current.
[d] Catalytic current.

and three ferrocenes:

13. 1,1'-dimethylferrocene
14. ferrocenecarboxylic acid
15. ferrocene

were examined.[19, 20] Cyclic voltammograms of these compounds were measured with a film-coated GOD-CPE (without mixed-in mediator) in solution (5% v/v ethanol) containing each mediator at 1 mmol dm^{-3} or in saturated solution of each mediator for some mediators of low solubility. Some representative results[20] are given in Table 10-3. The midpoint potential E_{mid} was determined by $E_{mid} = (E_{pa} + E_{pc})/2$, where E_{pa} and E_{pc} are the anodic and cathodic peak potentials, respectively. These midpoint potentials are considered equal to their standard redox potential $E_M^{0\prime}$, in the solution examined (pH 7.0). The peak separation ΔE_p ($= E_{pa} - E_{pc}$) can be considered as a measure of the reversibility of the electrode reaction of mediators: Compounds 8 and 11 gave reversible cyclic voltammograms, whereas compounds 2 and 6 gave irreversible cyclic voltammograms with large peak separations (Table 10-3). The catalytic current versus potential curves were measured using a film-coated GOD(0.5 μg/0.09 cm^2) – M(0.1%) – CPE in 0.2 mol dm^{-3} phosphate buffer (5% ethanol; pH 7) containing 0.125 mol dm^{-3} D-glucose. Two representative curves[20] are shown in Fig. 10-3. The half-wave potential $E_{1/2}$ of the catalytic currents shifted only slightly from E_{mid} for the currents obtained with reversible mediators, whereas it positive-shifted largely from E_{mid} for the currents obtained with irreversible mediators of large peak separations. Also, the catalytic currents obtained with irreversible mediators gave drawn-out gently sloped current-potential curves compared with those obtained with reversible mediators. These results are in line with the theoretical prediction as given by Eqs. (10-15) through (10-21) (note that θ will be about unity, for instance, in Fig. 10-3), though the experiments were done with film-coated electrodes and hence the measured current-potential curves should be corrected for the influence of the SM [see Eq. (10-23)] for accurate examination.

Table 10-3 also shows that the I'_{max} values,[19] which were determined according to Eq. (10-10) (see Fig. 10-4), depend strongly on the mediators and decrease with decreasing (less positive) E_{mid}, that is, E_M^0 of the mediators. Because the I'_{max} value is

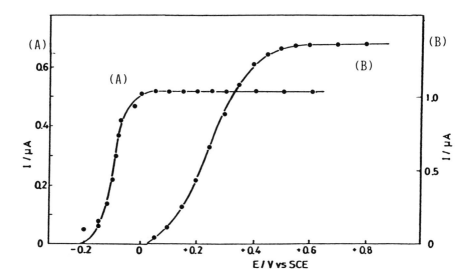

Figure 10-3 ■ Current versus potential curves obtained with a film- (dialysis membrane, 20 μm) coated GOD(0.5 μg/0.09 cm^2)-immobilized (curve A) 1,4-NQ(0.1% w/w)-mixed or (curve B) 1,4-BQ(0.1% w/w)-mixed CPE in 0.125 mol dm^{-3} D-glucose solution (5% v/v ethanol, 0.2 mol dm^{-3} phosphate buffer, pH 7.0).

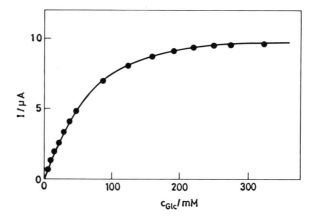

Figure 10-4 ■ Dependence of the limiting current, measured at 0.55 V versus SCE with a film- (20 μm) coated GOD(0.5Mg)-BQ(6%)-CPE, on the concentration of D-glucose in the same basal solution as Fig. 10-3.

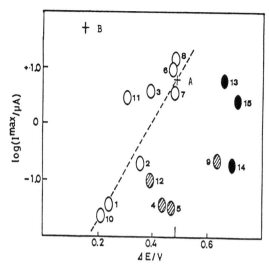

Figure 10-5 ■ Plot of $\log(I'_{max})$ versus ΔE for the 15 mediators examined (see text).

the measure of k_{cat}, which is defined by Eq. (10-5a), $k_{cat} = k_2 k_4/(k_2 + k_4)$, this result indicates that the rate of the half-reaction [Eq. (10-2)] is dependent on the standard potential of the mediators. Figure 10-5 shows the $\log(I'_{max})$ values[19] for the 15 mediators plotted against the difference of the redox potential, $\Delta E(M) = E_{mid, M} - E^0_{GOD'}$ $(= E^0_{M'} - E^0_{GOD'})$, where $E^0_{GOD'}$ $(= -0.41$ versus SCE at pH 7) is the redox potential of glucose oxidase. Also, in this figure are shown the I'_{max} value for oxygen [+marked by A on the abscissa of $\Delta E(O_2/H_2O_2) = 0.49$ V and the hypothetical (maximum) I'_{max} value corresponding for $k_{cat} = k_2$ when $k_4 \gg k_2$ in Eq. (10-5a) (+marked by B on the abscissa of $\Delta E(\text{D-gluconolactone}/\text{D-glucose}) = 0.14$ V, both estimated for this particular electrode. Figure 10-5 shows that in a series of p-quinone derivatives I_{max} increases with increasing $\Delta E(M)$, but (a) the structure factor that the mediator activity [the rate of the half reaction Eq. (10-2)] is greatly reduced by tetra-substitution in the benzoquinone ring [as seen by compounds 4 and 5 (and probably 9) compared with compounds 7 and 8] and (b) the mediator activity of a negatively charged compound is greatly reduced compared with its neutral homo-logue, as seen by compound 12 compared with 11. The negative charge effect is observed also for ferrocene derivatives, as seen by compound 14 compared with 15. The $\log(I'_{max})$ values of p-quinone derivatives, when the preceding structure factor is taken into account, are scattered on the line with the slope of $\log(I'_{max})/\Delta E(M) = 1/120$ mV (the inclined dashed line). Similar linear dependence of the rate of the half reaction [Eq. (10-2)] in solubilized glucose-oxidase reaction has been reported by Kulys and Cenas.[9] The mediator activity of ferrocenes is relatively low despite their high redox potentials. Further study is warranted to interpret this result as represent-ing the decrease of the reaction rates in a large ΔE region ("inverted" behavior), as predicted by theories of electron-transfer processes in condensed media. Finally the remarkably high rate constant of the reductive half reaction [Eq. (10-1)], as indicated by +B in Fig. 10-5, despite its relatively low value of ΔE, should be mentioned. This may reflect the substrate (here D-glucose) specificity of the enzyme.

10-4. Direct Electron Transfer

Only a limited number of enzymes are known to exchange electrons directly and rapidly (or reversibly) with the electrode. Heme proteins, such as cyt c, cyt c_3, and cyt b are known to exhibit reversible electron transfer at certain electrode surfaces and at what is called promoter-modified electrode surfaces (for review and recent progress, see refs. 21–24). Ferredoxins and their model compounds have also been shown to exchange electrons directly with the electrode (for review, see refs. 26–29). D-gluconate dehydrogenase (GADH) from bacterial membranes consists of three subunits, each containing FAD, di-heme, and (probably) the Fe — S cluster: FAD is the site to react with D-gluconate, whereas the site to react with ubiquinone, the electron acceptor in vivo, is supposed to be heme or the Fe — cluster.[30] This enzyme was immobilized on a carbon paste or graphite electrode by irreversible adsorption. Then the enzyme-modified electrode gave the catalytic oxidation current of D-gluconate without addition of external mediator,[31,32] as shown in Fig. 10-6. For comparison, two examples of the current-potential curves of the ubiquinone-mediated oxidation of D-gluconate at GADH-modified electrodes are shown[32] in Fig. 10-7. The difference between two categories of the current-potential curves in Figs. 10-6 and 10-7 is clear,

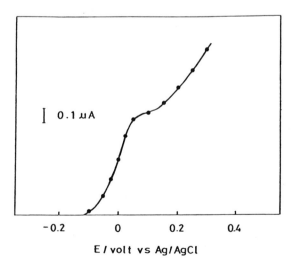

I 0.1 µA

−0.2   0   0.2   0.4

E / volt vs Ag/AgCl

Figure 10-6 ■ Current-potential curve obtained with a GADH-immobilized CPE in 5 mM D-gluconate in acetate buffer of pH 5.0 at 25°C. D-gluconate dehydrogenase (GADH) was immobilized by absorption on the electrode surface by a casting method (an enzyme solution was syringed and the solvent was evaporated), followed by washing with buffer solution.

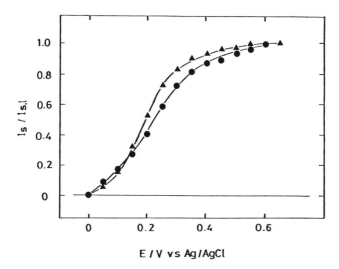

Figure 10-7 ▪ Current-potential curves obtained with GADH-immobilized ubiquinone (CoQ9)-mixed(1.2%) CPEs in the same solution as Fig. 10-6. Partially purified GADH, that is, GADH preparation containing membrane components (▲) and purified GADH mixed with cardiolipin at 1:50 in weight (●) were used.

indicating the different mechanisms of bioelectrocatalysis in these two cases. A model was proposed of the direct electron transfer of GADH adsorbed on electrode surfaces with their heme and/or ferredoxin subunits oriented toward the electrode surface[31, 32] (Fig. 10-8). Azurin, a blue copper protein, has also been shown to work as a mediator of enzyme-catalyzed electrochemical conversion of p-cresol into p-hydroxybenzaldehyde.[33] Also, it has been shown[22b] that cyt c_1 can mediate the (cathodic) electron transfer between the promotor-modified electrode and cytochrome c oxidase. Various metal enzymes, including heme and nonheme iron proteins, play important roles in biological electron transport systems. Those metal enzymes or macromolecular metal complexes appear promising to work as a mediator or bridge

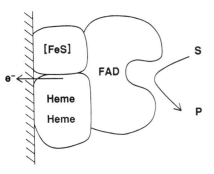

Figure 10-8 ▪ A proposed model of the adsorption of GADH at a direct-electron–transfer biocatalyst electrode.

for the electron transfer, eventually achieving the "direct" transfer of electrons between the electrode and the enzyme.

Recently Degani and Heller[34,35] showed that direct electrical communication between the electrode and the enzyme is possible through chemical modification of the enzyme: Ferrocenecarboxylic acid, a mediator, was covalently attached to glucose oxidase by an amide linkage (12 ferrocene/ferricenium centers per enzyme molecule) to form electron relays between the redox center of the enzyme and the periphery of the enzyme that may easily exchange electrons with the electrode. These authors also showed that the mediator-modified enzyme approach is useful for glucose oxidase with other mediators and also for other enzymes, such as D-amino acid oxidase. According to Aizawa, Yabuki, and Shinohara,[36] the electrochemically synthesized polypyrrole-enzyme (e.g., glucose oxidase) membrane is enzymatically active and electrically conductive; thus the biocatalysis at the polypyrrole-enzyme membrane electrode can be controlled electrochemically. Recently it has been reported[37] that mediator-modified flexible polymers, such as the methyl(ferrocenylethyl)siloxane homopolymer and the methyl(ferrocenylethyl)dimethylsiloxane copolymer can mediate the electron transfer between glucose oxidase and a carbon paste electrode. Also, bioelectrocatalysis with conductive organic salt (such as TTF-TCNQ and NMP-TCNQ) electrodes has been studied at enzyme- (such as glucose oxidase, acetylcholine, and choline oxidase) modified electrodes.[38,39]

10-5. Biocatalyst Electrodes as Biosensors

A variety of oxidoreductases including dehydrogenases as well as oxidases can be used for designing biosensors based on biocatalyst electrodes (for review, see refs. 5 and 40). The glucose-oxidase–immobilized carbon paste electrodes or porous electrodes with entrapped mediator, as previously stated, will be useful for applications as biosensors. Biosensors for D-gluconate based on ubiquinone- or benzoquinone-mediated gluconate-dehydrogenase–modified electrodes have been studied.[17,41] Biosensors based on oligosaccharide-dehydrogenase–modified electrodes with entrapped mediator (benzoquinone) have also been studied and tested for such applications as determination of maltose concentration[42] and α-amylase activity.[42,43] Ikeda, Shiraishi, and Senda[44] have also shown with peroxidase- and lactate-dehydrogenase–immobilized carbon paste electrodes that the addition of an ionic polymer (poly-L-lysine) into the EL is highly effective to entrap a soluble mediator (tetracyanoferric and -ferrous ions) in the EL.

NAD-dependent dehydrogenases can be used to construct a biocatalyst electrode where NAD works as a mediator (for electrooxidation) between the electrode and the enzyme. An alcohol dehydrogenase (NAD) electrode was constructed on a carbon paste electrode containing NAD.[14] Furthermore, in order to reduce the high overvoltage for oxidation of NADH, a diaphorase-immobilized carbon paste electrode containing vitamin K_3 as a mediator was studied.[45] This enzyme-modified electrode was able to oxidize NADH at much less positive applied potential than a bare carbon paste electrode. Thus biosensors for lactate, ethanol, glycerol, and glucose-6-phosphate were constructed on the basis of film-coated NAD-dependent dehydrogenase (for each substrate) and diaphorase co-immobilized carbon paste electrodes containing both NAD and a mediator (vitamin K_3).[45] Conductive organic salt electrodes[46, 47] as well as chemically modified-mediators[48, 49] appear also promising for constructing sensors based on NAD-dependent dehydrogenase electrodes.

We take again the glucose-oxidase–modified electrode with an entrapped mediator as the representative biocatalyst electrode on which a glucose sensor is to be constructed, and we discuss some important factors that determine the characteristics of the sensor.

10-5-1. Calibration Curves

When the concentration of the mediator entrapped in the EL is so large that $K_2'/^*C_M \ll 1$ [see Eq. (10-9)], the dependence of the current on the concentration of the substrate, that is, the analyte (here D-glucose) is determined by Eqs. (10-10) and (10-11). Eliminating ^lC_s from Eqs. (10-10) and (10-11) and taking that $^mC_s = {}^*C_s$ [see the text following Eq. (10-11)], we get the equation for a calibration curve for the sensor based on a film-coated enzyme-modified electrode:[16]

$$\frac{I_l}{nFA}\left[1 + \frac{k'_{cat}[E]l\beta}{K_1'P_m} + \frac{\beta}{K_1'}\left({}^*C_s - \frac{(I_l/nFA)}{P_m}\right)\right] = \frac{k'_{cat}[E]l\beta}{K_1'}{}^*C_s \quad (10\text{-}24)$$

Then it can be predicted that I_l increases first linearly with increasing $^*C_s$ but deviates downward from the linearity at large $^*C_s$. The linearity range extends to the concentration larger than K_1 owing to (a) the diffusion-enzyme reaction in the EL (that is, $K_1' > K_1$) and (b) the diffusion in the SM [Eq. (10-11)]. When P_m is relatively so small that, for example, $(k'_{cat}[E]l\beta/K_1'P_m) \gg 1$, Eq. (10-24) is reduced to

$$I_l = nFAP_m{}^*C_s \quad (10\text{-}25)$$

that is, the current turns out to be controlled solely by the permeability of the SM to the substrate. These predictions have been experimentally proved to be true with film-coated GOD-BQ-CPEs.[16] Representative calibration curves that were obtained with three different film-coated GOD-BQ-CPEs are shown in Fig. 10-9.[14, 16] These results indicate that the film-coated glucose oxidase electrodes can be designed to make glucose sensors for use from low to high concentration range by selection of P_m, $k'_{cat}[E]l$, and K_1' (and $K_2'/^*C_M$ if necessary). Designing a glucose sensor capable of monitoring the glucose level in blood directly (without dilution) is feasible.

Choice of the coating film (the SM) is important in designing film-coated enzyme electrode sensors. It not only is concerned with immobilization or entrapping of the enzyme but also protects the enzyme from undesirable contaminations. It must be permeable for the substrate (and product) but is preferably as impermeable as possible for the mediator. When P_m is so small that Eq. (10-25) is valid, the response of the enzyme electrode sensor becomes very stable, controlled solely by diffusion across the SM and independent of the reaction rate of the immobilized enzyme

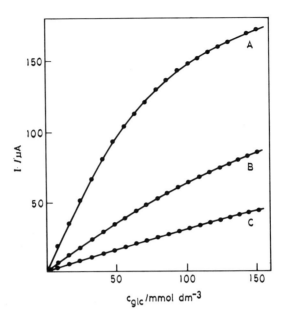

Figure 10-9 ■ Dependence of the limiting current obtained with three different film-coated GOD(180 μg)-BQ(30%)-CPEs on the concentration of D-glucose in the basal solution of 0.2 mol dm^{-3} acetate buffer (pH 5.0); A, nitrocellulose film of 50 μm thickness; B and C, dialysis membrane (Visking Co.) of 50 and 100 μm thickness, respectively. The current was measured at 0.5 V versus SCE at 25°C.

(hence, the mode of enzyme reaction, such as pH dependence, the lifetime of the immobilized enzyme), though at the cost of decreased sensitivity (the decreased slope of the calibration curve) and prolonged response time (see the following text).

10-5-2. Response Time

The response time of film-coated enzyme-modified electrode sensors is determined by (a) the diffusion-enzyme reaction process of the substrate and mediator in the EL and (b) the diffusion process of the substrate in the SM.[14] In most film-coated enzyme-modified electrode sensors the thickness of the EL l is much smaller than the thickness of the (effective) SM l_m. Then the response time is practically determined by the diffusion process of the substrate in the SM, which can be evaluated by the parameter

$$\tau_{SM} = l_m^2 / \pi^2 D_{SM} \tag{10-26}$$

where D_{SM} is the diffusion coefficient of S in the SM. Then, the transient phenomena, such as the flux of the substrate across the membrane, can be evaluated by

$$(M_t/M_{inf}) = 1 - 2[\exp(-t/\tau_{SM}) - \exp(-4t/\tau_{SM}) \cdots]$$

where M_t and M_{inf} represent the responses at the transient state of time t and at the stationary state (at infinite time), respectively. Thus, for example, the "response time" necessary for the response to reach 95% of the stationary value can be calculated to be about $3.7 \times \tau_{SM}$.

10-5-3. Enhanced Sensitivity

The sensitivity of the enzyme-modified electrode sensors can greatly be enhanced by "substrate amplification": The substrate of the first enzyme is regenerated by the second enzyme.[50-53] Thus, a biosensor for NADH (and NAD) has been designed on a glucose-6-P dehydrogenase and diaphorase co-immobilized carbon paste electrode containing vitamin K_3 as a mediator.[54] This sensor exhibited very high sensitivity to detect NADH (and NAD) in the presence of D-glucose-6-P, a reductant for enzymatic regeneration of NADH.

An amperometric immunoelectrode based on a glucose-oxidase–modified electrode with a ferrocene-drug complex as a mediator has been studied.[55] The principle of the assay is that binding of the ferrocene-drug complex by antibody inhibits its ability to act as a mediator in the glucose-oxidase electrode and thus the catalytic current is greatly reduced. This can be reversed by adding a nonlabeled drug (that is, analyte) that competes for the available antibody binding sites.

10-5-4. Effect of Oxygen

Biocatalyst electrodes with entrapped mediator give the current response in the absence of oxygen in analyte solution. This is in contrast to, for example, conventional D-glucose sensors now available on the market, which are mostly based on an oxidase-immobilized oxygen- (or hydrogen peroxide-) detecting electrode. Furthermore, when the entrapped mediator is present in excess over oxygen and $(K_2'/^*C_M)$ $\ll 1$ in Eq. (10-9) so that Eq. (10-10) is valid, the current becomes independent of the presence of oxygen (and other oxidants that can function as a mediator). Consequently, the current response becomes independent of either the absence or presence of oxygen in solution when the sensor is properly designed.[14, 18] However, when the redox property (redox potential and kinetics) of the selected mediator is such that the reduced form of the mediator will be oxidized by an oxidant Ox_1, which may be present in analyte solution (that is, $M_{red} + Ox_1 = M_{ox} + Red_1$; for example, the autooxidation of the mediator by oxygen) and that the resultant reductant Red_1 is not electrooxidizable at the electrode under the given applied potential, then the current response will be reduced accordingly.

Oxygen (O_2/H_2O_2) itself works satisfactorily as a mediator at an oxidase-immobilized platinum electrode. Thus, a film-coated GOD-PtGE-air (O_2) gave the current response to D-glucose in deaerated solution.[13]

10-5-5. Interference by Oxidizable Substances

Electrooxidizable substances, such as L-ascorbic acid and uric acid can interfere with the current response of a glucose-oxidase–modified electrode with an entrapped mediator. The interference can be eliminated by placing a filter in front of the electrode to prevent interfering substance from reaching the electrode surface. Physical (that is, permeability-selective) filters are commonly employed in designing glucose-oxidase–immobilized H_2O_2-detecting electrode sensors. Another device will be an electrochemical filter,[56,57] for instance, a gold minigrid electrode placed in front of the coating film, which when properly designed, can effectively eliminate by electrooxidation the interference by L-ascorbic acid at a film-coated GOD-BQ-CPE.[56]

10-6. Bioreactors and Biofuel Cells

Applications of the biocatalyst electrodes to organic synthesis have been discussed extensively.[58,59] Both electroenzymatic and electromicrobial reductions of organic compounds using NADH-dependent enzymes with a mediator(s) have been studied. Electrochemical regeneration of NADH from NAD by a bipyridine-rhodium(I) complex as an electron-transfer agent has been reported.[60] Bioelectrochemical conversion of p-cresol into p-hydroxybenzaldehyde has been studied.[32]

D-gluconate is commericially produced at present by two methods: the electrochemical method (for review, see ref. 61) and the microbial method (for review, see ref. 62). The first method is the electrocatalysis method in which D-glucose is oxidized to D-gluconate by $Br_2/2Br^-$ redox coupling as an electron-transfer catalyst on the electrode. The second method is the biocatalysis method in which D-glucose is biochemically oxidized to D-gluconate by oxygen using *A. niger* as a biocatalyst. Application of the biocatalyst electrode to convert D-glucose into D-gluconate suggests a third method: the bioelectrocatalysis method, which may have the advantage of both the specificity and mildness of biochemical processes and the energy saving and ease of control of electrochemical processes.[2] In view of the fact that the redox potential of $Br_2/2Br^-$ coupling is 1.06 V versus NHE and that of O_2/H_2O_2) is 0.29 V versus NHE (at pH 7.0), the replacement of the catalyst ($Br_2/2Br^-$) electrode by the biocatalyst (for example, GOD-O_2/H_2O_2) electrode would result in saving 0.79 V of applied voltage. Furthermore, constructing a fuel cell by combining the biocatalyst electrode as the anode with an O_2/H_2O electrode as the cathode should result theoretically in a gain of electric energy, in addition to the production of D-gluconate from D-glucose. A laccase-A–modified electrode[63] could be a biocatalyst electrode for use as the cathode. Electrochemical conversion of D-glucose simultaneously to D-gluconate and D-sorbitol by anodic and cathodic reactions, respectively, has been the subject of recent interest.[64] In this connection recent research[65] on biological conversion of D-glucose to D-gluconate and/or D-sorbitol may be of interest. Finally, the biocatalyst electrodes should be a key for constructing the hoped-for biomass fuel cells.[2]

Conclusion

The study on bioelectrocatalysis at enzyme-modified electrodes has shown a pathway to molecular devices that will have a broad area of applications, duplicating some of the biological activities and possibly even improving upon them. Of primary importance is the problem of interfacing between biological electron-transfer systems and electronic configurations. The electron transfer is conjugated with the molecular events. The electrical characteristics are dependent on the molecular environment. Research on interfacing microfabrication technology and advances in molecular engineering should develop a further way to advanced molecular devices that may ultimately lead to mimicking the very functions of biological systems.

We thank our colleagues and students for their help and enthusiasm in the development of this subject. It is also a pleasure to acknowledge the support by Grant-in-Aid for Scientific Research on Priority Area "Dynamic Interactions and Electronic Processes of Macromolecular Complexes" from the Ministry of Education, Science, and Culture.

References

1. M. R. Tarasevich, "Bioelectrocatalysis," in *Comprehensive Treatise of Electrochemistry*, Vol. 10, S. Srinivasan, Y. A. Chizmadzhev, J. O'M. Bockris, B. E. Conway, and E. Yeager, eds., Plenum, New York (1985), Chap. 4, pp. 231–295.
2. M. Senda and T. Ikeda, "Biopolymer-Modified Electrodes," in *"Koubunshi Hyomen no Kiso to Ouyou (Polymer Surfaces, Fundamentals and Applications)*, Vol. 2, Y. Ikada, ed., Kagaku Dojin, Kyoto (1986), pp. 201–226.
3. M. Senda, *Shokubai (Catalysis)*, **29**, 586–589 (1987).
4. M. Senda, T. Ikeda, and T. Osakai, *Proceedings of the JSPS-NUS Symposium on Analytical Chemistry*, March 20–23, 1988, National University of Singapore, Singapore (1989), pp. 1–18.
5. M. Senda and T. Ikeda, *Bioinstrumentation: Research, Developments and Application*, D. L. Wise, ed., M&N Toscano, New York (1989), pp. 189–210.
6. M. Senda, Abstract, *International Bio. Symposium*, Nagoya '88, March 10-12, 1988 (1988), pp. 65–73.
7. M. L. Fultz and R. A. Durst, *Anal. Chim. Acta*, **140**, 1–18 (1982).
8. J. M. Johnson, H. B. Halsall, and W. R. Heineman, *Anal. Biochem.*, **133**, 186–189 (1983).
9. J. J. Kulys and N. K. Cenas, *Biochim. Biophys. Acta*, **744**, 57–63 (1983).
10. T. Ikeda, I. Katasho, M. Kamei, and M. Senda, *Agric. Biol. Chem.*, **48**, 1969–1979 (1984).
11. A. E. G. Cass, G. Davis, G. D. Francis, H. A. O. Hill, W. J. Aston, I. J. Higgins, E. V. Plotkin, L. D. L. Scott, and A. P. F. Turner, *Anal. Chem.*, **56**, 667–671 (1984).
12. A. L. Crumblis, H. A. O. Hill, and D. J. Page, *J. Electroanal. Chem.*, **206**, 327–331 (1986).
13. M. Senda, T. Ikeda, H. Hiasa, and I. Katasho, *Nippon Kagaku Kaishi (J. Chem. Soc. Jpn.)*, **1987**, 358–365 (1987).
14. M. Senda, T. Ikeda, K. Miki, and H. Hiasa, *Anal. Sci.*, **2**, 501–506 (1986).
15. T. Ikeda, K. Miki, and M. Senda, *Anal. Sci.*, **4**, 133–138 (1988).
16. T. Ikeda, H. Hamada, and M. Senda, *Agric. Biol. Chem.*, **50**, 883–890 (1986).
17. T. Ikeda, K. Miki, F. Fushimi, and M. Senda, *Agric. Biol. Chem.*, **51**, 747–754 (1987).
18. T. Ikeda, H. Hamada, K. Miki, and M. Senda, *Agric. Biol. Chem.*, **49**, 541–543 (1985).
19. T. Ikeda, H. Hiasa, and M. Senda, *Redox Chemistry and Interfacial Behavior of Biological Molecules*, G. Dryhurst and K. Niki, eds., Plenum, New York (1988), pp. 193–201.
20. H. Hiasa, MS Thesis (Agricultural Chemistry), Kyoto University (1987).
21. D. J. Cohen, F. M. Hawkridge, H. N. Blount, and C. R. Hartzell, *Charge and Field Effects in Biosystems*, M. J. Allen and P. N. R. Usherwood, eds., Abacus Press, Kent, U.K. (1984), pp. 19–31.
22. (a) I. Taniguchi and K. Yasukouchi, *Hyoumen (Surfaces)*, **23**, 597–614 (1985); (b) I. Taniguchi, *Redox Chemistry and Interfacial Behavior of Biological Molecules*, G. Dryhurst and K. Niki, eds., Plenum, New York (1988), pp. 113–123.

23. F. A. Armstrong, H. A. O. Hill, and N. J. Walton, *Acc. Chem. Res.*, **21**, 407–413 (1988).
24. J. E. Frew and H. A. O. Hill, *Eur. J. Biochem.*, **172**, 261–269 (1988).
25. (a) K. Niki, Chapter 11 of this book, pp. 251–274 (1990); (b) Fan Ke-Jun, I. Satake, K. Ueda, H. Akutsu, and K. Niki, *Redox Chemistry and Interfacial Behavior of Biological Molecules*, G. Dryhurst and K. Niki, eds., Plenum, New York (1988), pp. 125–138.
26. T. Kakutani, K. Toriyama, T. Ikeda, and M. Senda, *Bull. Chem. Soc. Jpn.*, **53**, 947–950 (1980).
27. A. Nakamura and Ueyama, Chapter 13 of this book, pp. xxx–xxx (1990).
28. G. Dryhurst, K. M. Kadish, F. Scheller, and R. Renneberg, eds., *Biological Electrochemistry*, Vol. 1, Academic, New York (1982), p. 548.
29. H. Matsubara, Y. Katsube, and K. Wada, eds., *Iron-Sulfur Protein Research*, Japan Sci. Soc. Press, Tokyo (1987) p. 329.
30. K. Matsushita, E. Shinagawa, and M. Ameyama, *Methods in Enzymology*, **89**, 187–193 (1982).
31. T. Ikeda, F. Fushimi, K. Miki, and M. Senda, *Argic. Biol. Chem.*, **52**, 2655–2658 (1988).
32. F. Fushimi, MS Thesis (Agricultural Chemistry), Kyoto University (1989).
33. H. A. O. Hill, B. N. Oliver, D. J. Page, and D. J. Hopper, *J. Chem. Soc., Chem. Commun.*, **1985**, 1469–1471 (1985).
34. Y. Degani and A. Heller, *J. Phys. Chem.*, **91**, 1285–1289 (1987).
35. Y. Degani and A. Heller, *J. Am. Chem. Soc.*, **110**, 2615–2620 (1988).
36. M. Aizawa, S. Yabuki, and H. Shinohara, *Redox Chemistry and Interfacial Behavior of Biological Molecules*, G. Dryhurst and K. Niki, eds., Plenum, New York (1988), pp. 173–179.
37. N. K. Cenas and J. J. Kulys, *Bioelectrochem. Bioenergetics*, **8**, 103–113 (1981).
38. (a) W. J. Albery and P. N. Bartlett, *J. Electroanal. Chem.*, **194**, 211–222 (1985); (b) W. J. Albery, P. N. Bartlett, M. Bycroft, D. H. Carston, and B. J. Driscoll, *J. Electroanal. Chem.*, **218**, 119–126 (1987).
39. P. D. Hale and R. M. Wightman, *Mol. Cryst. Liq. Cryst.*, **160**, 269–279 (1988).
40. A. P. F. Turner, I. Karube, and G. S. Wilson, eds., *Biosensors, Fundamentals and Applications*, Oxford University Press, Oxford (1987), p. 770.
41. T. Ikeda, K. Miki, and M. Senda, *Argic. Biol. Chem.*, **52**, 1557–1563 (1988).
42. T. Ikeda, T. Shibata, and M. Senda, *J. Electroanal. Chem.*, **261**, 351–362 (1989).
43. H. Kinoshita, private communication.
44. T. Ikeda, T. Shiraishi, and M. Senda, *Agric. Biol. Chem.*, **52**, 3187–3188 (1988).
45. (a) K. Miki, T. Ikeda, and M. Senda, *Rev. Polarography (Kyoto)*, **33**, 20–21 (extended abstract) (1987); (b) K. Miki, S. Todoriki, T. Ikeda, and M. Senda, *Anal. Sci.*, **5**, 269–274 (1989).
46. J. J. Kulys, *Enzyme Microb. Technol.*, **3**, 344–352 (1981).
47. (a) W. J. Albery and P. N. Bartlett, *J. Chem. Soc., Chem. Commun.*, 234–236 (1984); (b) W. J. Albery, P. N. Bartlett, A. E. G. Cass, and K. W. Sim, *J. Electroanal. Chem.*, **218**, 127–134 (1987).
48. L. Gorton, *J. Chem. Soc., Faraday Trans.*, 1, **82**, 1245–1258 (1986), and references therein.
49. B. F. Y. Y. Hin and C. R. Lowe, *Anal. Chem.*, **59**, 2111–2115 (1987).
50. (a) F. Mizutani, Y. Shimura, and K. Tsuda, *Chem. Lett.*, 199–202 (1984); (b) F. Mizutani, T. Yamanaka, Y. Tanabe, and K. Tsuda, *Anal. Chim. Acta*, **177**, 153–166 (1985).
51. F. Schubert, D. Kirstein, K. Schroder, and F. Scheller, *Anal. Chim. Acta*, **169**, 391–396 (1985).
52. F. W. Scheller, D. Kirstein, L. Kirstein, F. Schubert, V. Wollenberger, B. Olsson, L. Gorton, and G. Johansson, *Phil. Trans. Roy. Soc. London*, **B316**, 85–94 (1987).
53. T. Wasa, N. Nakayama, Y. Matsumoto, and T. Yao, *Rev. Polarogr. (Kyoto)*, **31** 31–32 (extended abstract) (1985).

54. K. Miki, S. Todoriki, T. Ikeda, and M. Senda, *Denki Kagaku*, **58** (1990), in press.

55. K. D. Gleria, H. A. O. Hill, C. J. McNeil, and M. J. Green, *Anal. Chem.*, **58**, 1203–1205 (1986).

56. T. Ikeda, I Katasho, and M. Senda, *Anal. Sci.*, **1**, 455–457 (1985).

57. F. W. Scheller, F. Schubert, R. Rennenberg, H. G. Mueller, M. Jaenchen, and H. Weise, *Biosensors*, **1**, 135–160 (1985).

58. H. Simon, J. Bader, H. Gunther, S. Neumann, and J. Thanos, *Angew. Chem. Int. Ed. Engl.*, **24**, 539–533 (1985).

59. G. M. Whitesides and C. H. Wong, *Angew. Chem. Int. Ed. Engl.*, **24**, 617–638 (1985).

60. R. Wienkamp and E. Steckhan, *Angew. Chem. Int. Ed. Engl.*, **21**, 782–784 (1982).

61. R. Jansson, *Chem. & Eng. News*, Nov. 19, 43–57 (1984).

62. S. Teramoto, "Gluconate Fermentation. II" in *Microbe Technology*, Vol. 5, N. Tomoda, K. Sakaguchi, S. Yamada, and Y. Asai, eds., Kyoritsu, Tokyo (1964) pp. 51–68.

63. C. W. Lee, H. B. Gray, F. C. Anson, and B. G. Malmstrom, *J. Electroanal. Chem.*, **172**, 289–300 (1984).

64. K. Park, P. N. Pintauro, M. M. Baizer, and K. Nobe, *J. Electrochem. Soc.*, **132**, 1850–1855 (1985).

65. (a) M. Zachariou, R. K. Scopes, *J. Bacteriol.*, **167**, 863–869 (1986); (b) Y. Tani, V. Vongsuvanlert, *J. Ferm. Technol.*, **65**, 405–411 (1987).

Redox Behavior of Tetraheme Protein—Cytochrome c_3 [†]

Kejun Fan, Hideo Akutsu, and Katsumi Niki

Department of Physical Chemistry, Faculty of Engineering, Yokohama National University, Hodogaya-Ku, Yokohama 240, Japan

11-1. Cytochrome c_3

Cytochrome c_3 isolated from anaerobic sulfate-reducing bacteria *Desulfovibrio vulgaris* Miyazaki is a tetrahemoprotein.[1-6] It is an electron carrier from hydrogenase to ferredoxin in the respiratory chain in *Desulfovibrio*. Cytochrome c_3, which is reduced reversibly with molecular hydrogen by the catalytic action of hydrogenase, contains about 107 amino acid residues and has a molecular weight of approximately 14,000. The four hemes are covalently attached to the polypeptide chain through thioether linkages provided by cysteinyl residues in either a Cys-a-b-c-d-Cys-His or a Cys-a-b-Cys-His sequence. The axial ligands are two imidazoles of histidinyl residues.[7] X-ray crystallographic investigations of cytochrome c_3 from *D. vulgaris* Miyazaki was done with a 1.8 Å resolution in 1984.[8] All heme groups are exposed to the solvent, and the imidazole rings of the fifth and sixth ligands are approximately parallel to each other in hemes I, II, III. The imidazole rings of His-22, 25, the pyrrole ring D of heme III, and the aromatic side chain of Phe-20 are close as shown in Fig. 11-1. Moreover, Phe-20 is a strictly invariant residue in the six known cytochromes c_3 from *D. vulgaris* Miyazaki,[9] *D. vulgaris* Hildenborough,[10] *D. gigas*,[11] *D. desulfurican* Norway,[12] *D. Salexigens*,[13] and *D. desulfuricans*.[10] Cytochrome c_3 from *D. vulgaris* Miyazaki and *D. desulfuricans* Norway do not share a common precipitation antigenic determinant. In spite of the low degree of sequence homology, the relative heme arrangements of these cytochromes c_3 are quite similar.[8,12] The iron-to-iron distances of *D. vulgaris* Miyazaki cytochrome c_3 are almost identical with those of *D. desulfurican* Norway cytochrome c_3 as shown in Table 11-1. The inclination angles between the hemes and the positions of the side groups, such as propionic acid, ethylidene, and methyl groups are also similar. Cytochrome c_3 from *D. gigas* is an acidic protein with pI = 5.2,

---

[†]According to IUPAC convention, the formal potential defined by the Nernst equation should be used. In this article, however, the formal potential is referred to the redox potential because of its popularity in biological science.

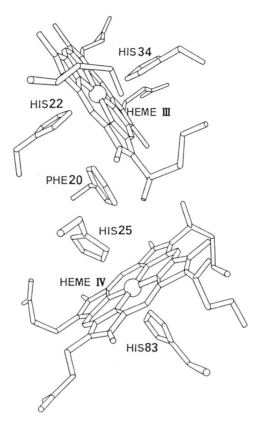

Figure 11-1 ▪ The configuration of hemes III and IV and intervening aromatic groups.

Table 11-1 ▪ Heme–Heme Angles of Cytochrome c_3 from *D. vulgaris* Miyazaki and *D. desulfurican* Norway (in parentheses)

| Heme | Iron-to-Iron Distances (Å) | | | |
	II (IV)	III (III)	IV (II)	I (I)
II	—	12.2	15.8	16.4
(IV)	—	(12.8)	(16.8)	(16.3)
III	88.7	—	11.0	17.8
(III)	(84)	—	(10.9)	(17.3)
IV	59.1	80.3	—	12.0
(II)	(62)	(90)	—	(12.7)
I	73.0	22.0	80.0	—
(I)	(55)	(35)	(89)	—

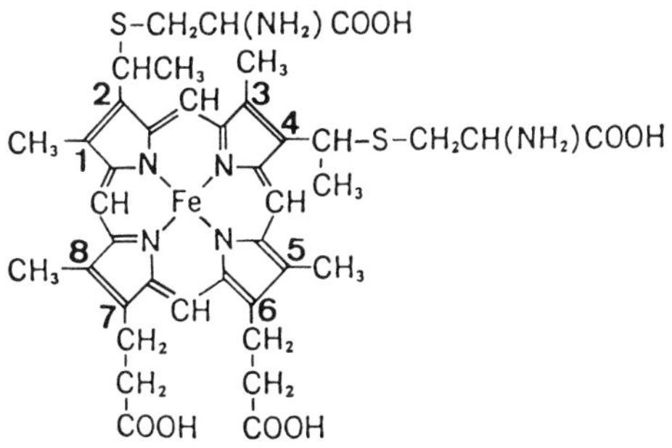

S-CH$_2$CH(NH$_2$) COOH

The molecular structure diagram of c-type heme showing:

CHCH$_3$ CH$_3$
2   CH 3
CH$_3$ 1   4 CH-S-CH$_2$CH(NH$_2$)COOH
         N   N   CH$_3$
         CH Fe CH CH$_3$
         N   N
CH$_3$ 8   5 CH$_3$
         7   CH   6
         CH$_2$ CH$_2$
         CH$_2$ CH$_2$
         COOH COOH

Figure 11-2 ▪ The molecular structure of c-type heme.

Table 11-2 ▪ Carbon–Carbon Distances (> 10 Å) between Heme Methyl Groups of Cytochrome c_3 from *D. vulgaris* Miyazaki Obtained in Crystallographic Studies

	Distances between Heme Methyl Groups in Positions			
	1–8	1–3	3–5	5–8
Heme I	5.31	8.06	8.04	> 10
Heme II	5.33	7.90	7.97	9.96
Heme III	5.25	7.92	8.05	9.88
Heme IV	5.36	8.06	7.98	9.94

whereas most other cytochromes c_3 are basic proteins with pI > 10. Figure 11-2 shows the molecular structure of C-type heme and Table 11-2 gives the carbon–carbon distances (> 10 Å) of four heme methyl groups.

The high density of heme groups in the cytochrome c_3 molecule is thought to be the major factor leading to the unique physicochemical properties of cytochrome c_3. For example, the redox potential of cytochrome c_3 was shown to be as low as -300 mV,[14,15] which is 560 mV more negative than the monohemoprotein cytochrome c. Mössbauer spectroscopy provided evidence for a strong heme–heme interaction at 4.2 K.[16] The electrical conduction of a cytochrome c_3 solid thin film containing a trace amount of hydrogenase as a catalyst was found to be very peculiar.[17] High conductivity was observed at low temperature and under high hydrogen gas pressure.[18] The difference in ionization potential between the feric and ferrous forms was as much as 77 KJ mol^{-1} compared with 29 KJ mol^{-1} for cytochrome c.[19]

11-2. Macroscopic Redox Potentials

Electrochemical studies have shown that cytochrome c_3 undergoes a reversible voltammetric response at various electrodes without surface modifiers.[20,21] The normal pulse and scan-reversal polarograms of cytochrome c_3 are shown in Fig. 11-3 in which the ratio of the limiting currents on the cathodic scan to the anodic scan is

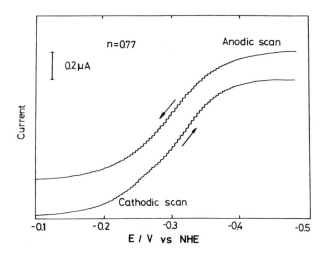

Figure 11-3 ▪ Normal pulse and scan-reversible polarograms of cytochrome c_3 from *D. vulgaris* Miyazaki at a dropping mercury electrode in 30 mM phosphate buffer solution (pH 7.0; 25°C).

almost unity and the half-wave potentials are identical in either direction.[14, 15] These results provide strong evidence that:

1. The electron-transfer reaction of cytochrome c_3 is reversible on a mercury electrode.
2. The diffusion coefficients of both oxidized and reduced forms of cytochrome c_3 are the same.
3. The rate constant of the redox reaction should exceed 0.01 cm s^{-1}.[14]

From the nernstian slope, the apparent number of electrons involved in the reduction of cytochrome c_3 was 0.77, suggesting that the redox reaction is not a single step. That is, the reduction is not a four-electron transfer process but four-consecutive one-electron transfer steps with small redox potential spacings. In this case, four macroscopic redox potentials ($E_i^{0'}$, $i = 1$–4) can be defined as:

$$E = E_i^{0'} + (RT/F)\ln(f_i/f_{i+1}) \qquad i = 1\text{–}4 \qquad (11\text{-}1)$$

where E, R, F, T, and f_i are the equilibrium potential, gas constant, Faraday constant, absolute temperature, and mole fraction of cytochrome c_3 in the i-electron reduced state, respectively.

Figure 11-4 gives the differential pulse polarograms (DPP) of cytochrome c_3 of *D. vulgaris* Miyazaki together with these from *D. vulgaris* Hildenborough, *D. desulfurican* Norway, and *D. gigas*. It can be predicted that the four hemes in cytochromes c_3 are nonequivalent because their DPP curves are not symmetrical. The DPP is more sensitive to small variations in the values of macroscopic redox potentials of cytochrome c_3 than the cyclic voltammogram, and the background signal is less of a problem for the DPP than for the cyclic voltammogram.[14, 15, 21] In addition, the nonlinear regression analysis of DPP was found to be much simpler than that of the

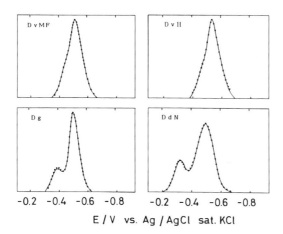

Figure 11-4 ■ Differential pulse polarograms for reduction of cytochromes c_3 from *D. vulgaris* Miyazaki, *D. vulgaris* Hildenborough, *D. gigas*, and *D. desulfurican* Norway. Circles represent the experimental data and solid lines are the nonlinear regression analysis results obtained in Eq. (11-2).

cyclic voltammogram. The theoretical equation [eq. (11-2)] was derived by Niki et al.[21] for the molecule with *n*-active sites, which undergoes a reversible *n*-consecutive one-electron transfer reactions at the dropping mercury electrode. The relation between the differential pulse polarographic current i_{DPP} and the macroscopic redox potentials E_p^0 for $p(p = 1-n)$ electrons redox reactions is given by

$$
i_{DPP} = (i_d)_{Cott}\left\{ \frac{\sum\limits_{p=1}^{n} p\exp\left[pF/RT\left(E_p^0 + \Delta E - E\right)\right]}{\sum\limits_{p=1}^{n} \exp\left[pF/RT\left(E_p^0 + \Delta E - E\right)\right] + 1} \right.
$$

$$
\left. - \frac{\sum\limits_{p=1}^{n} p\exp\left[pF/RT\left(E_p^0 - E\right)\right]}{\sum\limits_{p=1}^{n} \exp\left[pF/RT\left(E_p^0 - E\right)\right] + 1} \right\} \tag{11-2}
$$

where ΔE and E represent the potential modulation amplitude and electrode potential, respectively, and $(i_d)_{Cott}$ denotes the limiting diffusion current given by Cottrell equation for the one-electron–transfer process. The nonlinear regression analysis of DPP curve of cytochrome c_3 was made by using Eq. (11-2) and the simulated results are also shown in Fig. 11-4. The agreement between the experimental DPP and the simulated DPP is excellent and the results[22] are summarized in Table 11-3.

Bianco and Haladjian[23] evaluated the four macroscopic redox potentials of cytochrome c_3 from *D. desulfurican* Norway by using a graphical subtractive method. This method is only applicable to a system in which four different electrochemically active species with equal concentration are involved and all species undergo a

Table 11-3 ■ The four Macroscopic Redox Potentials (mV) of Cytochromes c_3
from *D. vulgaris* Miyazaki, *D. vulgaris* Hildenborough, *D. gigas*, and *D. desulfurican*
Norway Determined by Differential Pulse Polarography in 30 mM
Phosphate Buffer Solution (pH 7.0; 25 °C)

Cytochromes c_3	$E_1^{0'}$	$E_2^{0'}$	$E_3^{0'}$	$E_4^{0'}$
D. vulgaris Miyazaki	− 240	− 297	− 315	− 357
D. vulgaris Hildenborough	− 263	− 321	− 329	− 381
D. gigas	− 187	− 286	− 305	− 324
D. desulfurican Norway	− 126	− 246	− 290	− 339

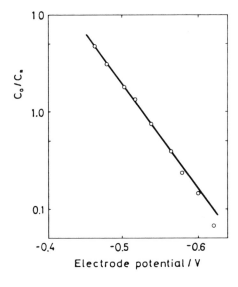

Figure 11-5 ■ The nernstian-type plot of *D. vulgaris* Miyazaki cytochrome c_3 obtained in DPP curves. The potential is referred to Ag/AgCl/KCl$_{(sat)}$. From ref. 14, with permission.

reversible one-electron–transfer reaction. That is, one cannot use the graphical substractive method to evaluate the redox potentials of the tetra-heme proteins in which a reversible four-consecutive one-electron–transfer reaction takes place.

In the potentiometric studies, the plot of $\log(C_0/C_R)$ versus the electrode potential produces a straight line as shown in Fig. 11-5, where C_0 and C_R represent the concentrations of the heme in ferric and ferrous forms in solution, respectively. These concentrations were determined from cathodic and anodic limiting currents in the NPP measurements.

11-3. NMR Studies

In the case of cytochrome c_3, besides the macroscopic redox potential, another type of redox potential should be examined, which reflects the redox behavior of each

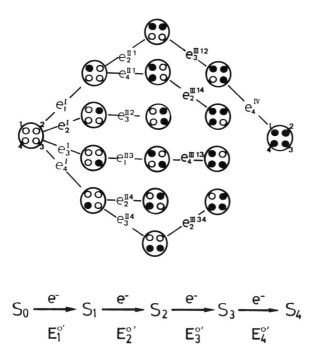

$$S_0 \xrightarrow{\ e^-\ } S_1 \xrightarrow{\ e^-\ } S_2 \xrightarrow{\ e^-\ } S_3 \xrightarrow{\ e^-\ } S_4$$
$$\quad E_1^{0'} \qquad E_2^{0'} \qquad E_3^{0'} \qquad E_4^{0'}$$

Figure 11-6 ■ A diagram of the five macroscopic oxidation states (S_i) and 16 redox molecular species. The numbers in S_0 or S_4 species denote the heme numbers. Open circles and closed circles represent the ferric and ferrous hemes in the cytochrome c_3 molecule, respectively. $E_i^{0'}$ ($i = 1$–4) is the macroscopic redox potential in the ith reduction step. e_i ($i = 1$–4) denotes the microscopic redox potential of heme i. The superscripts I, II, III, and IV indicate the reduction step, and 1, 2, 3, and 4 represent the heme kept in the reduced state in the reaction, respectively. One set of 15 independent microscopic redox potentials is represented here.

heme and is referred to as the microscopic redox potential. Because there are 16 microscopic oxidation species, 32 microscopic redox potentials (among them, 15 are independent) were defined. Both macroscopic and microscopic redox potentials are the fundamental parameters in understanding and describing the biological roles, the inter- and intramolecular electron-transfer processes of cytochrome c_3. The 16 microscopic species among the 5 macroscopic oxidation states and one set of 15 independent microscopic redox potentials are shown in Fig. 11-6, where e_i denotes the microscopic redox potential of heme i. The superscripts I, II, III, and IV indicate the first, second, third and fourth reduction steps, respectively, and 1, 2, 3, and 4 represent the hemes that kept the reduced state in the reaction.

In following sections, we will introduce a methodology to estimate the microscopic redox potentials as well as a direct determination of the macroscopic redox potentials of *D. vulgaris* Miyazaki cytochrome c_3 by using [^1]H NMR technique coupled with electrochemical measurements.[24] The advantage of the [^1]H NMR technique is that the

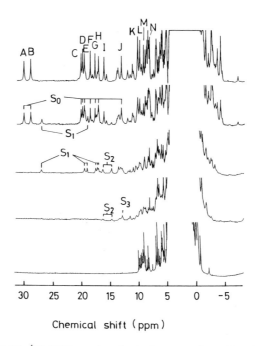

Chemical shift (ppm)

Figure 11-7 ■ 500 MHz ^{1}H NMR spectra of cytochrome c_3 from *D. vulgaris* Miyazaki, in a variety of redox stages at 30°C. The stage of oxidation was changed from fully oxidized (top) to fully reduced (bottom) by changing the hydrogen partial pressure in the presence of hydrogenase. The heme methyl signals are labeled on top. S_i is the macroscopic oxidation state. From ref. 27, with permission.

heme methyl proton is extremely sensitive to the redox states of heme iron[25, 26] and the chemical shifts of the heme methyl proton in the fully oxidized state S_0 occur in the spectral window away from the crowded region of 0–10 ppm that contains many overlapping resonances of polypeptide protons as shown in Fig. 11-7.[24, 25] Sixteen heme methyl resonances were expected to be seen in NMR measurements; however, only 14 methyl resonances were assigned in the measurements and they were labeled A, B, C, ..., N from the low field side. The top spectrum is fully oxidized and the bottom is fully reduced where heme methyl resonances disappeared in the low field region.[27]

Because a small amount of hydrogenase was involved in this sample, increasing the hydrogen gas partial pressure in an NMR tube can reduce cytochrome c_3. As reduction proceeds, several new methyl resonances appear at the expense of decreasing intensity in the fully oxidized signals. The resonances reached a maximum and then decreased, and another set of resonances appeared as shown in Fig. 11-7. Including the fully oxidized resonances, five sets of resonances were observed in the entire reduction (or oxidation) reaction of cytochrome c_3. The five resonance sets were assigned to cytochrome c_3 molecules in fully oxidized, 1-electron, 2-electron,

3-electron, and 4-electron reduced states, respectively, and they were represented as $S_i (i = 0–4)$. This observation is consistent with the four-consecutive one-electron redox mechanism deduced by DPP studies already described.

Although the intensities of signals changed dramatically along with the progress of the reduction of cytochrome c_3, the peak positions remained the same. On the basis of the observations, it can be concluded that the intermolecular electron exchange rate is slow with respect to the NMR time scale. The intramolecular electron exchange rate could be interpreted as fast or slow with respect to the NMR time scale. If the exchange rate is slow, a heme methyl group would give rise to from two to six sets (oxidized and reduced forms) of 16 signals in the intermediate oxidation stages, which originate from the microscopic redox species in each oxidation state. However, this was not the case for cytochrome c_3 because only one heme methyl signal was observed in each oxidation state. From this NMR behavior of cytochrome c_3, it can be concluded that the intramolecular electron exchange is fast with respect to the NMR time scale.

Because the peak intensity of the heme methyl group is proportional to the mole fraction of cytochrome c_3 in the distinct macroscopic oxidation state, mole fractions of cytochrome c_3 in five different states are easily estimated from NMR spectra. For

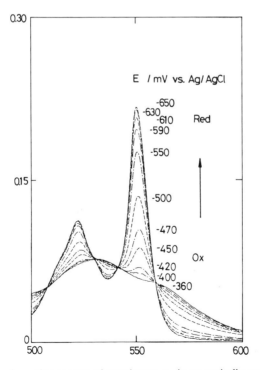

Figure 11-8 ■ The absorption spectra of cytochrome c_3 in an optically transparent thin-layer electrode (OTTLE) cell with various applied potentials at 30 °C.

example, cytochrome c_3 in states S_0 and S_1 were distinguished in the second spectrum in Fig. 11-6, because the summation of each fraction should be unity. The ratio of $S_0:S_1$ was obtained from the integrated intensities of any isolated signal belonging to S_0 or S_1. In order to get the information for the equilibrium potential corresponding to the same NMR sample, the absorption spectrum of the NMR sample was measured before and after each measurement. The corresponding relation between the reduction ratio (estimated from absorption spectrum) and equilibrium potential was established by introducing an absorbance-potential measurement with an optically transparent thin layer electrode (OTTLE). The absorption spectra of cytochrome c_3 measured in an OTTLE cell are presented in Fig. 11-8.

There are four isobestic points in the region of interest, suggesting that the spectral changes can be interpreted in terms of a two-oxidation-state (oxidized and reduced forms) model. Nernstian-type plots were made from OTTLE measurements as shown in Fig. 11-9. A good agreement was obtained in the plots between the forward and reverse potentiometric titrations. However, because the DPP curve is not symmetrical, the nernstian plot should not be linear. The straight line obtained here was caused by the small potential spacings of *D. vulgaris* Miyazaki cytochrome c_3. Another example, shown in Fig. 11-10, is the nernstian-type plot of cytochrome c_3 from *D. gigas*, in which, instead of a straight line, a curve was obtained because of the large potential spacing between $E_1^{0'}$ and $E_2^{0'}$.

Nevertheless, the nernstian-type plot of *D. vulgaris* Miyazaki can be used to link the mole fraction in each NMR spectrum to its equilibrium potential versus NHE

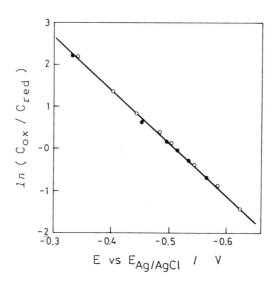

Figure 11-9 ■ Nernstian-type plots of cytochrome c_3 observed with an OTTLE. Closed and open circles are the plots for the oxidation and reduction processes, respectively, and the solid line is the least-squares fitting result, which overlaps with the theoretical line simulated from macroscopic redox potentials.

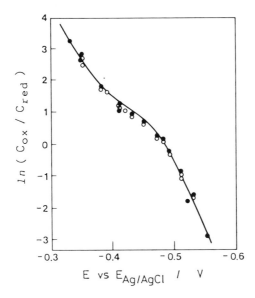

Figure 11-10 ■ Nernstian-type plot of cytochrome c_3 from *D. gigas* as a result of potentiometry by using an OTTLE. Open and closed circles have the same meaning as Fig. 11-9 and the solid line is the calculated curve from macroscopic redox potentials.

scale. Subsequently, the estimated mole fraction in the NMR spectrum was plotted as a function of the equilibrium potential as presented in Fig. 11-11. A nonlinear least-squares fitting was carried out on the original data and the four macroscopic redox potentials were obtained as -260, -312, -327, and -369 mV from the intersections of the best-fitted curves of f_{i-1} and f_i at $30\,°C$ and p^2H 7.0.[25] These values agreed quite well with those obtained from DPP curve analysis as shown in Table 11-3.

In summary, NMR measurement could distinguish the five macroscopic oxidation states as isolated signals. The mole fractions of each oxidation state could be determined by the use of these signals, and the X-axis of Fig. 11-10 is correlated to the electrode potentials with respect to the NHE scale by using OTTLE. The macroscopic redox potentials could be determined directly from these observations without any assumptions or any models for the electron-transfer process.

11-4. Analytical Processes for Estimation of Microscopic Redox Potentials

Because the intramolecular electron exchange rate is fast with respect to the NMR time scale, the heme methyl resonances in the intermediate oxidation states appeared in the region between the positions for the fully oxidized and fully reduced states. This experimental result indicates that the transferred electron is delocalized among the four hemes in the NMR time scale sense. Therefore, the chemical shifts of heme

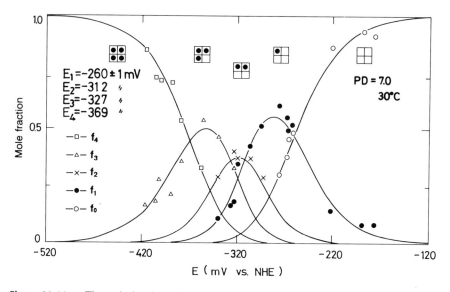

Figure 11-11 ■ The mole fractions of cytochrome c_3 in the five macroscopic oxidation states as a function of the equilibrium potential (E). Open circles, closed circles, crosses, triangles, and squares represent the observed mole fractions of cytochrome c_3 in the fully oxidized, one-electron, two-electron, three-electron, and fully reduced states, respectively. Solid curves were calculated with the best-fitting parameters obtained on nonlinear least-squares fitting. The four macroscopic formal potentials were determined as $E_1^{0'} = -260$, $E_2^{0'} = -312$, $E_3^{0'} = -327$, and $E_4^{0'} = -369$ mV versus NHE, respectively.

methyl groups included the information on the electron distribution probability at each heme, which is useful for determination of the microscopic redox potentials.[27, 28]

In order to determine the chemical shifts of each heme methyl group in the five macroscopic oxidation states, saturation transfer experiments were carried out at the condition where cytochromes c_3 belonging to several oxidation states coexist, and the obtained chemical shifts are summarized in Table 11-4.[27] The chemical shifts of the eight heme methyl groups (A–H) were identified for all oxidation states.

Here we would like to present an analysis method to estimate the 32 microscopic redox potentials from the chemical shifts in Table 11-4 and the derived formulations. We denote each heme as o (for oxidized) or r (for reduced). The microscopic redox potentials for the first reduction step e_i^1 ($i = 1$–4) can be defined as

$$E = e_1^1 + (RT/F)\ln([oooo]/[rooo]) \tag{11-3}$$

$$E = e_2^1 + (RT/F)\ln([oooo]/[oroo]) \tag{11-4}$$

$$E = e_3^1 + (RT/F)\ln([oooo]/[ooro]) \tag{11-5}$$

$$E = e_4^1 + (RT/F)\ln([oooo]/[ooor]) \tag{11-6}$$

Table 11-4 ▪ Chemical Shifts (ppm) of Heme Methyl Signals in the Five Macroscopic Oxidation States at p^2H 7.1 and 30 °C[a]

Signals	Oxidation States				
	S_0	S_1	S_2	S_3	S_4
A	30.51	12.11	10.81	7.57	3.23
B	29.31	26.97	13.06	4.90	3.25
C	20.51	19.83	16.37	6.66	3.15
D	20.24	19.30	14.75	6.03	3.16
E	19.90	16.88	13.46	13.02	3.47
F	18.98	17.70	9.73	4.16	3.26
G	18.07	17.26	8.83	3.78	3.29
H	17.53	7.60	6.77	5.45	3.18
I	16.45	7.35	6.60	5.20	
J	13.52	13.15	11.14		
K	10.62	6.13	5.50		
L	10.35	8.56			
M	9.62	9.25	5.64		
N	8.47	7.07			

[a]From ref. 27, with permission.

where [oroo], for example, stands for the concentration of the microscopic molecular species in which only heme 2 is reduced. From these equations, we have

$$e_1^I - e_2^I = (RT/F)\ln(R_1^I/R_2^I) \tag{11-7}$$

$$e_2^I - e_3^I = (RT/F)\ln(R_2^I/R_3^I) \tag{11-8}$$

$$e_3^I - e_4^I = (RT/F)\ln(R_3^I/R_4^I) \tag{11-9}$$

where R_i^I is the distribution probability of the electron introduced in the first reduction step into heme i. For example,

$$R_1^I = [rooo]/([rooo] + [oroo] + [ooro] + [ooor])$$

Equations (11-7) through (11-9) give the relationships between the microscopic redox potentials to be determined and R_i^j that are observable. Analogously, similar equations were obtained for the fourth reduction step as follows:

$$E = e_1^{IV} + (RT/F)\ln([orrr]/[rrrr]) \tag{11-10}$$

$$E = e_2^{IV} + (RT/F)\ln([rorr]/[rrrr]) \tag{11-11}$$

$$E = e_3^{IV} + (RT/F)\ln([rror]/[rrrr]) \tag{11-12}$$

$$E = e_4^{IV} + (RT/F)\ln([rrro]/[rrrr]) \tag{11-13}$$

and

$$e_1^{IV} - e_2^{IV} = (RT/F)\ln(R_2^{IV}/R_1^{IV}) \tag{11-14}$$

$$e_1^{IV} - e_3^{IV} = (RT/F)\ln(R_3^{IV}/R_1^{IV}) \tag{11-15}$$

$$e_1^{IV} - e_4^{IV} = (RT/F)\ln(R_4^{IV}/R_1^{IV}) \tag{11-16}$$

Such equations also can be deduced for the second and third reduction steps, respectively. Because $\sum_i R_i^j = 1$ and $\sum_j R_i^j = 1$, only 9 out of 16 R_i^j are independent. If we refer to the macroscopic redox potentials, which were determined previously, additional independent equations can be introduced. From Eqs. (11-1) and (11-2) through (11-6), we obtained

$$e_i^{I} = E_1^{0'} + (RT/F) \ln R_i^{I} \tag{11-17}$$

and similarly,

$$e_i^{IV} = E_4^{0'} - (RT/F) \ln R_i^{IV} \tag{11-18}$$

Consequently, we have 11 independent observations and corresponding equations when we use $E_1^{0'}$ and $E_4^{0'}$ in addition to the 9 independent R_i^j.

If we assume that the effect of the change in the oxidation state of heme i on the redox potential of heme j (which can be called an interacting potential between hemes i and j, I_{ij}) is not affected by the oxidation states of the other two hemes, any microscopic redox potential can be described with the four microscopic redox potentials and six interacting potentials. Thirty-two parameters can be reduced to 10. Thus, we can determine these parameters analytically because we have enough independent equations as shown in the preceding text.

The interacting potential I_{ij} can be defined[28] as

$$I_{ij} = e_i^{IIj} - e_i^{I} = e_i^{IIIjk} - e_i^{IIk} = e_i^{IV} \left(\text{or } e_i^{IVjkl} \right) - e_i^{IIIkl} \tag{11-19}$$

where i, j, k, and l denote heme numberings. It follows, for example, that

$$e_i^{IV} = e_i^{I} + I_{ij} + I_{ik} + I_{il} \tag{11-20}$$

and

$$I_{ij} = I_{ji} \tag{11-21}$$

Rearrangement of Eq. (11-14) gives

$$e_1^{IV} - e_2^{IV} = \left[e_1^{I} - e_2^{I} \right] + \left[I_{13} + I_{14} - I_{23} - I_{24} \right] \tag{11-22}$$

$$e_1^{IV} - e_3^{IV} = \left[e_1^{I} - e_3^{I} \right] + \left[I_{12} + I_{14} - I_{32} - I_{34} \right] \tag{11-23}$$

$$e_1^{IV} - e_4^{IV} = \left[e_1^{I} - e_4^{I} \right] + \left[I_{12} + I_{13} - I_{42} - I_{43} \right] \tag{11-24}$$

Substitution of Eqs. (11-14) through (11-16) into Eqs. (11-22) through (11-24) yields

$$\left[I_{13} + I_{14} - I_{23} - I_{24} \right] = (RT/F) \ln\left(\left(R_2^{I} R_2^{IV} \right) / \left(R_1^{I} R_1^{IV} \right) \right) \tag{11-25}$$

$$\left[I_{12} + I_{14} - I_{32} - I_{34} \right] = (RT/F) \ln\left(\left(R_3^{I} R_3^{IV} \right) / \left(R_1^{I} R_1^{IV} \right) \right) \tag{11-26}$$

$$\left[I_{12} + I_{13} - I_{42} - I_{43} \right] = (RT/F) \ln\left(\left(R_4^{I} R_4^{IV} \right) / \left(R_1^{I} R_1^{IV} \right) \right) \tag{11-27}$$

where Eqs. (11-7) through (11-9) were also used. Thus, we can obtain

$$I_{12} - I_{34} = (RT/2F) \ln\left(\left(R_3^{I} R_3^{IV} R_4^{I} R_4^{IV} \right) / \left(R_1^{I} R_1^{IV} R_2^{I} R_2^{IV} \right) \right) = a_1 \tag{11-28}$$

$$I_{13} - I_{24} = (RT/2F) \ln\left(\left(R_2^{I} R_2^{IV} R_4^{I} R_4^{IV} \right) / \left(R_1^{I} R_1^{IV} R_3^{I} R_3^{IV} \right) \right) = a_2 \tag{11-29}$$

$$I_{14} - I_{23} = (RT/2F) \ln\left(\left(R_2^{I} R_2^{IV} R_3^{I} R_3^{IV} \right) / \left(R_1^{I} R_1^{IV} R_4^{I} R_4^{IV} \right) \right) = a_3 \tag{11-30}$$

On the other hand, the electron-distribution probabilities in the second reduction step can be defined as

$$\frac{[\text{roor}] + [\text{roro}] + [\text{rroo}]}{[\text{orro}] + [\text{rroo}] + [\text{oror}]} = \frac{R_1^I + R_1^{II}}{R_2^I + R_2^{II}} \tag{11-31}$$

$$\frac{[\text{orro}] + [\text{rroo}] + [\text{oror}]}{[\text{roro}] + [\text{orro}] + [\text{oorr}]} = \frac{R_2^I + R_2^{II}}{R_3^I + R_3^{II}} \tag{11-32}$$

$$\frac{[\text{roro}] + [\text{orro}] + [\text{oorr}]}{[\text{roor}] + [\text{oror}] + [\text{oorr}]} = \frac{R_3^I + R_3^{II}}{R_4^I + R_4^{II}} \tag{11-33}$$

$$\frac{[\text{roor}] + [\text{oror}] + [\text{oorr}]}{[\text{roor}] + [\text{roro}] + [\text{rroo}]} = \frac{R_4^I + R_4^{II}}{R_1^I + R_1^{II}} \tag{11-34}$$

The right-hand terms are replaced by t_{12}, t_{23}, t_{34}, and t_{41}, and the left-hand terms by

$$[\text{roor}]/[\text{rroo}] = \exp\big((F/RT)\big(e_4^{II1} - e_2^{II1}\big)\big) \tag{11-35}$$

$$[\text{roro}]/[\text{rroo}] = \exp\big((F/RT)\big(e_3^{II1} - e_2^{II1}\big)\big) \tag{11-36}$$

$$[\text{orro}]/[\text{rroo}] = \exp\big((F/RT)\big(e_3^{II2} - e_1^{II2}\big)\big) \tag{11-37}$$

$$[\text{oror}]/[\text{rroo}] = \exp\big((F/RT)\big(e_4^{II2} - e_1^{II2}\big)\big) \tag{11-38}$$

$$[\text{rroo}]/[\text{orro}] = \exp\big((F/RT)\big(e_1^{II2} - e_3^{II2}\big)\big) \tag{11-39}$$

$$[\text{oror}]/[\text{orro}] = \exp\big((F/RT)\big(e_4^{II2} - e_3^{II2}\big)\big) \tag{11-40}$$

$$[\text{roro}]/[\text{orro}] = \exp\big((F/RT)\big(e_1^{II3} - e_2^{II3}\big)\big) \tag{11-41}$$

$$[\text{oorr}]/[\text{orro}] = \exp\big((F/RT)\big(e_4^{II3} - e_2^{II3}\big)\big) \tag{11-42}$$

$$[\text{roro}]/[\text{oorr}] = \exp\big((F/RT)\big(e_1^{II3} - e_4^{II3}\big)\big) \tag{11-43}$$

$$[\text{orro}]/[\text{oorr}] = \exp\big((F/RT)\big(e_2^{II3} - e_4^{II3}\big)\big) \tag{11-44}$$

$$[\text{roor}]/[\text{oorr}] = \exp\big((F/RT)\big(e_1^{II4} - e_3^{II4}\big)\big) \tag{11-45}$$

$$[\text{oror}]/[\text{oorr}] = \exp\big((F/RT)\big(e_2^{II4} - e_3^{II4}\big)\big) \tag{11-46}$$

$$[\text{oror}]/[\text{roor}] = \exp\big((F/RT)\big(e_2^{II4} - e_1^{II4}\big)\big) \tag{11-47}$$

$$[\text{roro}]/[\text{roor}] = \exp\big((F/RT)\big(e_3^{II1} - e_4^{II1}\big)\big) \tag{11-48}$$

Thus, we have

$$\frac{\exp\big((F/RT)\big(e_4^{II1} - e_2^{II1}\big)\big) + \exp\big((F/RT)\big(e_3^{II1} - e_2^{II1}\big)\big) + 1}{\exp\big((F/RT)\big(e_3^{II2} - e_1^{II2}\big)\big) + \exp\big((F/RT)\big(e_4^{II2} - e_1^{II2}\big)\big) + 1} = t_{12} \tag{11-49}$$

$$\frac{\exp\big((F/RT)\big(e_1^{II2} - e_3^{II2}\big)\big) + \exp\big((F/RT)\big(e_4^{II2} - e_3^{II2}\big)\big) + 1}{\exp\big((F/RT)\big(e_1^{II3} - e_2^{II3}\big)\big) + \exp\big((F/RT)\big(e_4^{II3} - e_2^{II3}\big)\big) + 1} = t_{23} \tag{11-50}$$

$$\frac{\exp\big((F/RT)\big(e_1^{II3} - e_4^{II3}\big)\big) + \exp\big((F/RT)\big(e_2^{II3} - e_4^{II3}\big)\big) + 1}{\exp\big((F/RT)\big(e_1^{II4} - e_3^{II4}\big)\big) + \exp\big((F/RT)\big(e_2^{II4} - e_3^{II4}\big)\big) + 1} = t_{34} \tag{11-51}$$

$$\frac{\exp\big((F/RT)\big(e_2^{II4} - e_1^{II4}\big)\big) + \exp\big((F/RT)\big(e_3^{II4} - e_1^{II4}\big)\big) + 1}{\exp\big((F/RT)\big(e_3^{II1} - e_4^{II1}\big)\big) + \exp\big((F/RT)\big(e_2^{II1} - e_4^{II1}\big)\big) + 1} = t_{41} \tag{11-52}$$

By using interacting potentials (I_{ij}) and the relationships of Eqs. (11-28) through (11-30), we can obtain the following equations:

$$(C_1 Z + C_2/X + 1)/(C_3 Z + C_4/X + 1) = t_{12} \tag{11-53}$$

$$(C_5/Z + C_6 Y + 1)/(C_7 Y + C_8/Z + 1) = t_{23} \tag{11-54}$$

$$(C_9/X + C_{10} Z + 1)/(C_{11} Z + C_{12}/X + 1) = t_{34} \tag{11-55}$$

$$(C_{13} Y + C_{14}/Z + 1)/(C_{15}/Z + C_{16} Y + 1) = t_{41} \tag{11-56}$$

where

$$X = \exp((F/RT)(I_{12} - I_{13})) \tag{11-57}$$

$$Y = \exp((F/RT)(I_{13} - I_{14})) \tag{11-58}$$

$$Z = \exp((F/RT)(I_{14} - I_{12})) \tag{11-59}$$

and

$$C_1 = \exp((F/RT)(e_4^1 - e_2^1)) \tag{11-60}$$

$$C_2 = \exp((F/RT)(e_3^1 - e_2^1)) \tag{11-61}$$

$$C_3 = \exp((F/RT)(e_3^1 - e_1^1 - a_3)) \tag{11-62}$$

$$C_4 = \exp((F/RT)(e_4^1 - e_1^1 - a_2)) \tag{11-63}$$

$$C_5 = \exp((F/RT)(e_1^1 - e_3^1 + a_3)) \tag{11-64}$$

$$C_6 = \exp((F/RT)(e_4^1 - e_3^1 - a_2 + a_3)) \tag{11-65}$$

$$C_7 = \exp((F/RT)(e_1^1 - e_2^1 + a_3)) \tag{11-66}$$

$$C_8 = \exp((F/RT)(e_4^1 - e_2^1 - a_1 + a_3)) \tag{11-67}$$

$$C_9 = \exp((F/RT)(e_1^1 - e_4^1 + a_1)) \tag{11-68}$$

$$C_{10} = \exp((F/RT)(e_2^1 - e_4^1 - a_3 + a_1)) \tag{11-69}$$

$$C_{11} = \exp((F/RT)(e_1^1 - e_3^1 + a_1)) \tag{11-70}$$

$$C_{12} = \exp((F/RT)(e_2^1 - e_3^1 - a_2 + a_1)) \tag{11-71}$$

$$C_{13} = \exp((F/RT)(e_2^1 - e_1^1 - a_2)) \tag{11-72}$$

$$C_{14} = \exp((F/RT)(e_3^1 - e_1^1 - a_1)) \tag{11-73}$$

$$C_{15} = \exp((F/RT)(e_2^1 - e_4^1)) \tag{11-74}$$

$$C_{16} = \exp((F/RT)(e_3^1 - e_4^1)) \tag{11-75}$$

Because $XYZ = 1$, Eqs. (11-53) through (11-56) can be rewritten simply as

$$l_i Y + m_i(1/Z) + n_i = 0 \qquad i = 1, 2, 3, \text{ or } 4 \tag{11-76}$$

where

$$\{l_1, m_1, n_1\} = \{(C_2 - C_4 t_{12}), (1 - t_{12}), (C_1 - C_3 t_{12})\}$$

$$\{l_2, m_2, n_2\} = \{(C_6 - C_7 t_{23}), (C_5 - C_8 t_{23}), (1 - t_{23})\}$$

$$\{l_3, m_3, n_3\} = \{(C_9 - C_{12} t_{34}), (1 - t_{34}), (C_{10} - C_{11} t_{34})\}$$

$$\{l_4, m_4, n_4\} = \{(C_{13} - C_{16} t_{41}), (C_{14} - C_{15} t_{41}), (1 - t_{41})\}$$

Equation (11-76) gives the relationship between the interacting potentials (I_{ij}) and electron-distribution probabilities (R_i^j). In our calculation, we used Eqs. (11-17) and (11-18) to obtain e_i^I and e_i^{IV}; then we used Eqs. (11-20) and (11-76) to obtain I_{ij}.

On the other hand, the electron-distribution probabilities can be estimated from the chemical shifts of the heme methyl groups for five macroscopic oxidation states. The total paramagnetic shift of heme methyl signal i of heme 1 in the fully oxidized state can be represented [27, 28] as

$$\delta_1^i = \Delta_{11}^i + \Delta_{12}^i + \Delta_{13}^i + \Delta_{14}^i \tag{11-77}$$

The intrinsic shift Δ_{jj}^i is dominated by the contact contribution. The extrinsic shifts Δ_{jk}^i ($j \neq k$), can be ascribed to the pseudocontact paramagnetic shifts. For the intermediate oxidation states (S_1, S_2, and S_3), the paramagnetic shifts can be given as follows, provided that conformational changes are negligible:

$$\delta_1^i(I) = \delta_1^i - \left[R_1^I \Delta_{11}^i + R_2^I \Delta_{12}^i + R_3^I \Delta_{13}^i + R_4^I \Delta_{14}^i \right] \tag{11-78}$$

$$\delta_1^i(II) = \delta_1^i - \left[\left(R_1^I + R_1^{II} \right) \Delta_{11}^i + \left(R_2^I + R_2^{II} \right) \Delta_{12}^i \right.$$
$$\left. + \left(R_3^I + R_3^{II} \right) \Delta_{13}^i + \left(R_4^I + R_4^{II} \right) \Delta_{14}^i \right] \tag{11-79}$$

$$\delta_1^i(III) = \delta_1^i - \left[\left(R_1^I + R_1^{II} + R_1^{III} \right) \Delta_{11}^i + \left(R_2^I + R_2^{II} + R_2^{III} \right) \Delta_{12}^i \right.$$
$$\left. + \left(R_3^I + R_3^{II} + R_3^{III} \right) \Delta_{13}^i + \left(R_4^I + R_4^{II} + R_4^{III} \right) \Delta_{14}^i \right] \tag{11-80}$$

Because Δ_{jj} is much larger than Δ_{jk} in general, we may ignore Δ_{jk} ($j \neq k$) in the first step of estimation of the electron distribution probabilities. This leads to the simplifications

$$\delta_1^i = \Delta_{11}^i \tag{11-81}$$

$$\delta_1^i(I) = \delta_1^i - R_1^I \Delta_{11}^i \tag{11-82}$$

$$\delta_1^i(II) = \delta_1^i - \left(R_1^I + R_1^{II} \right) \Delta_{11}^i \tag{11-83}$$

$$\delta_1^i(III) = \delta_1^i - \left(R_1^I + R_1^{II} + R_1^{III} \right) \Delta_{11}^i \tag{11-84}$$

Equations (11-81) through (11-84) provide the formulations to evaluated electron distribution probabilities from chemical shifts, and the results are summarized in Table 11-5.

In principle, methyl groups belonging to the same heme should have the same electron-distribution probability. Taking advantage of this property, 14 heme methyl signals were classified into four groups by using the largest electron-distribution probability in each signal except for heme methyl groups J, N, and L, and four hemes

Table 11-5 ■ **Electron-Distribution Probabilities (R_i^j) Calculated from Heme Methyl Chemical Shifts**[a]

Signals	R^{I}	R^{II}	R^{III}	R^{IV}	Heme
A	0.674	0.048	0.119	0.159	1
H	0.692	0.058	0.092	0.158	
I	0.689	0.057	0.106	0.145	
K	0.609	0.085			
B	0.090	0.534	0.313	0.063	2
F	0.081	0.507	0.345	0.057	
G	0.055	0.570	0.342	0.033	
M	0.058	0.567			
C	0.039	0.199	0.559	0.202	3
D	0.055	0.266	0.511	0.168	
J	0.036	0.191			
E	0.184	0.208	0.027	0.581	4
L	0.252	0.220			
N	0.268				

[a] Equations (11-81) through (11-84) were used for calculation. The average value for the observed chemical shifts in the fully reduced state (S_4) was used as the diamagnetic shift for signals I–N. By definition, R_i^j is the distribution probability at heme i of the electron introduced in the jth reduction step. Fourteen heme methyl groups were classified into four hemes according to the largest ^{j} except for signals J, L, and N: In the latter cases, R^{I} was used. The four hemes are numbered in the order of the major reduction taking place. From ref. 27, with permission.

are numbered in the order of the major reduction taking place. However, the electron-distribution probability for the same heme is scattered somewhat in Table 11-5, suggesting that the extrinsic paramagnetic shifts are not negligible.

11-5. Estimation of Microscopic Redox Potentials

Some important conclusions can be deduced from Table 11-5. The electron transferred in each reduction step is actually delocalized among the four hemes. Nevertheless, we can identify the site, keeping the major electron-distribution probability. Namely, hemes 1, 2, 3, and 4 mainly are reduced at the first, second, third, and fourth reduction steps, respectively. Although heme 4 had the second highest electron-distribution probability in the first reduction step, it was mainly reduced in the fourth step. This change clearly shows that the existence of the electron at heme 1 affects the redox potential of other hemes significantly and in different ways. It can be concluded that the interheme interaction plays an important role in the redox potential of each heme.

To minimize errors from neglecting the contribution from the extrinsic paramagnetic shifts, the calculated electron-distribution probability was averaged for the methyl groups belonging to the same heme. This was carried out using the methyl

Table 11-6 ▪ The Average Electron-Distribution Probabilities of Four Hemes in the Four Reduction Steps[a]

	R^I	R^{II}	R^{III}	R^{IV}	ΣR^i
Heme 1	0.683	0.053	0.105	0.159	1.000
Heme 2	0.075	0.537	0.336	0.051	1.000
Heme 3	0.047	0.233	0.535	0.185	1.000
Heme 4	0.184	0.208	0.027	0.581	1.000
Heme 4*	0.194	0.170	0.031	0.605	1.000
Σ heme i	0.989	1.031	1.003	0.976	
Σ heme i*	1.000	1.000	1.000	1.000	

[a] Resonances A–H were used in averaging because their chemical shifts were identified for the five oxidation states. The asterisk indicates the values obtained from $\Sigma^j_{ii} = 1$. From ref. 27, with permission.

groups, for which all chemical shifts in the five oxidation states could be identified (A–H). The averaged R are summarized[27] in Table 11-6. Because the summation either for each heme or for each reduction step is very close to unity, these values can be used in the calculation of the microscopic redox potentials for the first approximation.

In the first place, e_i^I and e_i^{IV} were calculated using Eqs. (11-17) and (11-18). The results[27] are given in parentheses in Table 11-7. Then, by a least-squares fitting with Eq. (11-76), we obtained

$$I_{13} - I_{14} = 1.1 \pm 0.7 \, (\text{mV}) \tag{11-85}$$

$$I_{14} - I_{12} = -33.4 \pm 0.9 \, (\text{mV}) \tag{11-86}$$

and Table 11-5 gives

$$I_{12} + I_{13} + I_{14} = e_1^{IV} - e_1^I = -50.9 \, (\text{mV}) \tag{11-87}$$

Table 11-7 ▪ The Microscopic Redox Potentials at the First and Fourth Reduction Steps (e_i^I and e_i^{IV}, respectively) and the Interacting Potentials (I_{ij})[a]

Microscopic Redox Potentials (mV)		Interacting Potentials (mV)		Macroscopic Potentials (mV)	
				Calculated	Observed
e_1^I −270.0(−270.0)	e_1^{IV} −320.9(−320.9)	I_{12}	4.8(4.9)	$E_1^{0'}$ −260(−260)	−260
e_2^I −327.6(−327.6)	e_2^{IV} −291.4(−291.4)	I_{13}	−21.0(−27.3)	$E_2^{0'}$ −310(−309)	−312
e_3^I −339.8(−339.8)	e_3^{IV} −324.9(−324.9)	I_{14}	−34.8(−28.5)	$E_3^{0'}$ −327(−329)	−327
e_4^I −302.8(−304.3)	e_4^{IV} −355.9(−354.8)	I_{23}	42.8(47.8)	$E_4^{0'}$ −369(−369)	−369
		I_{24}	−11.4(−16.5)		
		I_{34}	−6.9(−5.6)		

[a] The values in parentheses were calculated using the nonmodified electron-distribution probabilities of heme 4. Standard deviation is ±1 mV. From ref. 27, with permission.

Table 11-8 ■ The 32 Microscopic Redox Potentials (mV) of Cytochrome $c_3{}^a$

		Microscopic Redox pot.	Differences from e^I
Microscopic Redox Potentials of Heme 1			
e_1^I	$oooo + e = rooo$	$-270(-270)$	
e_1^{II2}	$oroo + e = rroo$	$-265(-265)$	$5(5)$
e_1^{II3}	$ooro + e = roro$	$-291(-297)$	$-21(-27)$
e_1^{II4}	$ooor + e = roor$	$-305(-298)$	$-35(-28)$
e_1^{III23}	$orro + e = rrro$	$-286(-292)$	$-16(-22)$
e_1^{III24}	$oror + e = rror$	$-300(-294)$	$-30(-24)$
e_1^{III34}	$oorr + e = rorr$	$-326(-326)$	$-56(-56)$
e_1^{IV}	$orrr + e = rrrr$	$-321(-321)$	$-51(-51)$
Microscopic Redox Potentials of Heme 2			
e_2^I	$oooo + e = oroo$	$-328(-328)$	
e_2^{II1}	$rooo + e = rroo$	$-323(-323)$	$5(5)$
e_2^{II3}	$ooro + e = orro$	$-285(-280)$	$43(48)$
e_2^{II4}	$ooor + e = oror$	$-339(-344)$	$-11(-16)$
e_2^{III23}	$roro + e = rrro$	$-280(-275)$	$48(53)$
e_2^{III34}	$oorr + e = orrr$	$-296(-296)$	$32(32)$
e_2^{III14}	$roor + e = rror$	$-334(-339)$	$-6(-11)$
e_2^{IV}	$rorr + e = rrrr$	$-291(-291)$	$37(37)$
Microscopic Redox Potentials of Heme 3			
e_3^I	$oooo + e = ooro$	$-340(-340)$	
e_3^{II1}	$rooo + e = roro$	$-361(-367)$	$-21(-27)$
e_3^{II2}	$oroo + e = orro$	$-297(-292)$	$43(48)$
e_3^{II4}	$ooor + e = oorr$	$-347(-345)$	$-7(-5)$
e_3^{III12}	$rroo + e = rrro$	$-318(-319)$	$22(21)$
e_3^{III14}	$roor + e = rorr$	$-368(-373)$	$-28(-33)$
e_3^{III24}	$oroo + e = orrr$	$-304(-298)$	$36(42)$
e_3^{IV}	$rror + e = rrrr$	$-325(-325)$	$15(15)$
Microscopic Redox Potentials of Heme 4			
e_4^I	$oooo + e = ooor$	$-303(-304)$	
e_4^{II1}	$rooo + e = roor$	$-338(-333)$	$-35(-29)$
e_4^{II2}	$oroo + e = oror$	$-314(-321)$	$-11(-17)$
e_4^{II3}	$ooro + e = oorr$	$-310(-310)$	$-7(-6)$
e_4^{III12}	$rroo + e = rror$	$-349(-349)$	$-46(-45)$
e_4^{III13}	$roro + e = rorr$	$-344(-338)$	$-41(-34)$
e_4^{III23}	$orro + e = orrr$	$-321(-326)$	$-18(-22)$
e_4^{IV}	$rrro + e = rrrr$	$-356(-355)$	$-53(-51)$

^{a}Parentheses indicate the results from the nonmodified electron-distribution probability values. o and r represent the ferric and ferrous heme, respectively.

These three equations gave I_{12}, I_{13}, and I_{14}. I_{23}, I_{24}, and I_{34} could be obtained with Eq. (11-20). The obtained interacting potentials also are summarized in parentheses in Table 11-7.

To check the reliability of the estimated microscopic redox potentials, the macroscopic redox potentials calculated from them were compared with observations. Only $E_2^{0'}$ and $E_3^{0'}$ are independent because the other two were used in the preceding calculation. The results showed there were -3 and $+2$ mV differences for $E_2^{0'}$ and $E_3^{0'}$, respectively.

As a second step of approximation, we modified the electron-distribution probabilities of heme 4 (R_4^j, $j = $ I, II, III, or IV). Because only one set of R values was obtained for heme 4, there was no averaging process. We modified these values to satisfy $\sum_i R_i = 1$, as shown by asterisks in Table 11-6. The microscopic redox potentials and interacting potentials obtained from the set of modified R values also are summarized in Table 11-6. These results show better agreement between the observed and calculated macroscopic redox potentials, indicating that this is a better set of R values. The 32 microscopic redox potentials calculated from the set of modified R values are shown in Table 11-8 together with their reaction equations and potential notations. From Table 11-8, it is clear that the redox potential of each heme is significantly changed, not only in value, but also in relative order, with the change of oxidation state. For example, the redox potential of heme 2 is the second lowest in the first reduction step. It operates as an electron donor among the four hemes. Yet, it becomes the heme with the highest potential in the third and fourth reduction steps and takes on the role of an electron acceptor. In contrast, the redox potential of heme 4 is changed in the opposite direction. This heme has the second highest potential in the first reduction step but becomes one with the lowest potential in the fourth reduction step. This strongly suggests that each heme plays a different role in the different oxidation states. Furthermore, because the microscopic redox potentials are distributed over a wide range (about 100 mV) and the interheme electron exchange within a molecule is rapid, cytochrome c_3 can provide different redox proteins or substrates with multielectron-transfer paths in a nonequilibrium state. Along with the protein–protein specific recognition, those characteristics could serve as the physicochemical base for the multifunctionality of this protein in a sulfate-reducing bacterium.

11-6. Partial Assignments of Hemes by NOE

The change of the redox potential of each heme was induced by interheme interactions. The remarkable feature of interacting potentials in Table 11-7 is the positive value of I_{23} ($= I_{32}$). This means that the presence of an electron at heme 2 (or 3) makes the reduction of heme 3 (or 2) much easier. If an electrostatic interaction is the dominant contribution to an interacting potential, only a negative one would be expected. Therefore, a positive interaction would be elucidated in terms of other factors, including structural parameters. For this purpose, it is important to make the corresponding relation of four hemes in NMR and crystal structure clear. Nuclear Overhauser effect measurement was carried out in the NMR experiment. NOE

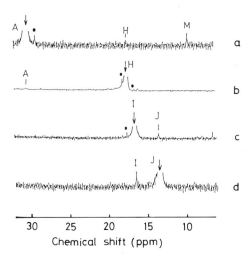

Figure 11-12 ■ NOE difference spectra of cytochrome c_3 in the fully oxidized state at (a) 19 °C and (b, c, d) 30 °C. The arrows indicate the irradiated positions. A closed circle denotes power spillage. In the case of (a), the measurement was carried out at a lower temperature because a single proton signal at around 9.6 ppm, which gives an NOE signal, overlapped with heme methyl signal M at 30 °C.

difference spectra with irradiation of the methyl proton at signals A, H, I, and J are presented in Fig. 11-12. On irradiation at the methyl proton at signal A, weak NOE signals were observed at the positions of heme methyl signal H. The NOE signal at about 9.6 ppm originates from a single-proton signal. The irradiation of the proton corresponding to signal H also gives a NOE signal at the position of signal A. As can be seen in Table 11-5, both signals A and H belong to heme 1. Because the intraheme NOE can be observed only between the methyl groups at 1 and 8 positions as shown in Table 11-2,[29] we can assign signals A and H to either 1 or 8 methyl groups of heme 1. On irradiation of proton at signal I, a strong NOE was observed at signal J and vice versa. Because there is no overlapping at signals I and J, and because they belong to hemes 1 and 3, respectively, we can conclude that this is the interheme NOE. The crystal structure of this protein shows that the shortest interheme methyl carbon distance is 0.417 nm (methyl 5 of heme I–methyl 1 of heme IV; Roman numbers are designations in the crystal structure) and the second shortest one is 0.534 nm (methyl 1 of heme II–methyl 5 of heme III).[8] The former is the only pair that provides a strong NOE signal. Because the NOE experiments showed that signal I is neither methyl 1 nor 8, signals I and J can be assigned to the methyl 5 of heme I and methyl 1 of heme IV in the crystal structure, respectively. This leads to the assignment of hemes 1 and 3 to I and IV in the crystal structure, respectively. This is consistent with that inferred on the basis of the crystal structure of heme groups and the results of electrochemistry[20, 21] but contradicts that by EPR,[30] which attributed heme 3 to heme II. The assignment of heme 1 is in accord with that for *D. desulfuricans* Norway cytochrome c_3 on the basis of chemical modification[31] and electrochemistry.

The NOE results show that the hemes with the highest and lowest potentials in the first reduction step (hemes 1 and 3) are the hemes I and IV in the crystal structure, respectively. Heme I has the least exposure to the solvent and is surrounded by the highest positive charge density among the four hemes.[8] A resonance Raman study showed the existence of vibrational exciton coupling among the four hemes of *D. vulgaris* Miyazaki cytochrome c_3.[32] It was suggested that repulsive forces play a significant role in the interactions in the fully oxidized state, whereas dispersion forces become more dominant in the fully reduced state. This was discussed in connection with the changes in the interiron distances induced by the reduction. Such change can be reflected in the positive interacting potentials. NMR results suggest, however, that the influence of the change is localized to two particular hemes. Such a specifically correlated pair would be expected to originate from a specific conformation of the two hemes in the three-dimensional structure.

It is known that the orientations of the four hemes in the crystal structures are almost identical for cytochrome c_3 from *D. vulgaris* Miyazaki and *D. desulfuricans* Norway, in spite of their low sequential homology. As previously mentioned, two hemes (hemes III and IV by the designation in *D. vulgaris* Miyazaki crystal structure) are located very close to each other at an almost right angle with an intervening phenylalanine, which is conserved in all cytochrome c_3 so far examined. A specific interaction between these hemes was anticipated by crystallographers. NOE results showed that heme 3 in the NMR spectrum is heme IV in the crystal structure, and the interacting potential I_{23} (or I_{32}) has a specific nature.[27] Although the assignment of heme 2 is not established yet, it is highly probable that the positive interheme interaction has something to do with the specific conformation of hemes III and IV in the crystal structure.

References

1. M. Ishimoto, J. Koyama, and Y. Nagai, *Bull Chem. Soc. Jpn.*, **27**, 564–565 (1954).
2. M. Ishimoto, J. Koyama, and Y. Nagai, *J. Biochem.* (*Tokyo*), **41**, 763–770 (1954).
3. M. Ishimoto, J. Koyama, T. Yagi, and M. Shiraki, *J. Biochem.* (*Tokyo*), **44**, 413–423 (1957).
4. J. R. Postgate, *Biochem. J.*, **56**, xi–xii (1954).
5. J. R. Postgate, *J. Gen. Microbiol.*, **14**, 545–572 (1956).
6. J. R. Postgate, *The Sulphate-Reducing Bacteria*, 2nd ed., 1976, Cambridge University Press, Cambridge, p. 76.
7. T. Yagi, H. Inokuchi, and K. Kimura, *Acc. Chem. Res.*, **16**, 2–7 (1983).
8. Y. Higuchi, M. Kusunoki, Y. Matsuura, N. Yasuoka, and M. Kakudo, *J. Mol. Biol.*, **172**, 109–139, (1984).
9. W. Shinkai, T. Hase, T. Yagi, and H. Matsubara, *J. Biochem.*, **87**, 1747–1756 (1980).
10. E. B. Trousil and L. L. Campbell, *J. Biol. Chem.*, **249**, 386 (1974).
11. R. P. Ambler, M. Bruschi, and J. Le Gall, *FEBS Lett.*, **5**, 115 (1969).
12. R. Haser, M. Pierrot, M. Frey, F. Payan, J. P. Astier, M. Bruschi, and J. Le Gall, *Nature*, **282**, 806–810 (1979).
13. R. P. Amber, *System Zool.*, **22**, 554 (1973).
14. K. Niki, T. Yagi, H. Inokuchi, and K. Kimura, *J. Am. Chem. Soc.*, **101**, 3335–3339 (1979);

k^0 has been determined to be 1.2 cm S^{-1}, T. Sagara, S. Nakajima, H. Akutsu, K. Niki and G. S. Wilson, *J. Electroanal. Chem.*, in press.

15. K. Niki, T. Yagi, H. Inokuchi, and K. Kimura, *J. Electrochem. Soc.*, **124**, 1889–1891 (1977).
16. K. Ono, K. Kimura, T. Yagi, and H. Inokuchi, *J. Chem. Phys.*, **63**, 1640–1642 (1975).
17. K. Kimura, Y. Nakahara, T. Yagi, and H. Inokuchi, *J. Chem. Phys.*, **70**, 3317–3323 (1979).
18. Y. Nakahara, K. Kimura, H. Inokuchi, and T. Yagi, *Chem. Phys. Lett.*, **73**, 31–34 (1980).
19. N. Sato, K. Kimura, H. Inokuchi, and T. Yagi, *Chem. Phys. Lett.*, **73**, 35–35 (1980).
20. K. Niki, K. Kawasaki, N. Nishimura, Y. Higuchi, N. Yasuoka, and M. Kakudo, *J. Electroanal. Chem.*, **168**, 275–286 (1984).
21. K. Niki, Y. Kobayashi, and H. Matsuda, *J. Electroanal. Chem.*, **178**, 333–341 (1984).
22. K. Niki, K. Kawasaki, Y. Kimura, Y. Higuchi, and N. Yasuoka, *Langmuir*, **3**, 982–986 (1987).
23. P. Bianco and J. Halajian, *Electrochimica Acta*, **26**, 1001 (1981).
24. K. Fan, H. Akutsu, K. Niki, N. Higuchi, and Y. Kyogoku, *J. Chem. Soc. Jpn.*, 512–517 (1988) (in Japanese).
25. K. Fan, H. Akutsu, K. Niki, N. Higuchi, and Y. Kyogoku, *J. Electroanal. Chem.*, **278**, 295–306 (1990).
26. J. D. Satterlee, *Annual Reports on NMR Spectroscopy*, Academic Press, London (1986), p. 79.
27. K. Fan, H. Akutsu, Y. Kyogoku, and K. Niki, *Biochemistry*, **29**, 2257–2263 (1990).
28. H. Santos, J. J. G. Moura, J. Le Gall, and A. V. Xavier, *Eur. J. Biochem.*, **141**, 283–296 (1984).
29. S. Ramaprasad, R. D. Johnson, and G. N. La Mar, *J. Am. Chem. Soc.*, **106**, 3632–3635 (1984).
30. J. P. Gayda, T. Yagi, H. Benosman, and P. Bertrand, *FEBS Lett.*, **217,** 57–61 (1987).
31. A. Dolla, C. Cambillau, P. Bianco, J. Haladjian, and M. Bruschi, *Biochem. Biophys. Res. Comm.*, **147**, 818–823 (1987).
32. A. L. Verma, K. Kimura, A. Nakamura, T. Yagi, and T. Kitagawa, *J. Am. Chem. Soc.*, **110**, 6617–6623 (1988).

Electrochemical Pulse Analysis of Multielectron Transfer in Electroactive Macromolecular Complexes

Noboru Oyama

Department of Applied Chemistry, Tokyo University of Agriculture and Technology, Koganei, Tokyo 184, Japan

Takeo Ohsaka

Department of Electronic Chemistry, Graduate School at Nagatsuta, Tokyo Institute of Technology, Nagatsuta, Midori-ku, Yokohama 227, Japan

12-1. Introduction

By introducing polymer complexes as thin films on electrode surfaces, polymer-modified electrodes (also referred to as polymer-coated electrodes) can be prepared and an electron-transfer reaction in the resulting polymeric domain can be quantitatively elucidated by various electrochemical procedures. During the past decade since the appearance of polymer-modified electrodes, much attention has been directed toward the investigation of electron-transfer processes at electroactive thin polymer film-coated electrodes from the viewpoints of possible applications, based on their interesting properties [e.g., electron-transfer catalysis (mediation), redox conductivity, electrochromic properties, and photoelectrochemical response], as well as fundamental studies on the mechanism and kinetics of electron transfer.[1] At polymer-modified electrodes where electroactive species (or sites) are confined in the polymeric matrix the overall electrode reaction consists of a "heterogeneous" electron-transfer process between the electrode and the electroactive species as well as a "homogeneous" charge-transport process within the polymer film, as schematically illustrated in Fig. 12-1. The latter process was believed to occur via physical diffusion of electroactive

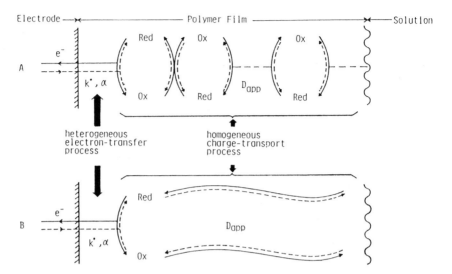

Figure 12-1 ■ Schematic depiction of the heterogeneous electron-transfer and homogeneous charge-transport processes at electrodes coated with polymer films containing electroactive species. The illustration represents two typical cases of the charge-transport process within electroactive films: charge is transported through polymer films (A) by electron transfer via an electron-hopping reaction between the electroactive species (or sites) and (B) by physical diffusion of the electroactive species themselves. k^0 and α are the standard rate constant and the transfer coefficient, respectively, of the heterogeneous electron-transfer process, and D_{app} is the apparent diffusion coefficient for the diffusion-like charge-transport process within the polymer films. Ox and red represent the oxidized and reduced species (or sites), respectively, in the polymer films. Note that the heterogeneous electron-transfer and homogeneous charge-transport processes are associated with charge-compensating counterion motion. —— anodic process; --- cathodic process.

species itself and/or electron self-exchange (electron hopping) between adjacent pairs of oxidized and reduced reactants that serves to shuttle electrons from the electrode to reactants that are located away from the electrode surface. Substantially, both processes are followed by the motion of a charge-compensating counterion that is necessarily coupled to electron transfer for charge neutrality, the motion of solvent molecules for solvation, the segmental motion of polymer chains, and other processes. By the interplay of these processes the electrochemical response of polymer films can be greatly complicated compared with the normal diffusion process in solution in the electrode reaction at bare electrodes. As a result, the voltammetric responses obtained are much more difficult to interpret and, in fact, a great deal of effort has been focused on unravelling these processes.[1] The heterogeneous electron-transfer process between the electrode and the electroactive species confined in the polymeric domain has also been successfully examined by conventional electrochemical methods.

In this chapter, we will present an application of electrochemical pulse methods to an analysis of the electron-transfer process in thin films of electroactive polymer complexes coated on electrodes. Moreover, some typical examples concerning the kinetic and thermodynamic studies of both heterogeneous and homogeneous electron-transfer processes at polymer-modified electrodes, which have been recently conducted in our laboratory, will be presented. Kinetic profiles of electron transfer in the macromolecular domain are characterized. The examples we discuss will be concerned not only with multielectron-transfer processes, but will serve to elucidate some of the results that have been obtained for multielectron-transfer processes.

12-2. Electrochemical Pulse Methods

Thus far a variety of electrochemical methods, including pulse (or step) methods of various types [potential-step chronoamperometry (PSCA), potential-step chronocoulometry (PSCC), chronopotentiometry (CP), normal pulse voltammetry (NPV), etc.], cyclic voltammetry, alternating current (AC) impedance method, rotating disk (and ring) electrode voltammetry, microstructured electrode-based methods (bilayer electrode, array electrode, etc.), and others (e.g., spectroelectrochemical method) have been widely applied to study electron-transfer processes at polymer-modified

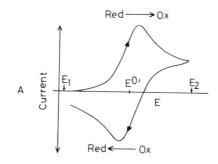

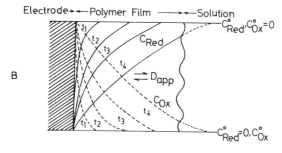

Figure 12-2 ■ (A) Cyclic voltammogram for a reversible electrode reaction red $\rightleftarrows$ ox + ne. $E^{0\prime}$ is the formal redox potential of red/ox couple. (B) A schematic time-dependent profile of the concentrations of fixed oxidized (C_{ox}, ---) and reduced (C_{red}, ——) sites in the films. The electrode potential is stepped from the potential (e.g., E_1) at which no oxidation of red occurs substantially to the potential (e.g., E_2) at which the oxidation of red is diffusion controlled. $t_1 < t_2 < t_3 < t_4$. C_{red}^0 is the bulk concentration of red confined initially in the film and C_{ox}^0 is the bulk concentration of the corresponding ox.

electrodes.[1] Here, among these, an application of PSCA, PSCC, CP, and NPV to an analysis of electron-transfer processes will be presented. These pulse (or step) methods have proved to be useful in the study of the kinetics of heterogeneous and homogeneous electron-transfer processes at polymer-modified electrodes.[1] The point to be noted is that the thickness of polymer films at polymer-coated electrodes is usually less than ca. 10 μm. For this reason, (a) the measurements should be confined to times sufficiently short to ensure that semiinfinite linear diffusion prevails or (b) in the analysis of the data obtained at a long time scale the contribution of finite diffusion must be considered. An apparent diffusion coefficient (D_{app}) usually has been employed as a kinetic parameter that characterizes the charge transport within polymeric films, because, in most cases, the homogeneous charge-transport processes have been found to obey Fick's diffusion law relations.[1,2]

A schematic depiction of time-dependent profiles of the concentrations of fixed oxidized (C_{ox}) and reduced (C_{red}) sites in the film is shown in Fig. 12-2. These profiles depend on the parameter $D_{app}t/l^2$, where t is the experimental time scale and l is the polymer film thickness. If $D_{app}t/l^2 \ll 1$ (or $\sqrt{D_{app}t} \ll l$, where $\sqrt{D_{app}t}$ approximately represents the thickness of the diffusion layer), a semiinfinite electrochemical charge diffusion condition prevails. At such a time scale, the conventional electrochemical analysis procedures are applicable. On the other hand, if $D_{app}t/l^2 > 1$, a finite diffusion must be considered.

For the case of PSCA, PSCC, and CP (Fig. 12-3), the finite diffusion relationships have been derived for a simple electrode reaction[3,4]:

$$\text{red} \rightleftharpoons \text{ox} + ne \tag{12-1}$$

in which the reduced half (red) of the redox couple is incorporated in the film of thickness l:

PSCA: $$i = \frac{nFAC_{red}^0 D_{app}^{a1/2}}{(\pi t)^{1/2}}\left[1 + 2\sum_{m=1}^{\infty}(-1)^m \exp\left(\frac{-m^2 l^2}{D_{app}^a t}\right)\right] \tag{12-2}$$

PSCC:

$$Q = \frac{2nFAC_{red}^0 D_{app}^{a1/2} t^{1/2}}{\pi^{1/2}}$$

$$\times \left\{1 + 2\sum_{m=1}^{\infty}(-1)^m\left[\exp\left(\frac{-m^2 l^2}{D_{app}^a t}\right) - \frac{ml\pi^{1/2}}{(D_{app}^a t)^{1/2}}\text{erfc}\left(\frac{ml}{(D_{app}^a t)^{1/2}}\right)\right]\right\} \tag{12-3}$$

CP:

$$i' = \frac{nFA(\pi D_{app}^a)^{1/2} C_{red}^0}{2\tau^{1/2}}$$

$$\times \left[1 + 2\sum_{m=1}^{\infty}\left\{\exp\left(\frac{-m^2 l^2}{D_{app}^a \tau}\right) - \frac{ml\pi^{1/2}}{(D_{app}^a \tau)^{1/2}}\text{erfc}\left(\frac{ml}{(D_{app}^a \tau)^{1/2}}\right)\right\}\right]^{-1} \tag{12-4}$$

where i is the current, Q is the charge, i' is the constant current, A is the electrode area, τ is the transition time, C_{red}^0 is the initial concentration of red in the film, D_{app}^a is the apparent diffusion coefficient for the anodic process, n is the number of electrons involved in the heterogeneous electron-transfer reaction, and F is the

Controlled
Variable

Dependent
Variable

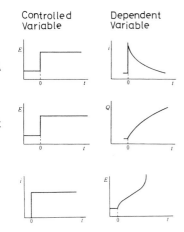

Figure 12-3 ■ Typical step wave forms applied in PSCA, PSCC, and CP and their response curves.

Faraday constant. In the case of the reduction, D_{app}^{a} and C_{red}^{0} in Eqs. (12-2)–(12-4) are replaced by the apparent diffusion coefficient for the cathodic process (D_{app}^{c}) and the concentration of ox in the film (C_{ox}^{0}), respectively. The leading terms on the right-hand side of Eqs. (12-2) through (12-4) correspond to the response that is obtained when the diffusion is semiinfinite: When $D_{app}^{a}t \ll l^{2}$, the second terms in the square bracket on the right-hand side of Eqs. (12-2) through (12-4) are negligible. The first terms of Eqs. (12-2) and (12-3) are well known as the Cottrell equation. The succeeding infinite series allows for the effects of finite diffusion within the film. PSCA and PSCC have been employed to estimate D_{app} for most films because of their experimental ease. In addition, IR drop losses may be prevented by addition of an appropriate voltage to ensure that the reactant at the electrode–film interface is consumed rapidly. Under these conditions film charge propagation is rate-limiting and a diffusionally limited response is observed.

NPV is also useful in the estimation of D_{app},[5–10] as can be seen from the fact that PSCA and NPV are based on the same principle.[11] Figure 12-4 shows the principle of this technique. The electrode is held at an initial potential (E_i), at which negligible electrolysis occurs. After a fixed waiting period τ' (0.5 ~ 180 s), the potential is changed abruptly to value E for a period about 1–100 ms in duration. The potential pulse is ended by a return to the initial value E_i. The current is sampled at a time τ_s near the end of the pulse and a signal proportional to this sampled value is presented as a constant output on a recorder until the sample taken in the next pulse step replaces it. The whole cycle is repeated. The step potential is made a few millivolts more extreme with additional cycle (i.e., $|E_{s,i} - E_{s,i+1}|$ is usually 2 ~ 20 mV). The output is a plot of sampled current versus step potential E and it takes the form schematically represented in Fig. 12-4(C).

At sufficiently short times when the depletion of the reactant has not reached the film–solution interface, that is, the thickness of the diffusion layer is much shorter than that of the polymer film coated on electrode, the normal pulse voltammetric Cottrell equation is obeyed:

$$(i_d)_{Cott} = nFAD_{app}^{a1/2}C_{red}^{0}/(\pi\tau_s)^{1/2} \tag{12-5}$$

where $(i_d)_{Cott}$ is the limiting current and τ_s is the sampling time. Equation (12-5) is

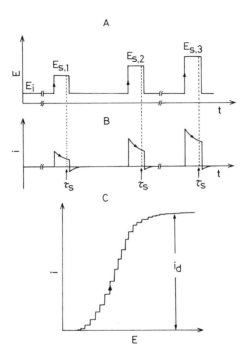

Figure 12-4 ▪ (A) Potential program of NPV. (B) Current response of NPV. (C) Normal pulse voltammogram (*i-E* curve). i_d represents the diffusion-controlled current.

equivalent to the conventional Cottrell equation that is expressed by the first term on the right-hand side of Eq. (12-2). The $(i_d)_{Cott}$ value can be easily obtained from the plateau portion of the voltammogram [see Fig. 12-4(C)]. A plot of $(i_d)_{Cott}$ versus τ_s, sometimes called a normal pulse voltammetric Cottrell plot, is linear as expected from Eq. (12-5) and thus the slope yields D^a_{app} with known n, A, and C^0_{red}.

NPV has also been successfully applied for the kinetic study of the heterogeneous electron-transfer process of various types of polymer-modified electrodes.[5–10] This process has been found to obey the conventional Butler–Volmer equation, and thus the relevant kinetic parameters (standard rate constant k^0 and transfer coefficient α) have been estimated from the analysis of the normal pulse voltammogram by using the current-potential relationship[12]

$$E = E^r_{1/2} \pm \frac{RT}{\alpha nF} \ln\left(\frac{4k^0\sqrt{\tau_s}}{\sqrt{3}\sqrt{D}}\right) \mp \frac{RT}{\alpha nF} \ln\left\{x\left(\frac{1.75 + x^2\left(1 + \exp(\mp\xi)^2\right)}{1 - x(\exp(\mp\xi))}\right)^{1/2}\right\}$$

$$(12\text{-}6)$$

with

$$x = i/(i_d)_{Cott} \qquad\qquad (12\text{-}7)$$

$$\xi = (nF/RT)\left(E - E^r_{1/2}\right) \qquad\qquad (12\text{-}8)$$

$$D = \left(D^a_{app}\right)^{\alpha a}\left(D^c_{app}\right)^{\alpha c} \qquad\qquad (12\text{-}9)$$

where the upper and lower signs $\pm$ or $\mp$ of Eq. (12-6) correspond to the reduction and oxidation, respectively, E is the electrode potential, i is the normal pulse voltammetric current, $E_{1/2}^r$ is the voltammetric reversible half-wave potential, and α denotes cathodic and anodic transfer coefficients (α_c and α_a denote the cathodic and anodic reactions, respectively). D_{app}^a and D_{app}^c are the apparent diffusion coefficients for anodic and cathodic processes, respectively, F is the Faraday constant, R is the gas constant, and T is the absolute temperature. $(i_d)_{Cott}$ is expressed by Eq. (12-5). Equation (12-6) is applicable for the case where the infinite diffusion conditions prevail. Positive feedback techniques are usually used to minimize the effects of film and solution resistance on the determination of k^0. The parameters (k^0 and α) can be determined from the intercept and slope of the plots of the logarithm of the third term on the right-hand side of Eq. (12-6) versus the electrode potential E, respectively. The facility to measure k^0, α, and D_{app} in one experiment and the apparent precision of the data indicate that this technique should be applied more widely.

12-3. Polyviologens

12-3-1. Pendant Viologen Polymers

NPV of PMV Films. Poly(styrene-*co*-chloromethylstyrene) pendant viologens (PMV) with various loadings ($x = 3 \sim 34\%$) of viologen sites are chemically stable and their electrochemical behavior is uncomplicated as long as the electrode potential is cycled within the range where the valence of the viologen site is changed between dication and monocation. In addition, their electron self-exchange reaction is relatively fast. Thus, in analogy with other polyviologen-coated electrodes, the PMV film-coated electrode is suitable as a model system for studying the electron transfer in a polymeric matrix[8, 9, 13]

PMV

Figure 12-5 shows typical normal pulse voltammograms for the one-electron reduction of the viologen dication (V^{2+}) to the corresponding radical monocation $V^{\cdot+}$ as PMV coated on basal-plane pyrolytic graphite (BPG) electrodes at various

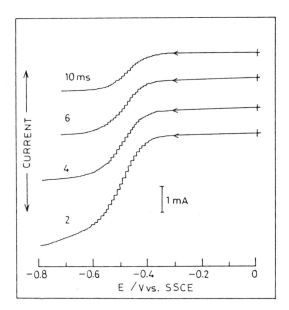

Figure 12-5 ■ Typical normal pulse voltammograms for the one-electron reduction of the viologen dication as PMV ($x = 18$, $y = 32$, $z = 50$) coated on BPG electrode at various sampling times in a 0.2 M KCl solution (pH 3.0). $C_{V^{2+}}^{obsd}$: 2.1×10^{-4} mol cm^{-3}; electrodes area: 0.17 cm^2. Sampling times (ms) are given on each voltammogram.

sampling times in a 0.2 M KCl solution (pH 3.0). These S-shaped voltammograms are similar to those observed for solution-phase redox species at an uncoated electrode and for other electroactive polymer-coated electrodes.[5-7,10] Plots of the cathodic limiting current (i_{lim}) of these normal pulse voltammograms against the inverse square root of the sampling time (τ_s) were found to be linear [see Fig. 12-6(A)], as expected for the diffusion-controlled limiting current. Thus, the values of D_{app} for the charge-transport process within the PMV films were obtained from the slopes of the i_{lim} versus $\tau_s^{-1/2}$ plots by using the normal pulse voltammetric Cottrell equation (12-5).

Figure 12-6(B) shows the typical examples of the plots of

$$\ln\left(x\left[\{1.75 + x^2(1 + \exp(\xi))^2\}/\{1 - x(1 + \exp(\xi))\}\right]^{1/2}\right) \quad \text{versus} \quad E$$

for the reduction of V^{2+} to $V^{\cdot+}$ as PMV coated on BPG electrodes at various sampling times. These plots correspond to the normal pulse voltammograms shown in Fig. 12-5. The plots gave the straight lines, the slopes of which were constant at the different sampling times ranging from 2 to 10 ms. Further, the potential E^* at the intersection of these straight lines with the abscissa shifted to more negative values with a decrease in τ. Thus, according to Eq. (12-6), from the slopes of the straight lines and the E^* potentials shown in Fig. 12-6(B), the values of α_c and k^0 were estimated by using the known values of $E_{1/2}^r$, D, and τ. The values of $E_{1/2}^r$ were

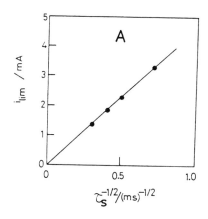

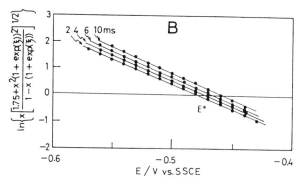

Figure 12-6 ■ (A) Plots of limiting current i_{lim} versus (sampling time)$^{-1/2}$ for the normal pulse voltammograms shown in Fig. 12-5. (B) Plots of $\ln(x[\{1.75 + x^2(1 + \exp(\xi))^2\}/\{1 - x(1 + \exp(\xi))\}]^{1/2})$ versus E for the normal pulse voltammograms shown in Fig. 12-5. Sampling times (ms) are indicated on each straight line.

estimated as the average of the anodic and cathodic peak potentials of the cyclic voltammograms for the oxidation-reduction of the $V^{2+}/V^{\cdot+}$ redox couple in PMV coatings on BPG electrodes in a supporting electrolytic solution.

The data for D_{app}, k^0, and α_c thus obtained are summarized in Fig. 12-7. Both k^0 and D_{app} increased with an increase in $C_{V^{2+}}^{obsd}$. For a series of PMV films with $z = 59\%$, as $C_{V^{2+}}^{obsd}$ were increased from 6.7×10^{-5} to 3.3×10^{-4} mol cm^{-3}, D_{app} increased from 2.5×10^{-11} to 3.9×10^{-10} cm^2 s^{-1} and k^0 increased from 3.0×10^{-5} to 8.9×10^{-5} cm s^{-1}. On the other hand, the α_c value, which is one of the kinetic parameters of the heterogeneous electron-transfer reaction, was almost independent of $C_{V^{2+}}^{obsd}$ and equals 0.42 ± 0.04. In the case of the PMV $(x = 34, y = 7,$ and $z = 59)$-Nafion and -poly(p-styrenesulfonate) (PSS) polymeric intermolecular complexes where the constant amount of PMV and the various amounts of Nafion (or PSS) were blended and coated on BPG electrode surfaces, the increase in k^0 and D_{app} with

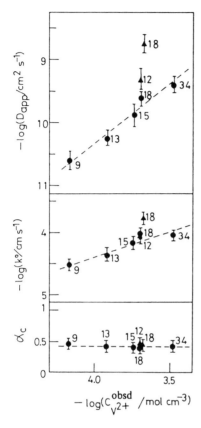

Figure 12-7 ▪ Dependences of D_{app}, k^0, and α_c on $C_{V^{2+}}^{obsd}$. The amount of each of the PMVs coated on electrodes was kept constant (2.9×10^{-5} g cm^{-2}). The fractions of styrene groups in PMVs (z) are ●, 59% and ▲, 50%.

increasing $C_{V^{2+}}^{obsd}$ has been observed.[9] When x is the same, the values of k^0 and D_{app} for the PMV films with different zs are different. For example, the values of k^0 and D_{app} for the PMV with $x = 18$, $y = 32$, and $z = 50$ are about 2 and 7 times larger than those for the PMV with $x = 18$, $y = 23$, and $z = 59$. These results suggest that the rates of the heterogeneous electron-transfer reaction at the electrode–PMV film interfaces and the homogeneous charge-transport process within the PMV films depend on the structure of the PMVs and more particularly on the morphology of the PMV films swollen in the supporting electrolytic solutions. As found for other redox polymer-coated electrode systems,[5-7,9,14] it was also found from Fig. 12-7 that there is a linear relationship between log k^0 and log D_{app}.

Charge-Transport Mechanism in PMV Films. Charge (electron or ion) transport through the solvent-swollen polymer films undergoing electrolysis [i.e., oxidation or reduction of incorporated electroactive species (or sites)] is generally believed to occur via an electron-hopping process between redox species and/or the physical diffusion of redox species themselves (which are temporarily confined in polymer domains).[3-10,13-26] Both processes will require, for charge neutrality, concurrent uptake of counterions into the polymer matrix or expulsion of co-ions initially present in the film as an ion pair. Thus, the charge transport will be determined by the intrinsic electron-transfer process between adjacent redox species, the physical diffusion of electroactive species themselves, the charge-compensating counterion motion that is necessarily coupled to electron transfer, the motion of solvent, and/or the segmental motion of polymeric chain. The charge-transport rate, which will be determined by only the slower of these various processes, is characterized by an effective diffusion coefficient for charge transport D_{app}, and thus the values of D_{app} for a number of redox polymer films have been measured.[1-10, 13-24] These previous data demonstrate that redox couples (as sites or species) incorporated in polymer and polyelectrolyte coatings on electrodes exhibit a much wider range of effective diffusion coefficients (over several orders of magnitude, 10^{-6}–10^{-14} cm^2 s^{-1}) than they do when dissolved in solutions.

With fixed-site electroactive polymers in a supporting electrolytic solution, electrons are reasonably considered to be transported through redox polymers by hopping between redox sites.[15] As pointed out by Buttry and Anson[19] and others,[8,9,14,17,27] in this situation the Dahms–Ruff electron-hopping charge-transport mechanism,[28,29] which was originally developed for the charge transport in solutions, can be applied. Because the electron-transfer sites are fixed, the effective charge-transport diffusion coefficient D_{app} is composed of contributions from actual diffusion governed by the diffusion coefficient D_0 and electron transfer (electron self-exchange) and is expressed by the equation[19]

$$D_{app} = D_0 + \frac{\pi}{4} k_{ex} \delta^2 C \qquad (12\text{-}10)$$

where k_{ex} is the second-order electron-exchange rate constant for the redox couple, δ is the distance between the sites when the electron transfer occurs, and C is the concentration of exchange sites (the sum of the concentrations of the oxidized and reduced forms of the redox couple). Rapid charge transport is therefore favored by proper alignment and high density of the exchange sites in the polymer matrix. The concentration dependence [such as that expected from Eq. (12-10)] of D_{app} has not been observed many times. Well behaved examples for such a concentration dependence of D_{app} are the cases of the transition-metal bipyridyl redox polymer films,[27] the Co(bpy)$_3^{2+/+}$ (bpy = 2,2'-bipyridine) confined in Nafion films,[19] and the polymeric intermolecular complexes composed of viologen polymer and Nafion or poly(p-styrenesulfonate).[9] In most cases where the redox species are electrostatically held in a polyelectrolyte film carrying the opposite charge, the increasing C introduces an additional effect: As C is increased, the degree of the electrostatic cross-linking between the redox sites (ions) and the opposite charged sites of polymer films increases, and as a result, the physical diffusion of the redox ion itself, the charge-compensating counterion (or co-ion) motion that is necessarily coupled to electron transfer, the motion of solvent, and/or the segmental motion of the polymer chain become slower. This results in the decrease of D_{app} with increasing C. In these cases it is considered that the electron self-exchange reactions still occur. However, the rate-determining step of the overall charge-transport process is not the electron-exchange process, but one of the other processes mentioned previously.

Thus, we must vary the redox site concentration without the extraneous influences on D_{app} caused by replacing the redox site with a dissimilar diluent site, because solvent swelling, cross-linking, and other aspects of internal polymer structure (as already mentioned) can thereby also be altered. The PMV system examined here was chosen to modify this condition. As expected, the D_{app} increased with an increase in C, although the obtained concentration dependence of D_{app} is not that expected from the Dahms–Ruff equation (see Fig. 12-7)[19,28,29] (i.e., the dependence of D_{app} on $C_{V^{2+}}^{obsd}$ was not completely linear). This finding demonstrates the significant contribution of electron self-exchange between viologen redox couples confined in the polymer chain to the overall charge transport in the PMV films.

The discrepancy between the obtained C dependence of D_{app} and that predicted from Eq. (12-10) probably arises because of the additional factors introduced by changing x in a series of the PMV films used. The changing x is not thought to lead only to the change of the concentration of the viologen site in PMV films. The viologen sites are more hydrophilic than the styrene and chloromethylstyrene moieties in PMV polymers. Thus, the larger x, the larger the degree of swelling of the PMV films. As the PMV films become more swollen, the motions of the PMV polymer chain itself, solvent, and counterion are considered to become easier. As a result of easier movement of the PMV polymer chains, the rate of the viologen site self-diffusive motion may increase. The rate of the self-diffusive motion of the viologen site is expressed as D_0 in Eq. (12-10). The increased swelling of the PMV films may also result in the increase in k_{ex}. On the basis of these considerations, it may be thought that D_0 and k_{ex} in Eq. (12-10) are functions of C. This reasoning seems to explain reasonably the actual observed C dependence of D_{app}.

The k_{ex} value can be estimated from the slope of the straight line drawn in the D_{app} versus C plot as a first approximation by assuming that D_{app} is proportional to C. The obtained values of k_{ex} are $(0.4–2) \times 10^5$ M^{-1} s^{-1} with the assumption of $\delta = 10$ Å. These values are much smaller than the electron self-exchange rate constant $(8 \times 10^6$ M^{-1} s$^{-1})$ of the monomeric V^{2+}/V$^{\cdot+}$ redox couple in the ordinary solution.[30]

Temperature Dependences of k^0, D_{app}, and the Formal Redox Potential. The temperature dependences of k^0, D_{app}, and the formal redox potential ($E^{0\prime}$) of the viologen dication/monocation radical redox couple (V^{2+}/V$^{\cdot+}$) were examined using a nonisothermal electrochemical cell.[13] The relevant activation parameters (i.e., the enthalpy ($\Delta H_{diffusion}^{\ddagger}$), entropy ($\Delta S_{diffusion}^{\ddagger}$), and free energy ($\Delta G_{diffusion}^{\ddagger}$) of activation for the diffusional charge-transport processes within the PMV films, and the enthalpy $[(\Delta H_{\ddagger}^{\circ})_{real}, (\Delta H_{\ddagger}^{\circ})_{ideal}]$ and entropy $[(\Delta S_{\ddagger}^{\circ})_{real}, (\Delta S_{\ddagger}^{\circ})_{ideal}]$ of activation for the heterogeneous electron-transfer reactions at the electrode/PMV film interfaces) and the reaction entropy (ΔS_{rc}°) of the V^{2+}/V$^{\cdot+}$ couple can be evaluated using the relations

$$\log D_{app} = \log D_0 - \frac{E_a}{2.303R} \frac{1}{T} \tag{12-11}$$

$$\log k^0 = \log \left\{ Z_{el} k \rho \exp \left[\frac{(\Delta S_{\ddagger}^{\circ})_{real}}{R} \right] \right\} - \frac{(\Delta H_{\ddagger}^{\circ})_{real}}{2.303R} \frac{1}{T} \tag{12-12}$$

$$\Delta S_{rc}^{\circ} \approx F \left(\frac{dE^{0\prime}}{dT} \right) \tag{12-13}$$

where E_a is the experimental energy of activation from which the activation enthalpy $(\Delta H_{diffusion}^{\ddagger})$ can be estimated by using $E_a = \Delta H_{diffusion}^{\ddagger} + RT$ and the preexponential factor (D_0) for the physical diffusion, according to Eyring,[31] can be written as $D_0 = e\lambda^2(kT/h)\exp(\Delta S_{diffusion}^{\ddagger}/R)$ (k is the Boltzman constant. h is the Planck constant, λ is the average distance between the equilibrium positions in the diffusion process), Z_{el} is the heterogeneous collision frequency, the temperature dependence of which was assumed to be only of minor importance quantitatively, κ and ρ are constants expected to be close to unity, and $(\Delta H_{\ddagger}^{\circ})_{real}$ and $(\Delta S_{\ddagger}^{\circ})_{real}$ represent the enthalpic and entropic barriers at the standard potential that remains after correction for the enthalpic and entropic driving forces, ΔH_{rc}° $(= T\Delta S_{rc}^{\circ})$ and ΔS_{rc}°, respectively, for the electrode reaction.

Further, the so-called ideal parameters $(\Delta H_{\ddagger}^{\circ})_{ideal}$ and $(\Delta S_{\ddagger}^{\circ})_{ideal}$, which are derived from the temperature dependence of the rate constant measured at a constant Galvani potential difference and equal the enthalpic and entropic barriers to electron transfer at the particular electrode potential at which they are evaluated (e.g., at the standard potential), can be related to $(\Delta H_{\ddagger}^{\circ})_{real}$ and $(\Delta S_{\ddagger}^{\circ})_{real}$, respectively, as[32]

$$\left(\Delta H_{\ddagger}^{\circ}\right)_{ideal} = \left(\Delta H_{\ddagger}^{\circ}\right)_{real} + \alpha T\Delta S_{rc}^{\circ} \tag{12-14}$$

$$\left(\Delta S_{\ddagger}^{\circ}\right)_{ideal} = \left(\Delta S_{\ddagger}^{\circ}\right)_{real} + \alpha \Delta S_{rc}^{\circ} \tag{12-15}$$

The results are summarized in Tables 12-1 and 12-2. The "reaction entropy" ΔS_{rc}° is an important parameter in considering the factors influencing the kinetics of electron-transfer processes, because it provides a sensitive monitor of the changes in solvent polarization that are necessary in order for electron transfer to occur. The positive values of ΔS_{rc}° were obtained for the $V^{2+}/V^{\cdot+}$ couple as the PMVs (see Table 12-2). This may suggest that the net solvent ordering in the vicinity of the viologen site is more extensive in the dicationic (V^{2+} state) than in the monocationic state ($V^{\cdot+}$ state) in qualitative agreement with the expectations from an electrostatic treatment such as the Born model.[33] From Table 12-2, we can also see that the ΔS_{rc}° values are almost the same (5.0 ± 0.5 cal mol^{-1} K^{-1}) for three kinds of PMVs with $x = 13, 15$, and 34%. Thus, the relative differences between the net solvent ordering in the vicinity of the oxidized site and that in the vicinity of the reduced site are considered to be almost constant independent of x.

The activation parameters for the homogeneous charge-transport process within the PMV films appear in Table 12-2. It is obvious that there are substantial differences in $\Delta H_{diffusion}^{\ddagger}$ and $\Delta S_{diffusion}^{\ddagger}$ for this series of PMVs. We have no information about the correct value of the mean jump distance per unit diffusion (λ). Here, assuming that $\lambda = 1$ and 10 Å, the values of $\Delta S_{diffusion}^{\ddagger}$ are estimated (different investigators have used values[34] ranging from 1 to 5 Å for λ). The negative values of $\Delta S_{diffusion}^{\ddagger}$ are to be noted. This fact may suggest that a substantial increase in the net solvent ordering occurs during the formation of the transition state in the diffusional charge-transport process. The $\Delta S_{diffusion}^{\ddagger}$ values for the PMVs are largely negative compared with those obtained for the protonated poly(4-vinylpyridine) films containing multiply charged anionic metal complexes[7] (Table 12-2). An important observation is that as x is increased, both $\Delta H_{diffusion}^{\ddagger}$ and $\Delta S_{diffusion}^{\ddagger}$ increase; in other words, $\Delta H_{diffusion}^{\ddagger}$ and $\Delta S_{diffusion}^{\ddagger}$ linearly correlate. This fact may give evidence of a common mechanism for the homogeneous charge-transport process within the PMV films with

Table 12-1 ■ Thermodynamic Parameters for the Heterogeneous Electron-Transfer Reaction of the $V^{2+}/V^{\cdot+}$ Couple as PMVs

$PMV(x, y, z)$	$C^{obsd}_{V^{2+}}$ (mol cm^{-3})	$(\Delta H^{o}_{\ddagger})_{real}$ (kcal mol^{-1})	$(\Delta S^{o}_{\ddagger})_{real}$ [a] (cal mol^{-1} K^{-1})	ΔS^{o}_{rc} [b] (cal mol^{-1} K^{-1})	$(\Delta H^{o}_{\ddagger})_{ideal}$ [c] (kcal mol^{-1})	$(\Delta S^{o}_{\ddagger})_{ideal}$ [d] (cal mol^{-1} K^{-1})
PMV(34, 7, 59)	3.3×10^{-4}	6.4	−15	5.3	7.1	−13
PMV(15, 26, 59)	1.8×10^{-4}	5.4	−19	5.0	6.0	−17
PMV(13, 28, 59)	1.2×10^{-4}	4.2	−24	4.5	4.7	−22

[a] Calculated by assuming that $\kappa\rho = 1$ and $Z_{el} = 1 \times 10^4$ cm s^{-1}.
[b] Reaction entropy of the $V^{2+}/V^{\cdot+}$ couple as PMVs.
[c] Calculated from Eq. (12-14) by using $\alpha_c = 0.42$ at 298 K.
[d] Calculated from Eq. (12-15) by using $\alpha_c = 0.42$.

Table 12-2 ■ Thermodynamic Parameters for the Diffusional Charge-Transport Process within Redox Polymer Films

Redox Species	$C^{obsd}_{V^{2+}}$ [a] (mol cm^{-3})	E_a (kcal mol^{-1})	$\log D_0$ (cm^2 s^{-1})	$\Delta H^{\ddagger}_{diffusion}$ (kcal mol^{-1})	$\Delta S^{\ddagger}_{diffusion}$ (cal mol^{-1} K^{-1})	$\Delta G^{\ddagger}_{diffusion}$ (kcal mol^{-1})
PMV(34, 7, 59)	3.3×10^{-4}	6.9	−4.4	6.3	-7.5^b -17^c	8.5^b
PMV(15, 26, 59)	1.8×10^{-4}	5.5	−5.8	4.9	-14^b -23^c	9.1^b
PMV(13, 28, 59)	1.2×10^{-4}	4.0	−7.3	3.4	-21^b -30^c	9.7^b
$W(CN)_8^{4-,d}$	2.2×10^{-4}	7.3 ± 0.7	−2.64	6.7 ± 0.7	0.61^b -8.5^c	6.5^b
$IrCl_6^{3-,d}$	4.5×10^{-5}	7.5 ± 0.5	−0.63	6.9 ± 0.5	9.8^b 0.65^c	4.0^b
$IrCl_6^{3-,d}$	4.8×10^{-4}	7.8 ± 0.5	−1.82	7.2 ± 0.5	4.3^b -4.8^c	5.9^b

[a] Concentration of electroactive species (or site) in polymer films on electrodes.
[b] Calculated by assuming $\lambda = 1$ Å (at 298 K).
[c] Calculated by assuming $\lambda = 10$ Å (at 298 K).
[d] Coating film: protonated poly(4-vinylpyridine) film; supporting electrolyte: 0.2 M CF$_3$COONa-CF$_3$COOH (pH 1.5).

different xs. The slope of this linear plot yielded an isokinetic temperature of about $-63°C$. At this temperature charge transport within all three PMVs occurs at the same rate. Above $-63°C$, $\Delta S^{\ddagger}_{diffusion}$ more strongly influences the rate, whereas $\Delta H^{\ddagger}_{diffusion}$ would dominate below $-63°C$. Thus the charge-transport rate is governed more entropically than enthalpically at room temperature.

From Table 12-1 we can clearly see that as x is increased, both $(\Delta H^{\circ}_{\ddagger})_{ideal}$ and $(\Delta S^{\circ}_{\ddagger})_{ideal}$ increase. Moreover, there is a linear relationship between $(\Delta H^{\circ}_{\ddagger})_{ideal}$ and $(\Delta S^{\circ}_{\ddagger})_{ideal}$ as in the above-mentioned $\Delta H^{\ddagger}_{diffusion}$-$\Delta S^{\ddagger}_{diffusion}$ relation. Both isokinetic plots give almost the same slopes and thus the same isokinetic temperature (ca. $-63°C$). This may be related to the fact that there is a linear relationship between $\log k^{0}$ and $\log D_{app}$. Also, $(\Delta H^{\circ}_{\ddagger})_{real}$ and $(\Delta S^{\circ}_{\ddagger})_{real}$ were found to correlate linearly. Negative values of $(\Delta S^{\circ}_{\ddagger})_{ideal}$ [and $(\Delta S^{\circ}_{\ddagger})_{real}$] are remarkable and may be attributable to the reasons common to the negative $\Delta S^{\ddagger}_{diffusion}$ values.

12-3-2. Polyalkyleneviologen/Poly(p-Styrenesulfonic Acid) Polymer Complexes

One- and Two-Electron Transfer Processes. The polymer complexes composed of poly(alkyleneviologen) (PAV) and poly(p-styrenesulfonic acid) (PSS), where the viologen units are joined by a varying number of methylene groups, are suitable as a simple model for the analysis of multielectron transfer reactions through polymer films coated on electrodes, because two couples of well defined reversible redox waves for the viologen dication/monocation radical $V^{2+}/V^{\cdot+}$ and the monocation radical/neutral species $V^{\cdot+}/V^{0}$ couples are observed at the PAV/PSS film-coated electrodes and two one-electron reductions of V^{2+} to $V^{\cdot+}$ and $V^{\cdot+}$ to V^{0} as well as two-electron reduction of V^{2+} to V^{0} can be easily examined by arbitrarily controlling the step potential in PSCA and PSCC.[35,36] The polymer complexes are highly cross-linked by the formation of a complex via electrostatic interactions between cationic viologen groups of PAV and anionic SO_3^- groups of PSS and this makes the polymer complexes of this type insoluble in aqueous solutions, though PAV or PSS itself is considerably water soluble.

m : 3 − 8

PAV/PSS

Figure 12-8(A) shows a typical cyclic voltammogram for the PAV($m = 8$)/PSS film coated on BPG electrode in 0.5 M Na_2SO_4 solution. Two couples of well defined reversible redox waves are observed: The first reduction wave corresponds to the

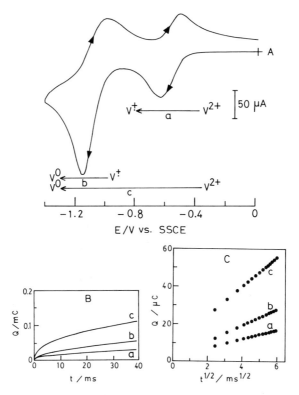

Figure 12-8 ■ (A) Cyclic voltammogram for the PAV($m = 8$)/PSS film coated on BPG electrode in 0.5 M Na$_2$SO$_4$ solution. Scan rate: 100 mV s^{-1}; concentration of viologen site (V^{2+}) in the film: 4.5×10^{-4} mol cm^{-3}; electrode area: 0.2 cm^2. (B) Typical potential-step chronocoulometric responses (Q-t curves) for (a) $V^{2+} \rightarrow V^{\cdot+}$, (b) $V^{\cdot+} \rightarrow V^0$, and (c) $V^{2+} \rightarrow V^0$ processes of the PAV($m = 8$)/PSS film used in (A). (C) Chronocoulometric Cottrell plots for the data in (B).

$V^{2+} \rightarrow V^{\cdot+}$ process and the second one the $V^{\cdot+} \rightarrow V^0$ process. The potential-step chronocoulometric responses (Q-t curves and Q-$t^{1/2}$ plots) for the $V^{2+} \rightarrow V^{\cdot+}$, $V^{\cdot+} \rightarrow V^0$, and $V^{2+} \rightarrow V^0$ processes are shown in Fig. 8(B) and (C). The linearity of Q-$t^{1/2}$ plots indicates that the charge-transport processes for these three reduction processes can be treated as Fickian diffusion processes and, thus, from the slopes of these plots the D_{app} values can be estimated (Table 12-3). For a given PAV/PSS system, the D_{app} values for $V^{2+} \rightarrow V^{\cdot+}$, $V^{\cdot+} \rightarrow V^0$, and $V^{2+} \rightarrow V^0$ processes decreased in the order

$$D_{app}(V^{\cdot+} \rightarrow V^0) > D_{app}(V^{2+} \rightarrow V^0) > D_{app}(V^{2+} \rightarrow V^{\cdot+})$$

suggesting the different electron self-exchange reaction rates between viologen sites in these processes.

Based on the relative values of D_{app} obtained for the three reduction processes, a possible mechanism of the charge-transport process for the two-electron reduction

Table 12-3 ■ **Apparent Diffusion Coefficients for the Charge-Transport Processes of PAV/PSS Films Coated on Electrodes in 0.5 M Na$_2$SO$_4$ Solution**

m	$D_{app}(V^{2+} \rightarrow V^{\cdot +})$ $(cm^2\ s^{-1})$	$D_{app}(V^{2+} \rightarrow V^{0})$ $(cm^2\ s^{-1})$	$D_{app}(V^{\cdot +} \rightarrow V^{0})$ $(cm^2\ s^{-1})$
4	$(2.0 \pm 0.3) \times 10^{-11}$	$(1.2 \pm 0.3) \times 10^{-9}$	$(3.0 \pm 0.4) \times 10^{-9}$
6	$(3.8 \pm 0.6) \times 10^{-11}$	$(3.5 \pm 0.4) \times 10^{-10}$	$(8.9 \pm 0.8) \times 10^{-10}$
7	$(5.3 \pm 2.1) \times 10^{-10}$	$(1.4 \pm 0.2) \times 10^{-9}$	$(1.8 \pm 0.2) \times 10^{-9}$
8	$(3.5 \pm 0.5) \times 10^{-11}$	$(1.2 \pm 0.2) \times 10^{-10}$	$(1.6 \pm 0.4) \times 10^{-10}$

within the PAV/PSS films may be proposed. When V^0 species are electrochemically generated in the complex films, the conproportionation reaction can take place:

$$V^{2+} + V^0 \rightleftharpoons 2V^{\cdot +} \tag{12-16}$$

The conproportionation constants (K_{conp}) for the PAV/PSS complexes can be estimated from the separation (ΔE_p) between the formal redox potentials of $V^{2+}/V^{\cdot +}$ and $V^{\cdot +}/V^0$ couples by using the equation

$$\Delta E_p = (RT/F)\ln K_{conp} \tag{12-17}$$

For the PAV($m = 8$)/PSS film, K_{conp} can be estimated as 4×10^8 from $\Delta E_p = 500$ mV, and other PAV/PSS films have similar values. Such large values of K_{conp} may suggest an ECE mechanism for the two-electron reduction as shown in Fig. 12-9. Thus, electrons are considered to be transported through the polymeric domain by one-electron hopping between $V^{\cdot +}$ and V^0 rather than by two-electron hopping between V^{2+} and V^0.

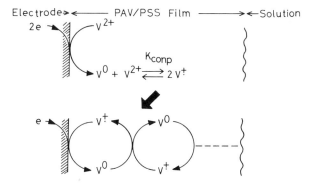

Figure 12-9 ■ A possible reaction scheme for the two-electron reduction process ($V^{2+} \rightarrow V^0$) of the PAV/PSS films.

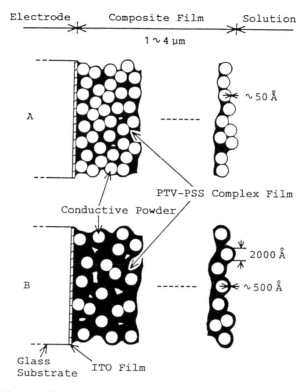

Electrode   Composite Film   Solution

$1 \sim 4$ μm

A

~ 50 Å

PTV–PSS Complex Film

Conductive Powder

B

2000 Å

~ 500 Å

Glass Substrate   ITO Film

Figure 12-10 ■ A schematic cross-sectional view of PAV($m = 4$)/PSS/conductive powder composite film-coated ITO electrodes. Conductive powder content x: (A) 86 wt%; (B) 37 wt%.

Effects of Introducing Conductive Powder into PAV/PSS Film on Electron Transfer. Figure 12-10 shows a schematic cross-sectional view of PAV/($m = 4$)/PSS/ conductive powder composite film-coated ITO electrodes.[37] By using the known average diameter of the conductive powder, 0.2 μm, and on the assumption that the PAV($m = 4$)/PSS complex films coat the surfaces of the conductive powders uniformly, the thicknesses of the PAV($m = 4$)/PSS complex films on the conductive powder were estimated from the amounts of the polymer and the powder blended. For example, the film thicknesses on the powders in the films of $x = 86$ and 37% were ~ 50 and 500 Å, respectively. In both cases the actually measured thickness of the composite films on the electrodes was ~ 4 μm. Figure 12-10 is based on these calculations and measured data.

From the dependence of the peak current i_p on the potential scan rate v for the composite films of various powder contents x, it was found that at a given v, i_p increases with increasing x and that the n values in the $i_p \propto v^n$ relation is equal to 0.5 at $x = 0$ and it increases almost linearly from 0.53 to 0.94 with x increasing from 37 to 86%.[37] In this experiment, composite films with almost the same thickness (4.0 ± 0.7

μm) were used. It is apparent from these results that as x is increased, the cyclic voltammetric behavior of the composite films comes close to that expected for a surface-confined redox species although the films used correspond to several hundreds of monolayers [estimated from the surface concentrations (~ 0.038–0.08 μmol/cm^2) of viologen sites in the films and a monolayer coverage of about 10^{-10} mol/cm^2]. Moreover, we found that the chronoamperometric Cottrell plots are not linear and are curved more for films of higher x. This suggests that the charge-transport processes within these films do not simply obey Fick's diffusion laws.

A probable explanation for these results is that as the conductive powder content x is increased, the effective electrode area and/or the apparent charge-transport rate might increase. The effective electrode area might increase with increasing x if the conductive powders make contact directly with the electrode and, in addition, are in contact with each other, because the powders are electronically conductive. Murray et al. reported that incorporation of small carbon particles in poly(N-vinylbenzyl-N'-methyl-4,4'-bipyridinum hexafluorophosphate) film raises the apparent diffusion coefficient.[38] They explained this by inserting a "roughness factor" into the Cottrell equation to represent the increased area of electrode/polymer interface or the decreased distance over which the charge has to diffuse.

As x is increased, the films can be expected to become more porous. As result, the rates of the charge-compensating counterion motion, the motion of solvent, and the segmental motion of polymeric chains might increase and, consequently, larger i_p values are obtained. For the films examined here, it was found that the concentrations C of the electroactive viologen site in the films decrease slightly with increasing x: C was 0.19, 0.17, 0.17, 0.22, 0.083, and 0.075 mol l^{-1} for the films of $x = 0$, 67, 75, 83, 91, and 95 wt%, respectively. This may be attributed to the increasing volume occupied by the conductive powder with increasing x. Such a decrease of C is considered to lead to a decrease in the contribution of the intrinsic electron-transfer reaction between the viologen sites to the overall charge-transport rate. However, the increase in x also leads to an alternative effect, as previously mentioned (i.e., the increases in the effective electrode area, the film porosity, and the contribution of "electronic conduction" to the overall charge transport). Consequently, as x increases, larger i_p values are obtained.

12-4. Polymeric Electrolytes Containing Electroactive Species

12-4-1. Anionic Perfluoro Polymer/Cationic Complex Systems

Perfluoro polysulfonate and polycarboxylate coatings have been extensively empolyed as electrode modifiers in preparing polymer-modified electrodes because of their chemical stability, strong adhesion to electrode surface, good swelling in aqueous solutions, water insolubility, reasonable cation-exchange capacity, and specific polymer domain.[14, 17, 19, 39] A heterogeneous electron-transfer process of [Os(bpy)$_3$]$^{2+/3+}$ (bpy: 2,2'-bipyridine) and [(trimethylammonio)methyl]ferrocene/ferricinium (TAF$^{+/2+}$) redox couples, which are confined in perfluoro polycarboxylate and polysulfonate polyelectrolyte coatings (abbreviated as PFPC and Nafion®, respectively) on graphite electrodes, at the electrode–film interfaces as well as homogeneous charge-transport process within the coatings are presented[40]:

Nafion   PFPC

$[Os(bpy)_3]^{2+}$ TAF$^+$

Figure 12-11 shows pH dependences of D_{app} and molar concentration (C_p^0) of the metal complexes incorporated in the coatings for $[Os(bpy)_3]^{2+}$-PFPC and TAF$^+$-PFPC systems. The volume concentration (C_p^0) of the $[Os(bpy)_3]^{2+}$ incorporated into the PFPC film increases from 1.4×10^{-5} to 3.2×10^{-4} mol cm^{-3} with increasing pH from 1 to 5.5, whereas C_p^0 of the incorporated TAF$^+$ is almost independent of pH in the range of pH ca. 2–5.5 and is ca. 2×10^{-4} mol cm^{-3}. This result for the $[Os(bpy)_3]^{2+}$ complex is explained by the equilibrium competition of protons and $[Os(bpy)_3]^{2+}$ ions for counterionic sites within the PFPC film. That is, at pH values as low as 1, protons compete quite successfully with $[Os(bpy)_3]^{2+}$ ions for counterionic sites within the PFPC, and with increasing pH $[Os(bpy)_3]^{2+}$ ions become dominantly trapped. The shape of the plot for the $[Os(bpy)_3]^{2+}$ resembles that of a typical pH titration curve. Thus, the effective pK_a of the carboxylic acid group in the PFPC is estimated to be ca. 2.9 from the pH at which the incorporation reaches one-half of its maximum value. This value is different from the 2.2 obtained from the incorporation experiments[39] of $[Ru(NH_3)_6]^{3+}$ into the PFPC and also from the pK_a of 1.9 measured directly for a similar perfluoro carboxylate polyelectrolyte.[41] As generally well known, the effective pK_a of polyelectrolytes varies with the degree of protonation and with the ionic strength.[42] These are considered as main reasons for the different effective pK_as.

On the other hand, the incorporation of TAF$^+$ complex changes slightly with pH, especially at pH > ca. 2. In this case, the effective pK_a, which is estimated from the C_p^0 versus pH curve by the same procedure as that used for the pK_a determination of the PFPC polyelectrolyte from the $[Os(bpy)_3]^{2+}$ incorporation experiments, is ca. 1, being clearly different from that obtained from the $[Os(bpy)_3]^{2+}$ incorporation experiments.

These different pH dependences of the incorporation of $[Os(bpy)_3]^{2+}$ and TAF$^+$ complexes into the PFPC film should be noted, because this result can be considered to suggest different interactions (between the complexes and the PFPC film) by which these complexes are incorporated into the film and confined in it. In the case of the

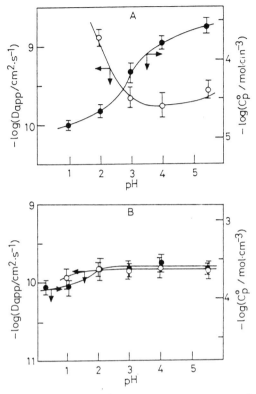

Figure 12-11 ■ pH dependences of C_p^0 and D_{app} for (A) [Os(bpy)$_3$]$^{2+}$-PFPC and (B) TAF$^+$-PFPC systems. The amount of the PFPC coatings is 4.2×10^{-5} g cm^{-2}. Supporting electrolyte: 0.2 M CF$_3$COONa. The values of D_{app} were estimated by PSCA and PSCC.

Nafion [or poly(p-styrenesulfonate)] film, such a pH dependence of the incorporation of the same complexes was not observed in the examined range of pH 1–5.5. In the case of the [Os(bpy)$_3$]$^{2+}$ complex, the incorporation can be considered to be mainly due to "electrostatic interactions"[3,5–7,18,19,22,39,42–47] between anionic carboxylic sites in the PFPC and cationic [Os(bpy)$_3$]$^{2+}$ complexes. On the other hand, in the case of the TAF$^+$ complex, there probably may be a minor contribution of electrostatic interactions to the incorporation. This is deducible from the fact that C_p^0 scarcely depends on pH. In this case, the incorporation seems to be mainly due to hydrophobic interaction.[19,39,48]

The data shown in Fig. 12-12 demonstrate the dependences of D_{app} on the redox complex confined in the film, its concentration (C_p^0), and the film confining it. These observed decreases of D_{app}s with increasing C_p^0 may be attributed to the effects of electrostatic (or hydrophilic) cross-linking[3,5–7,18,19,22,39,42–47,49] of the films by the electrostatic (or hydrophilic) interactions between the redox ions and the polyelectrolytes containing them (as previously mentioned) and/or of single-file

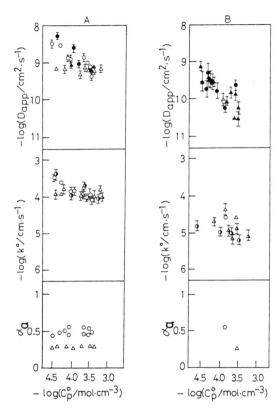

Figure 12-12 ▪ Dependences of D_{app}, k^0, and α_a on C_p^0. Coating films: (A) Nafion (thickness 4.3×10^{-5} cm); (B) PFPC (thickness 2.1×10^{-5} cm); supporting electrolyte: 0.2 M CF_3COONa + 0.1 M CH_3COONa/CH_3COOH (pH 5.5). ●, ◑, ○, $[Os(bpy)_3]^{2+}$; ▲, ◭, △, TAF^+. Methods used for the measurements: ●, ▲, PSCA and PSCC; ○, ◭, AC impedance method; ○, △, NPV.

diffusion.[19,39,50] An increase of the electrostatic cross-linking with increasing C_p^0 causes a decrease in the diffusion rate of the redox ion itself as well as decreases in the rates of the charge-compensating counterion motion, which is necessarily coupled to electron transfer, the motion of solvent, and/or the segmental motion of the polymeric chain. Also, the diffusing species that must move between more or less fixed sites within a polymeric matrix are considered to have their rate of motion limited by the decreasing availability of sites as C_p^0 of the diffusing species increases. These result in decrease of the overall rate of the charge transport with an increase in C_p^0, and thus the decreased D_{app}s are observed.

It appears that k^0 depends, though slightly, on C_p^0 (Fig. 12-12). The transfer coefficients were almost independent of the polymer films: The anodic transfer coefficients (α_a) for the oxidation of the $[Os(bpy)_3]^{2+}$ complex confined in the Nafion and PFPC films are 0.48 and 0.51, respectively, and those for the oxidation of the TAF^+ complex confined in the Nafion and PFPC films are 0.29 and 0.24, respectively.

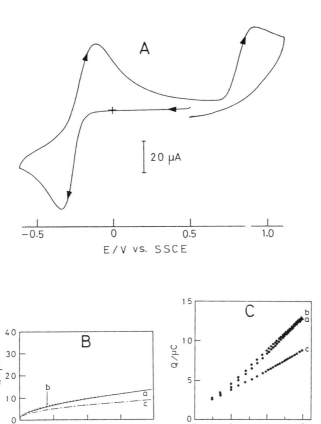

Figure 12-13 ▪ (A) A typical cyclic voltammogram for **1** confined in the CPFP film in 0.2 M CF$_3$COONa solution (pH 1.0) containing 0.1 mM of **1**. Electrode substrate: ITO 0.25 cm^2; CPFP film thickness 1.0 μm; concentration of **1** in the CPFP film: 0.46 M; scan rate: 50 mV s^{-1}. (B) Potential-step chronocoulometric responses for (a) the reduction of **1** to **2**, and the oxidations of (b) **2** to **1** and (c) **1** to **3** of ARS confined in the CPFP film coated on ITO electrode (0.25 cm^2) in 0.2 M CF$_3$COONa solution (pH 1.0) containing 0.01 mM ARS. Concentration of ARS in the CPFP film (thickness 1.0 μm): 35 mM. The electrode potentials were stepped from (a) 0.2 to -0.6, (b) -0.6 to 0.2, and (c) 0.2 to 1.3 V versus SSCE. (C) Chronocoulometric Cottrell plots for the data in (B).

12-4-2. Cationic Perfluoro Polymer/Alizarine Red S System

A variety of redox polymer film systems have been ever used as models for the fundamental elucidation of charge-transfer process within thin films on electrodes.[1,2] Among these, a system where a molecule with two separate electroactive centers that have different redox potentials is incorporated in a polyelectrolyte film on electrodes is of particular interest [as originally pointed out by Tsou and Anson[51] for the heterobinuclear metal complex $(NH_3)_5RuC_5H_4N$—$CH_2NHC(=O)$—$CpFeCp$ (C_5H_4N = 4-pyridyl, Cp = cyclopentadienide or substituted cyclopentadienide) within

Nafion coatings] because one could obtain new information on charge transport that is not obtained or is different from that obtained for cases where a kind of redox species is incorporated in a film or two kinds of redox species are simultaneously incorporated in it. The charge-transport kinetics for the oxidation and reduction of the two separate electroactive centers-containing molecule/thin polymer film system, that is, 9,10-dihydro-3,4-dihydroxy-9,10-dioxo-2-anthracenesulfonate **1** (which is often called alizarine red S dye and is abbreviated as ARS)-incorporating perfluoropolymer (CPFP) film system is presented.[52,53] ARS contains two separate electroactive centers, that is, 3,4-dihydroxy and 9,10-dioxo groups, and the electrode reactions of 3,4-dihydroxy and 9,10-dioxo groups are irreversible and reversible, respectively. Thus, if the electron exchange between electroactive centers contributes significantly to the overall charge-transport rates, we could observe different charge-transport rates for the oxidation (or reduction) process of these two kinds of the electroactive centers. The electroinactive functional group (i.e., SO_3^- group) serves as a counteranion site of the fixed cationic quaternized ammonio groups of CPFP and the net charge on ARS does not change before and after the oxidation or reduction of the electroactive centers.

Alizarine Red S

CPFP
$n/m = 6.5$

Figure 12-13(A) shows a typical cyclic voltammogram for **1** confined in the CPFP film on an ITO electrode in 0.2 M CF_3COONa solution (pH 1.0) containing 0.1 mM of **1**. In this case, the concentration (C_p) of **1** incorporated in the CPFP film was 0.46 M, and it was about 3 orders of magnitude larger than the concentration (0.1 mM) of **1** in the bathing solution in which the CPFP film-coated electrode was soaked, indicating that **1** can be concentrated into the CPFP film by an electrostatic interaction between the sulfonyl group of **1** and the quaternized ammonium site of the CPFP film. The reversible oxidation-reduction response at ca. -0.25 V versus a sodium chloride saturated calomel electrode (SSCE) corresponds to the two-electron reduction of **1** to 1,2-dihydroxyanthracene-9,10-diol-3-sulfonate **2**, and the oxidation of **2** to **1** (process I), whereas the irreversible response (anodic peak) at ca. 0.8 V versus SSCE corresponds to the oxidation of **1** to 1,2-dioxoanthracene-9,10-dione-3-sulfonate **3** (process II). These reactions are $2 H^+$-$2e^-$ reactions. The mechanism of these electrode reactions can be represented by the scheme Eq. 12-18

2   1   3   (12-18)

Potential-step chronoamperometric and chronocoulometric experiments were carried out in 0.2 M CF_3COONa solution (pH 1.0) in the presence of 0.1 mM of **1** (in order to hold C_p constant) within the experimental time scale (typically 2–40 ms) satisfying the conditions that the diffusion layer thickness is much less than the film thickness (1.0 μm). In Fig. 13(B) and (C) a typical example of potential-step chronocoulometric Cottrell plots is shown for the oxidations of **2** to **1** and **1** to **3** and the reduction of **1** to **2**. The point to be noted in this figure is that the slopes for the **1** to **2** and **2** to **1** processes are almost the same, but they are significantly larger than that for the **1** to **3** process. From these slopes, the D_{app} values for the **1** to **2**, **2** to **1**, and **1** to **3** processes [abbreviated as $D_{app}^{rev}(1 \rightarrow 2)$, $D_{app}^{rev}(2 \rightarrow 1)$, and $D_{app}^{irrev}(1 \rightarrow 3)$, respectively] were estimated to be $(1.5 \pm 0.2) \times 10^{-10}$, $(1.6 \pm 0.1) \times 10^{-10}$, and $(6.1 \pm 0.1) \times 10^{-11}$ cm^2 s^{-1}, respectively.

It is necessary for an understanding of the charge-transport mechanism to compare the D_{app} values with the diffusion coefficients D_s' corresponding to the motion of supporting electrolyte ions (i.e., CF_3COO^- and Na^+) that is necessarily (for charge neutrality) coupled to electron transfer. The D_s' values can be estimated approximately from the diffusion coefficient κD_s (κ is the distribution coefficient of an ion between solution and film) for film permeation of a dissolved redox species. For this purpose, I^- and $[Co(tpy)_2]^{2+}$ (tpy: 2,2',2''-terpyridine) are suitable, because these redox ions are electroactive in such a potential region (ca. 0 ~ 0.7 V versus SSCE) that the redox response of ARS is not substantially observed. The κD_s values were determined from the Koutecky–Levich plot of the steady-state current-potential curves obtained for the oxidation of $[Co(tpy)_2]^{2+}$ (or I^-) at a rotating-disk BPG electrode coated with the CPFP film incorporating ARS (**1**) in 0.2 M CF_3COONa solution (pH 1.0) containing **1** and $[Co(tpy)_2]^{2+}$ (or I^-) (Table 12-4). Note that the oxidation of I^- and $[Co(tpy)_2]^{2+}$ is not thermodynamically mediated by ARS incorporated in the CPFP film, as seen from the comparison of formal redox potentials $E^{0\prime}$ of ARS, $[Co(tpy)_2]^{2+/3+}$, and I_2/I^- couples, that is, the $E^{0\prime}$s of ARS (for process I), $[Co(tpy)_2]^{2+/3+}$, and I_2/I^- couples are -0.27, 0.035, and 0.44 V versus SSCE, respectively, in 0.2 M CF_3COONa aqueous solution (pH 1.0). The oxidation of I^- might be mediated by the I_2/I^- couple present within the CPFP film, because it is an anion-exchanging film and thus the trapping of I^- in the film is possible. However, this possibility may be ignored from the fact that after the above-mentioned rotating-disk voltammetric experiments, the electrode was transferred into a 0.2 M CF_3COONa

Table 12-4 ■ Comparison of D_{app}s for ARS with κD_ss for $[Co(tpy)_2]^{2+}$ and I^{-} [a]

C_p/M	D_{app}^{rev} [b] (cm^2 s^{-1})	D_{app}^{irrev} [b] (cm^2 s^{-1})	κD_s (cm^2 s^{-1})	
			$[Co(tpy)_2]^{2+}$ [c]	I^- [c]
0.035	$(5.6 \pm 0.5) \times 10^{-9}$ (red) $(5.7 \pm 0.5) \times 10^{-9}$ (ox)	$(2.0 \pm 0.2) \times 10^{-9}$ (ox)	$(3.7 \pm 0.6) \times 10^{-8}$	$(6.3 \pm 0.5) \times 10^{-6}$
0.46	$(1.5 \pm 0.2) \times 10^{-10}$ (red) $(1.6 \pm 0.1) \times 10^{-10}$ (ox)	$(6.1 \pm 0.1) \times 10^{-11}$ (ox)	$(2.6 \pm 0.4) \times 10^{-8}$	$(1.2 \pm 0.1) \times 10^{-6}$

[a]Supporting electrolyte: 0.2 M CF_3COONa (pH 1.0); CPFP film thickness: 1.0 μm.
[b]Ox: oxidation process; red: reduction process.
[c]The D_{soln} values of $[Co(tpy)_2]^{2+}$ and I^- in 0.2 M CF_3COONa solutions (pH 1.0) were estimated to be $(3.8 \pm 0.2) \times 10^{-6}$ and $(2.4 \pm 0.2) \times 10^{-5}$ cm^2 s^{-1}, respectively.

solution (pH 1.0) containing no ARS and I^- and the cyclic voltammetric response was examined as soon as possible. Consequently, we observed a well defined redox response of ARS, but no redox response corresponding to I_2/I^- couple was obtained. $[Co(tpy)_2]^{2+}$ is not confined in the polymer matrix.

As can be seen from Table 12-4, the diffusion coefficients for processes I and II and for the physical diffusion of I^- and $[Co(tpy)_2]^{2+}$ in the film are in the order: $\kappa D_s \gg D_{app}^{rev}(1 \to 2) \sim D_{app}^{rev}(2 \to 1) > D_{app}^{irrev}(1 \to 3)$, where $D_{app}^{rev}(1 \to 2)$, $D_{app}^{rev}(2 \to 1)$, and $D_{app}^{irrev}(1 \to 3)$ correspond to **1** to **2**, **2** to **1**, and **1** to **3** processes, respectively. The diffusion coefficients D_s' for the diffusion of supporting electrolyte ions used (Na^+ and CF_3COO^-) in the CPFP film have been unknown. The mean ionic diameters a of Na^+ and CF_3COO^- ions can be assumed to be between those of I^- and $[Co(tpy)_2]^{2+}$ ions: as of I^-, Na^+, CF_3COO^-, and $[Co(tpy)_2]^{2+}$ are roughly considered to be ca. 3, 4.5, 5, and >6 Å, respectively.[54] Thus, D_ss for I^- and $[Co(tpy)_2]^{2+}$ ions were obtained as a rough measure of D_s's for Na^+ and CF_3COO^- ions. Based on these considerations and the obtained values of κD_s, D_{app}^{rev}, and D_{app}^{irrev}, it can be expected that $D_s' \gg D_{app}^{rev} > D_{app}^{irrev}$. This suggests that the rate of the charge transport in the film is not substantially controlled by motion of the counterion (supporting electrolyte ion) in the polymer matrix that is coupled to electron transfer and molecular motion of ARS itself undergoing electrolysis.

The different values of D_{app}^{rev} and D_{app}^{irrev} seem to be instructive in understanding the charge-transport mechanism. In general, the D_{app} value can include contributions from both molecular motion (physical diffusion of reactant in film) and electron exchange between reactants, which will be represented as D_0 and D_{ex} (in units of centimeters squared per second), respectively[8, 19, 28, 29]:

$$D_{app} = D_0 + D_{ex} \qquad (12\text{-}19)$$

Equation (12-19) is identical to Eq. (12-10) and D_{ex} is expressed by

$$D_{ex} = (\pi/4)k_{ex}\delta^2 C_p \qquad (12\text{-}20)$$

where k_{ex} is the second-order electron-exchange rate constant for the redox couple, δ is the distance between the reactants when the electron transfer occurs, and C_p is the concentration of reactants (the sum of the concentrations of the oxidized and reduced forms of the redox couple). In **1** to **2** and **1** to **3** processes, the contribution (D_0) from molecular motion should be the same, because the same molecule takes part in both processes. Thus, based on Eq. (12-19) the difference in $D_{app}^{rev}(1 \to 2)$ and $D_{app}^{irrev}(1 \to 3)$ at a given C_p may be considered to originate from that in D_{ex}. As can be readily seen from Eq. (12-20), the degree of the contribution from electron exchange to the overall charge-transport rate depends on the electron self-exchange rate constant between electroactive reactants, that is, homogeneous electron self-exchange rate constant k_{ex}: The larger k_{ex} is, the larger is D_{ex}, and thus the larger is the contribution of electron exchange. Alternatively, k_{ex} can be correlated to the heterogeneous electron-transfer rate constant[55] k^0, that is, in a qualitative sense to the reversibility of the electrode reaction: For the redox system with large k_{ex}, the k^0 value is large. As seen from Fig. 13(A), process I is reversible, whereas process II is irreversible. Thus, the k^0 value for

Table 12-5 ■ **Kinetic Parameters of the Heterogeneous Electron-Transfer Process at the Electrode/Film Interfaces and the Homogeneous Charge-Transport Process in the Films for the Reduction Process of $Fe(CN)_6^{3-}$ Confined in Poly(N,N-Dialkyl Substituted Aniline) Films on BPG Electrodes[a]**

Film	$10^4 C$ (mol cm^{-3})	$10^9 D_{app}$ (cm^2 s^{-1})	$10^5 k^0$ (cm s^{-1})	α_c	$k^0 D_{app}^{-1/2}$ (s$^{-1/2}$)	Method
PDMA	2.5	1.6 ± 0.3	8.9 ± 1.8	0.31 ± 0.02	2.2	NPV
		2.5 ± 0.2				PSCA, PSCC
PMEA	2.7	2.8 ± 0.3	10 ± 5	0.31 ± 0.05	1.9	NPV
		5.8 ± 0.4				PSCA
		4.2 ± 0.2				PSCC
PDEA	2.3	11 ± 2	22 ± 5	0.26 ± 0.02	2.1	NPV
		20 ± 3				PSCA
		17 ± 2				PSCC
PVPH$^{+\,b}$	2.5	7.9	44	0.30	5.0	NPV
c		5000^d	150	0.48	0.67^e	NPV

[a] Supporting electrolyte: 0.2 M CF_3COONa-CF_3COOH (pH 1.0).
[b] Protonated poly(4-vinylpyridine).
[c] For the reduction of $Fe(CN)_6^{3-}$ at a bare BPG electrode in 0.2 M CF_3COONa-CF_3COOH solution (pH 1.0) containing 2 mM $Fe(CN)_6^{3-}$.
[d] Diffusion coefficient of $Fe(CN)_6^{3-}$ in the solution D_{soln}.
[e] Value of $k^0 D_{soln}^{-1/2}$.

process I should be larger than that for process II. Based on these considerations, the difference in D_{app} for processes I and II can be attributed to different contributions of electron exchange between electroactive centers to the overall charge-transport rates.

12-4-3. Poly(N,N-Dialkylaniline)/Anionic Metal Complex Systems

The electropolymerization of N,N-dialkylaniline derivatives such as N,N-dimethylaniline and N,N-diethylaniline leads to "ionene polymers" with quaternary ammonium sites in the polymeric backbone, and the resulting polymer films (the thicknesses are typically $10^{-5} \sim 10^{-3}$ cm) are electroinactive and insoluble in the organic solvents commonly used and H_2O.[43,44,56,57] In addition, these cationic polymer films have been found to have an anion-exchange character over a wide range of pH, irrespective of the pH in the solution.[57] Thus, these electropolymerized poly(N,N-dialkylaniline) films are promising as a new polymeric polyelectrolyte for the chemical modification of electrode surfaces.

Tables 12-5 and 12-6 summarize the kinetic parameters of the heterogeneous electron-transfer process at the electrode/film interfaces and of the homogeneous charge-transport process in the films for poly(N,N-dialkylaniline) film/multiply charged anionic metal complex systems. The data in Table 12-5 demonstrate that the D_{app}s are different for the different films incorporating $Fe(CN)_6^{3-}$ ions at almost the same concentrations of $Fe(CN)_6^{3-}$ in the films. The D_{app}s increase in the order: PDMA < PMEA < PVPH$^+$ < PDEA, where PVPH$^+$ is the protonated poly(4-vinyl-

Table 12-6 ▪ Kinetic Parameters of the Homogeneous Charge-Transport Process in the Films and of the Heterogeneous Electron-Transfer Process at the Electrode / Film Interfaces for the PDMA / Multiply Charged Anionic Metal Complex Systems[a]

Metal Complexes	Electrode Process[b]	$10^4 C$ (mol cm^{-3})	$10^{10} D_{app}$ (cm^2 s^{-1})	$10^5 k^0$ (cm s^{-1})	α_c	$k^0 D_{app}^{-1/2}$ (s$^{-1/2}$)	Method
W(CN)$_8^{4-/3-}$	ox	4.2	5.3 ± 1.1				PSCA
	ox	4.2	4.1 ± 0.8				PSCC
	ox	3.7	3.6 ± 0.8	7.0 ± 2.9	0.56 ± 0.03	3.7	NPV
	ox[d]	3.7	200	67	0.67	4.7	NPV
Mo(CN)$_8^{4-/3-}$	ox	3.2	6.8 ± 1.3	18 ± 3	0.33	6.9	NPV
	ox[d]	3.2	140	24	0.50	2.0	NPV
	ox[e]		f	50000	0.4	225	GDP[c]
Fe(CN)$_6^{4-/3-}$	ox	2.5	16 ± 3	8.9 ± 1.5	0.31 ± 0.62	2.2	NPV
	ox[d]	2.5	79	44	0.30	5.0	NPV
	red[g]		50000	150	0.48	0.67	NPV
	red[h]		f	1000		4.5	
Ru(CN)$_6^{4-/3-}$	ox	1.8	3.2 ± 0.7				PSCA
Os(CN)$_6^{4-/3-}$	ox	3.1	2.8 ± 0.6				PSCA
	ox	2.9	5.4 ± 1.1				NPV[i]
	red	2.7	3.0 ± 0.6				PSCA
IrCl$_6^{3-/2-}$	red	5.0	11 ± 3				PSCA
	red	5.0	9.2 ± 1.8				PSCC
	red[d]	5.0	79	37	0.67	4.2	NPV
	red[e]		f	50000	0.5	224	GDP[c]

[a] Supporting electrolyte: 0.2 M CF$_3$COONa-CF$_3$COOH (pH 1.0).
[b] Ox: oxidation precess; red: reduction process.
[c] GDP: galvanostatic double-pulse method.
[d] Coating film: protonated poly(4-vinylpyridine), PVPH$^+$.
[e] For the electrode reaction at a bare platinum electrode in solution.
[f] The diffusion coefficient in solution was taken as 5×10^{-6} cm^2 s^{-1}.
[g] For the electrode reaction at a bare BPG electrode in solution.
[h] For the electrode reaction at a bare gold electrode in solution.
[i] The electrode process was reversible in the time scale of the experiments (1–10 ms) and thu no kinetic parameters of the heterogeneous electron-transfer process were obtained.

pyridine). The D_{app} for the PDEA film is about 10 times larger than that for the PDMA film. This may be ascribed to a different extent of the electrostatic cross-linkings of the PDMA and PDEA films by Fe(CN)$_6^{3-}$ [and Fe(CN)$_6^{4-}$] ions. The Fe(CN)$_6^{3-}$ ions are more strongly confined electrostatically in the PDMA film than in the PDEA film, primarily reflecting the different steric bulkiness of the methyl and ethyl groups on the N atoms of the PDMA and PDEA films, respectively. Thus, the PDEA film is considered to be cross-linked to a smaller degree and thus to have less tight morphology than the PDMA film under conditions of the same concentrations C of the reactant confined in the films. As a result, the motions of Fe(CN)$_6^{3-}$ ions, counterions, solvents, polymeric chains, and so on, in the PDEA film are easier than those in the PDMA film. The order in the magnitude of the D_{app}s obtained for the PDMA, PMEA, and PDEA films and the different D_{app}s obtained for the PVPH$^+$ film and the poly(N,N-dialkyl substituted aniline) films may also be explained on the basis of the same idea as that previously mentioned:

PDMA R_1, R_2: CH_3
PMEA R_1: CH_3, R_2: C_2H_5
PDEA R_1, R_2: C_2H_5

PVPH$^+$

Furthermore, note that the diffusion coefficient D_{soln} and k^0 values for redox species in solution are typically 10^2–10^5 and 10–10^3 times higher than those D_{app} and k^0 for the same species in a polymer matrix, respectively (Table 12-6). The degree of decrease in D_{app} and k^0 for the polymers compared with those in solution changes with the polymers. However, the values of $k^0/D_{app}^{1/2}$ for the polymers are ca. 1 order of magnitude (i.e., 2–7) irrespective of the kind of polymer films used here. The relationship between k^0 and D_{app} is not predicted by the usual models for homogeneous electron-transfer reactions, when the reaction rate is not near diffusion control.

The smaller k^0s in the polymers may be explained by assuming that the active area of the electrode at which the heterogeneous electron-transfer reaction occurs is much smaller than the geometric electrode area. As discussed by Amatore, Savéant, and Tessier[58] for cyclic voltammetry, when the electron-transfer reaction occurs at a partially blocked electrode involving microscopic inhomogeneities, that is, involving numerous active sites of small size compared to the diffusion-layer dimensions, the electrochemical response is the same as that at a bare electrode, but with a decreased apparent standard rate constant of electron transfer. The value of k^0 measured under these conditions is the product of $(1 - \theta)$ and k^0 at the active sites, where θ is the surface coverage. If the actual k^0 at the active sites is the same as k^0 at the bare electrode, then this suggests that ca. 71 and 94% of the electrode surfaces are blocked in the PVPH$^+$-Fe(CN)$_6^{4-/3-}$ and the PDMA-Fe(CN)$_6^{4-/3-}$ systems, respectively. Thus, if the polymer film also causes a diminution in D_{app}, an apparent relation between the apparent k^0 and D_{app} can be found. In these cases, according to the treatment of Amatore, Savéant, and Tessier,[58] the distance $(2R_0)$ between the centers of two adjacent active sites and the size of the active site R_a are estimated to be not more than 3.4×10^{-6} and 9.3×10^{-7} cm, respectively, for the PVPH$^+$-Fe(CN)$_6^{4-/3-}$ system and not more than 7.0×10^{-7} and 8.5×10^{-8} cm, respectively, for the PDMA-Fe(CN)$_6^{4-/3-}$ system. The time scale of the NPV experiments in this study was typically 1–10 ms and so the thickness of the diffusion layer δ is estimated to be 5.0×10^{-6} to 1.6×10^{-5} cm using $D_{app} = 7.9 \times 10^{-9}$ cm^2 s^{-1} for the PVPH$^+$-Fe(CN)$_6^{4-/3-}$ system and 2.2×10^{-6} to 7.1×10^{-6} cm using $D_{app} = 1.6 \times 10^{-9}$ cm^2 s^{-1} for the PDMA-Fe(CN)$_6^{4-/3-}$ system. Thus, it is concluded that R_0s for these cases are smaller than the δs and that the diffusion layers of the various active sites intermingle. However, according to the same idea, we cannot explain the dependence of k^0 on C at the constant coating of the polymer, because we have no direct, nonelectrochemical evidence that changing C causes a variation of effective surface coverage.

Uncompensated resistance effects in the polymer films may be considered as another reason for the smaller k^0s. In this study, the conventional positive feedback techniques were employed to compensate the resistances associated with the polymer coatings as much as possible. Using these techniques, the residual uncompensated resistance can be decreased, though not completely, below typically ca. several Ω. Then the error in the values of k^0 is estimated to be at most $\pm 20\%$, which is almost comparable to the errors involved in the other measurements (e.g., the thickness measurement of the coating film) of the experiments. Thus, we expected that positive feedback techniques with correction for uncompensated resistance would minimize the effects of resistance on the determination of k^0.

The cathodic transfer coefficients α_c for the reduction of the $Fe(CN)_6^{3-}$ complex confined in four kinds of films (i.e., PDMA, PMEA, PDEA, and PVPH$^+$) on BPG electrodes are almost the same (0.26–0.31), but seem to be significantly different from the value 0.48 obtained for the reduction of the same complex at a bare BPG electrode in solution (see Table 12-5). The α_c values for the $W(CN)_8^{4-/3-}$ and $Mo(CN)_8^{4-/3-}$ couples appear to be different for the different films (Table 12-6). Further, it seems likely that the α_c (α_a) values for the $Fe(CN)_6^{4-/3-}$ and $Mo(CN)_8^{4-/3-}$ couples confined in the PDMA film are almost independent of C. In general, it is well known that the transfer coefficient is an important quantity characterizing the elementary process of an electrochemical reaction and offers useful information about the transition configuration during an electron-transfer process. The one thing to be noted from the present and previous data[5-9, 14, 40] is that the dependences of k^0 on C (or polymer films) are not always observed together with the dependences of α on C (or polymer films).

12-5. Electropolymerized Polymers

Electropolymerization has recently become of great interest from the viewpoints of the synthesis of new organic polymers and the chemical modification of electrode surfaces with a wide variety of functions (electrocatalysis, electrochromic property, redox activity, etc.). In order to elucidate these functions and further develop new devices based on electropolymerized film-coated electrodes with the desired characteristics of any particular application, it is necessary to clarify the charge- (ion and electron) transport processes on the polymer film-coated electrodes.

Figure 12-14 shows typical cyclic voltammetric responses of some electroactive films in 0.2 M $NaClO_4$ aqueous solution (pH 1.0). NPV, PSCA, and PSCC have been applied for the quantitative analysis of the overall electrode reactions of these films[10, 59-62] (Table 12-7). The results demonstrate that the electrode reactions for all these films can be analyzed in the same manner as those of a solution-phase redox species at an uncoated electrode: The heterogeneous electron-transfer process at the electrode/film interface obeys the conventional Butler–Volmer equation and the homogeneous charge-transport process within the film obeys Fick's law relations. The electropolymerized films examined are well swollen in an aqueous solution and an excess of supporting electrolytic ions are present within the films. Thus, these films are not considered to be completely electronically conductive materials under the experimental conditions used in this study (i.e., in an aqueous solution), so that the

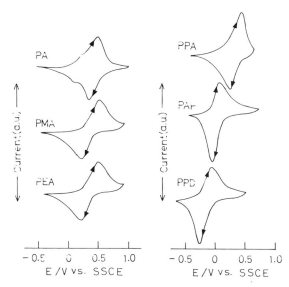

Figure 12-14 ■ Typical cyclic voltammetric responses of electroactive electropolymerized films in 0.2 M NaClO$_4$ aqueous solution (pH 1.0). PA: polyaniline; PMA: poly(N-methylaniline); PEA: poly(N-ethyaniline); PPA: poly(1-pyrenamine); PAP: poly(o-aminophenol); PPD: poly(o-phenylenediamine).

electrical double layer may be formed at the electrode/film interface and not at the film/solution interface, that is, a significant electrical field gradient is thought to be present only at the electrode/film interface where electron transfers to or from the electrode occur. Also, an electron transfer [e.g., via an electron hopping (or self-exchange) reaction] within the films should in fact be considered to occur. For this reason, the electroactive sites are fixed covalently in the polymeric backbones, so that the sites that are remote from the electrode surface cannot be oxidized and reduced unless successive electron transfers between the neighboring electroactive sites occur. Such an electron-transport mechanism in a thin film on electrodes via electron self-exchange reactions between adjacent oxidized and reduced polymer sites has been termed redox conduction.[63] Note that the charge-compensating counterion motion is necessarily coupled to the electron transfer. Thus, electronic and ionic conduction can be considered to occur concurrently during the redox reaction of these films.

The D_{app} value is a measure of the charge-transport rate within the film during the redox reaction of the polymer. As can be seen from Table 12-7, the values of D_{app}^a and D_{app}^c for PPD, PMA, PEA, and PA films are almost the same ($\sim 10^{-8}$ cm^2 s^{-1}). These D_{app} values are ca. 100 times larger than those of PPA, PAP, and polypyrrole films.[77] The D_{app}s depend not only on the kind of film but also on the solution (e.g., supporting electrolyte, solvent, and pH) used for their measurement.[62] For example, both D_{app}^a and D_{app}^c for PPD film decrease with increasing pH and the degree of the

Table 12-7 ■ Kinetic Parameters for the Electrode Reaction of Electropolymerized Films[a]

Film	$10^3 C^b$ (mol cm^{-3})	$10^8 (D_{app}^c)$ (cm^2 s^{-1})	$E^{0'}$ V vs. SSCE [c]	$10^4 k^0$ (cm s^{-1})	$\alpha_a (\alpha_c)$	Medium
PPD	0.78 ± 0.08	1.1 ± 0.1 (A)	−0.13 ± 0.01	5.8 ± 0.6	0.83 ± 0.03 (A)	0.2 M NaClO$_4$ + HClO$_4$ (pH 1.0)
		2.9 ± 0.2 (C)			0.23 ± 0.03 (C)	
PMA	1.1 ± 0.1	1.2 ± 0.1 (A)	0.37 ± 0.01	4.2 ± 0.8	0.86 ± 0.02 (A)	0.2 M NaClO$_4$ + HClO$_4$ (pH 1.0)
		1.5 ± 0.1 (C)			0.13 ± 0.02 (C)	
PEA	0.33 ± 0.03	4.2 ± 2.0 (A)	0.37 ± 0.01	5.1 ± 2.2	0.84 ± 0.02 (A)	0.2 M NaClO$_4$ + HClO$_4$ (pH 1.0)
		2.3 ± 1.1 (C)			0.16 ± 0.02 (C)	
PA	0.73 ± 0.07	0.94 ± 0.18 (C)	0.39 ± 0.01	2.5 ± 0.9	0.26 ± 0.05 (C)	0.2 M CF$_3$CCONa + CF$_3$COOH (pH 1.5)
PPA	0.97 ± 0.09	0.015 ± 0.005 (A)	0.24 ± 0.01	0.19 ± 0.04	0.67 ± 0.03 (A)	0.2 M NaClO$_4$ + HClO$_4$ (pH 1.0)
PAP	4.5 ± 0.5	0.013 ± 0.002 (A)	0.044 ± 0.002	1.1 ± 0.3	0.75 ± 0.02 (A)	0.2 M NaClO$_4$ + HClO$_4$ (pH 1.0)
		0.026 ± 0.003 (C)			0.21 ± 0.03 (C)	

[a] A and C are used for the anodic and cathodic processes, respectively.
[b] Concentration of electroactive sites in each film.
[c] The values of $E^{0'}$ were estimated as the average of the anodic and cathodic peak potentials of the cyclic voltammograms for the oxidation-reduction reaction of each film.

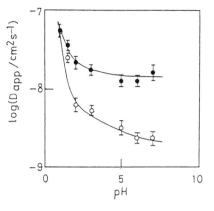

Figure 12-15 ∎ pH dependences of D_{app}^a and D_{app}^c of PPD film. ● for the oxidation process (D_{app}^a); ○ for the reduction process (D_{app}^c). Supporting electrolyte: 0.2 M NaCl. At pH < 2, no buffer solutions were used and at pH 3–7 a citric acid-disodium hydrogen phosphate buffer (20 mM) was used.

decrease is different for each (Fig. 12-15). The heterogeneous electron-transfer reaction of PPD in aqueous solutions may be written as

$$\text{(structure)} + n(2H^+ + 2e) \rightleftharpoons \text{(structure)}$$

$$(12\text{-}21)$$

The kinetic parameters of the heterogeneous electron-transfer process of the films are the basic and important data characterizing their electrochemical properties. The k^0 values are in the range of ca. $10^{-5} \sim 10^{-4}$ cm s^{-1}. The sum of α_a and α_c for a given film is almost equal to 1, as expected from the theory for a simple heterogeneous electron-transfer reaction.

Recently, electrochemical techniques (cyclic voltammetry, AC impedance method, etc.) other than the above-mentioned pulse methods also have been used to understand the charge-transport processes of conductive electropolymerized polymers. Feldberg[64] has expressed the anodic voltammogram of polythiophene as a sum of a faradaic surface wave due to the oxidation of the polymer and a capacitive current due to the charging of the double layer in the oxidized polymer. More recently, it has been suggested that in the case of polypyrrole film its anodic voltammogram could be decomposed into the bell-shaped curve and the sigmoid.[65] The former curve is ascribed to the oxidation of the polypyrrole film, whereas the latter may be ascribed to the charging current of the oxidized polypyrrole. The oxidized conductive film works as an electrode for oxidizing the reduced polymer. Then, the interface between the oxidized and the reduced films moves from substrate electrode up to the thickness of the polymer film.

A small amplitude current-pulse method and an AC impedance method have also been used for kinetic examination of the charge-transport processes at conductive polymer film-coated electrodes.[66,67,68] Because, unlike the conventional large amplitude pulse methods (e.g., PSCA and PSCC) for evaluating charge transport in thin films on electrode surfaces, these are small amplitude methods, the electrochemical, chemical, and morphological properties of the films are not perturbed by the analysis.[40,69] These methods should be more widely applied for evaluation of charge-transport processes in conductive polymer films on electrode surfaces.

12-6. Organized Macromolecules

Organized macromolecules such as Langmuir–Blodgett films and bilayer membranes have been receiving considerable interest in the fields of molecular devices and biomimetic chemistry. Synthetic bilayer membranes are two-dimensional molecular "organizates" and can thus be expected to play their unique role as organized media in chemical and/or physical processes. With this in mind we prepared the polyion complex films by electrostatically fixing PAVs on the regularly oriented anionic bilayer membrane $(2C_{12}SO_3)^{70,71}$:

$$
\begin{array}{c}
\quad\quad\;\; O \\
\quad\quad\;\; \| \\
C_{12}H_{25}O-C-CH_2 \\
\quad\quad\quad\quad\quad | \\
C_{12}H_{25}O-C-CH-SO_3^- \; Na^+ \\
\quad\quad\quad\;\; \| \\
\quad\quad\quad\;\; O
\end{array}
$$

$$2C_{12}SO_3$$

The absorption spectra of the reduced viologen polymers and the redox potentials as well as the charge-transport rates in the resulting polyion complex films have been found to strongly depend on the number m of the spacer methylene groups of PAVs: Blue shift in the absorption spectra due to the intramolecular association of the cation radicals and the dimer radical formation is observed for the odd polymers, whereas in the even polymers such a dimerization is not observed. The odd polymers are more easily reduced than the even polymers. The odd-even m dependence of the apparent diffusion coefficient D_{app} for the one-electron reduction of V^{2+} to $V^{\cdot+}$ is also observed[71] (Fig. 12-16). These observed odd-even effects seem to be associated with the fixed conformation of the viologen polymers ordered on the regularly charged surfaces of the bilayer membranes.[72] That is to say, a face-to-face orientation of the neighboring viologen groups within the linear polymer is probable for the odd polymers, whereas a head-to-tail orientation is probable for the even polymers. The face-to-face orientation is favorable for the intramolecular association, and thus dimerization of the cation radicals is dominant in the odd polymers. On the other hand, in the even polymers, radical–radical interaction is suppressed by the head-to-tail orientation. Organized macromolecules, as demonstrated by the above-mentioned synthetic bilayer membranes, can be expected as new polymeric matrices for the molecular ordering, and a systematic study concerning the electron-transfer

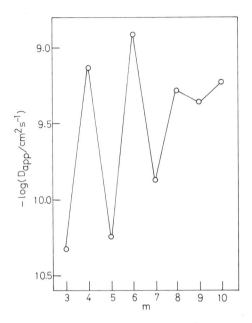

Figure 12-16 ■ Alkyl chain dependence of D_{app} for the $V^{2+} \rightarrow V^{\cdot+}$ process of the PAVs fixed on anionic bilayer membrane ($2C_{12}SO_3$).

process in such polymeric domains should thus be conducted for further clarification of this process.

12-7. A New Technique for Elucidation of Charge Transport in Thin Films: In Situ Quartz Crystal Microbalance Technique

As mentioned in the preceding sections, it has been generally understood that the overall rate of charge transport can be controlled by various processes, namely, by molecular motion of the reactant itself, electron hopping between adjacent localized redox sites, diffusion of counterions and solvents, and/or segmental motion of polymer chains. However, the separation of these processes from each other is a difficult problem, and much speculation can be found in the literature concerning this question. Recently, applications of the quartz crystal microbalance (QCM) technique for the study of the motion of counter- and co-ions and solvents in polymer matrices during the redox reaction of electroactive sites have been demonstrated by Buttry and co-workers for poly(vinylferrocene) (PVF) or polyaniline film-coated electrode systems[73,74] and more recently by other investigators for some electroactive polymer films.[75,76] This technique can be used to probe mass changes of the electrode surface that may be electrochemically induced and that have two significant attributes: ability to make the mass determination in situ, in conjunction with the electrochemical

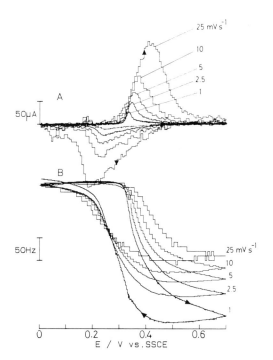

Figure 12-17 ■ (A) Cyclic voltammograms and (B) frequency-potential curves for PVF film-coated quartz crystal electrode in 0.2 M NaClO$_4$ solution. The amount of PVF film on the electrode is 17.3 μg cm^{-2}.

measurements, and excellent sensitivity, being capable of measuring mass changes corresponding to submonolayer adsorption and desorption (of the order of nanograms).

Figure 12-17 show typical cyclic voltammograms and frequency-potential curves for PVF film-coated quartz crystal electrode in 0.2 M NaClO$_4$ aqueous solution.[77] The frequency decreases during the oxidation (as a result of the insertion of anions into the film) and then reaches a constant value following the cessation of faradaic current flow, whereas on the return scan the frequency increases gradually and attains its original value (as a result of the release of the anions inserted during the anodic process). The mass change of the film can be calculated using the Sauerbrey equation[78]:

$$\Delta f = -C_f \Delta m \tag{12-22}$$

where Δf is the frequency change, Δm is the mass change per centimeter squared of the film, and C_f is the proportionality constant of the crystal.[79] Comparison of the charge and frequency changes for the oxidation of the film indicated that charge compensation (based on electroneutrality) is achieved by ClO$_4^-$ insertion with essentially no accompanying solvent. In addition, the frequency change for a given film has been found to largely depend on the film morphology (porous or dense films), the nature and concentration of supporting electrolyte, the kind of solvent, and so forth. The in situ QCM technique is a powerful tool for the study of electrochemical reactions that are accompanied by mass changes on the electrode surface as well as within thin films on the electrode surface and thus should become a widely used technique for unraveling charge-transport processes in polymeric films.

12-8. Conclusions

There have been many interesting developments in attempting to unravel the electron-transfer processes on polymer-coated electrodes, that is, heterogeneous electron-transfer processes at electrode/film interfaces and homogeneous charge-(ion and electron) transport processes within films. Electrochemical pulse (and step) methods such as PSCA, PSCC, and NPV have proved to be useful in a quantitative analysis of these processes. These advances have opened up whole new areas of investigation, that combine modern aspects of electrochemistry, polymer chemistry, organic chemistry, inorganic chemistry, biochemistry, physics, and so forth, and should prove very fruitful in future study.

Acknowledgments. Our work was generously supported by a Grant-in-Aid for Scientific Research on Priority Areas: Dynamic Interactions and Electronic Processes of Macromolecular Complexes from the Ministry of Education, Science and Culture, Japan.

References

1. Reviews: L. R. Faulkner, *Chem. Eng. News*, **62**, 28 (1984); R. W. Murray, in *Electroanalytical Chemistry*, A. J. Bard, ed., Marcel Dekker (1984), Vol. 13, p. 191; A. R. Hillman, in *Electrochemical Science and Technology*, R. G. Linford, ed., Elsevier Applied Science (1987), p. 103; H. D. Abruna, in *Electroresponsive Molecular and Polymeric Systems*, T. A. Skotheim, ed., Marcel Dekker (1988), Vol. 1, p. 97; M. Kaneko and D. Wöhrle, *Advances in Polymer Sciences*, **84**, Springer, Berlin (1988), p. 141.
2. N. Oyama and T. Ohsaka, in *Molecular Design of Electrode Surfaces*, R. W. Murray, ed., Wiley, New York, 1990.
3. N. Oyama, S. Yamaguchi, Y. Nishiki, K. Tokuda, H. Matsuda, and F. C. Anson, *J. Electroanal. Chem.*, **139**, 371 (1982).
4. P. Daum, J. R. Lenhard, D. Rolison, and R. W. Murray, *J. Am. Chem. Soc.*, **102**, 4649 (1980).
5. N. Oyama, T. Ohsaka, M. Kaneko, K. Sato, and H. Matsuda, *J. Am. Chem. Soc.*, **105**, 6003 (1983).
6. K. Sato, S. Yamaguchi, H. Matsuda, T. Ohsaka, and N. Oyama, *Bull. Chem. Soc. Jpn.*, **56**, 2004 (1983).
7. N. Oyama, T. Ohsaka, and T. Ushirogouchi, *J. Phys. Chem.*, **88**, 5274 (1984).
8. N. Oyama, T. Ohsaka, H. Yamamoto, and M. Kaneko, *J. Phys. Chem.*, **90**, 3850 (1986).
9. T. Ohsaka, N. Oyama, K. Sato, and H. Matsuda, *J. Electrochem. Soc.*, **132**, 1871 (1985).
10. T. Ohsaka, S. Kunimura, and N. Oyama, *Electrochim. Acta*, **33**, 639 (1988).

11. For example, A. J. Bard and L. R. Faulkner, *Electrochemical Methods, Fundamentals and Applications*, Wiley, New York (1980), Chap. 5.
12. H. Matsuda, *Bull. Chem. Soc. Jpn.*, **53**, 3439 (1980).
13. T. Ohsaka, H. Yamamoto, and N. Oyama, *J. Phys. Chem.*, **91**, 3775 (1987).
14. T. Ohsaka, H. Yamamoto, M. Kaneko, A. Yamada, M. Nakamura, S. Nakamura, and N. Oyama, *Bull. Chem. Soc. Jpn.*, **57**, 1844 (1984).
15. F. B. Kaufman and E. M. Engler, *J. Am. Chem. Soc.*, **101**, 547 (1979).
16. P. J. Peerce and A. J. Bard, *J. Electroanal. Chem.*, **112**, 97 (1980).
17. H. S. White, J. Leddy, and A. J. Bard, *J. Am. Chem. Soc.*, **104**, 4811 (1982).
18. N. Oyama and F. C. Anson, *J. Electrochem. Soc.*, **127**, 640 (1980).
19. D. A. Buttry and F. C. Anson, *J. Am. Chem. Soc.*, **105**, 685 (1983).
20. J. Facci and R. W. Murray, *J. Phys. Chem.*, **85**, 2870 (1981).
21. R. H. Schmehl and R. W. Murray, *J. Electroanal. Chem.*, **152**, 97 (1983).
22. K. Kuo and R. W. Murray, *J. Electroanal. Chem.*, **131**, 37 (1982).
23. M. Majda and L. R. Faulkner, *J. Electroanal. Chem.*, **169**, 77 (1984).
24. K. Doblhofer, W. Durr, and M. Jauch, *Electrochim. Acta*, **27**, 677 (1982).
25. C. P. Andrieux and J. M. Savéant, *J. Electroanal. Chem.*, **111**, 377 (1980).
26. E. Laviron, *J. Electroanal. Chem.*, **112**, 1 (1980).
27. J. Facci, R. H. Schmehl, and R. W. Murray, *J. Am. Chem. Soc.*, **104**, 4959 (1982).
28. I. Ruff, *Electrochim. Acta*, **15**, 1059 (1970).
29. H. Dahms, *J. Phys. Chem.*, **72**, 362 (1968).
30. J. C. Curtis, B. P. Sullivan, and T. J. Meyer, *Inorg. Chem.*, **19**, 3833 (1980).
31. H. Eyring, *J. Chem. Phys.*, **4**, 283 (1936).
32. J. T. Hupp and M. J. Weaver, *J. Electroanal. Chem.*, **145**, 43 (1983).
33. R. M. Noyes, *J. Am. Chem. Soc.*, **84**, 513 (1962).
34. S. Glasstone, K. J. Laidler, and E. Eyring, *The Theory of Rate Processes*, McGraw-Hill, New York (1941), p. 525.
35. T. Ohsaka, M. Nakanishi, O. Hatozaki, and N. Oyama, *Electrochim. Acta*, **35**, 63 (1990).
36. O. Hatozaki, T. Ohsaka, and N. Oyama, unpublished.
37. K. Nomura, K. Hirayama, T. Ohsaka, M. Nakanishi, O. Hatozaki, and N. Oyama, *J. Macromol. Sci.-Chem.*, **A26**, 593 (1989).
38. P. Burgmayer and R. W. Murray, *J. Electroanal. Chem.*, **135**, 335 (1982).
39. Y.-M. Tsou and F. C. Anson, *J. Electrochem. Soc.*, **131**, 595 (1984).
40. N. Oyama, T. Ohsaka, T. Ushirogouchi, S. Sanpei, and S. Nakamura, *Bull. Chem. Soc. Jpn.*, **61**, 3103 (1988).
41. Z. Twardowski, H. L. Yeager, and B. O'Dell, *J. Electrochem. Soc.*, **129**, 328 (1982).
42. H. Zumbrunnen and F. C. Anson, *J. Electroanal. Chem.*, **152**, 111 (1983).
43. N. Oyama, T. Ohsaka, and M. Nakanishi, *J. Macromol. Sci.-Chem.*, **A24**, 375 (1987).
44. T. Ohsaka, T. Okajima, and N. Oyama, *J. Electroanal. Chem.*, **215**, 191 (1986).
45. N. Oyama and F. C. Anson, *Anal. Chem.*, **52**, 1192 (1980).
46. F. C. Anson, T. Ohsaka, and J. M. Savéant, *J. Phys. Chem.*, **87**, 640 (1983).
47. N. Oyama, T. Shimomura, K. Shigehara, and F. C. Anson, *J. Electroanal. Chem.*, **112**, 271 (1980).
48. N. Oyama, T. Ohsaka, K. Sato, and H. Yamamoto, *Anal. Chem.*, **55**, 1429 (1983).
49. N. Oyama, T. Ohsaka, and T. Okajima, *Anal. Chem.*, **58**, 979 (1986).
50. K. Heckmann, in *Biomembranes*, A. L. Manson, ed., Plenum, New York (1972), Vol. 3, p. 127.
51. Y. M. Tsou and F. C. Anson, *J. Phys. Chem.*, **89**, 3818 (1985).
52. T. Ohsaka, N. Oyama, Y. Takahira, and S. Nakamura, *J. Electroanal. Chem.*, **247**, 339 (1988).
53. T. Ohsaka, Y. Takahira, O. Hatozaki, and N. Oyama, *Bull. Chem. Soc. Jpn.*, **62**, 1023 (1989).

54. J. Kielland, *J. Am. Chem. Soc.*, **59**, 1675 (1937).
55. R. A. Marcus, *Electrochim. Acta*, **13**, 995 (1968).
56. N. Oyama, T. Ohsaka, and T. Shimizu, *Anal. Chem.*, **57**, 1526 (1985).
57. T. Ohsaka, T. Okajima, and N. Oyama, *J. Electroanal. Chem.*, **200**, 159 (1986).
58. C. Amatore, J. M. Savéant, and D. Tessier, *J. Electroanal. Chem.*, **146**, 37 (1983).
59. N. Oyama, K. Hirabayashi, and T. Ohsaka, *Bull. Chem. Soc. Jpn.*, **59**, 2071 (1986).
60. N. Oyama and T. Ohsaka, *Synth. Met.*, **18**, 375 (1987).
61. K. Chiba, T. Ohsaka, and N. Oyama, *J. Electroanal. Chem.*, **217**, 239 (1987).
62. N. Oyama, T. Ohsaka, K. Chiba, and K. Takahashi, *Bull. Chem. Soc.*, **61**, 1095 (1988).
63. P. G. Pickup, W. Kutner, C. R. Leidner, and R. W. Murray, *J. Am. Chem. Soc.*, **106**, 1991 (1984).
64. S. W. Feldberg, *J. Am. Chem. Soc.*, **106**, 4671 (1984).
65. K. Aoki, Y. Tezuka, K. Shinozaki, and H. Sato, *Denki Kagaku*, **57**, 397 (1989).
66. R. M. Penner, L. S. Van Dyke, and C. R. Martin, *J. Phys. Chem.*, in press.
67. I. Rubinstein, E. Sabatani and J. Rishpon, *J. Electrochem. Soc.*, **134**, 3078 (1987).
68. T. Ohsaka, M. Hirata and N. Oyama, *Kobunshi Ronbunsyu*, in press (1990).
69. T. Ohsaka, T. Ushirogouchi and N. Oyama, *Bull. Chem. Soc. Jpn.*, **58**, 3252 (1985).
70. M. Shimomura, K. Utsugi, O. Hatozaki, and N. Oyama, *Langmiur*, in press.
71. O. Hatozaki, T. Ohsaka, M. Shimomura, and N. Oyama, unpublished.
72. M. Shimomura, K. Utsugi, and K. Okuyama, *J. Chem. Soc. Chem. Commun.*, 1805 (1986).
73. P. T. Varineau and D. A. Buttry, *J. Phys. Chem.*, **91**, 1292 (1987).
74. D. Orata and D. A. Buttry, *J. Am. Chem. Soc.*, **109**, 3574 (1987).
75. B. J. Feldman and O. R. Melroy, *J. Electroanal. Chem.*, **234**, 213 (1987).
76. M. R. Deakin and H. Byrd, *Anal. Chem.*, **61**, 290 (1989).
77. N. Oyama, M. Mizunuma and T. Ohsaka, unpublished.
78. G. Sauerbery, *Z. Phys.*, **155**, 206 (1959).
79. E. M. Genies, G. Biden, and A. F. Diaz, *J. Electroanal. Chem.*, **149**, 101 (1983).

Catalytic Activity and Multielectron Transfer of Metal Clusters

Akira Nakamura and Norikazu Ueyama

Department of Macromolecular Science, Faculty of Science, Osaka University, Toyonaka, Osaka 560, Japan

Biological reactions of carbon dioxide, dinitrogen, water, and hydrocarbon compounds require multielectron transfer during catalytic redox reaction by metalloenzymes. Recently various spectroscopic studies of the metalloenzymes have indicated that the electronic properties of the metal clusters involved are crucially important for better understanding of these catalytic reactions. Magnetic interaction between the active site and the surrounding metal centers has also been observed in many of the metalloenzymes.[1] Many active studies on the details of these electronic and magnetic interactions are now in progress worldwide.

First in this chapter, the recent studies of metalloenzymes, such as hydrogenase, nitrogenase, and aconitase having iron–sulfur clusters as suppliers of multielectrons, will be introduced. Next, the structural model complexes for the active site and their catalytic model reactions will be mentioned.

[4Fe—4S] clusters in adjacent proteins function as effective electron-transfer mediators between proteins with a suitable slope of redox potentials. The [4Fe—4S] model complexes are found to exhibit four redox couples as shown[2] in Scheme 1. Native bacterial ferredoxins employ the $3-/2-$ redox couple at -0.3 to -0.5 V versus NHE in the biological system.[3] High potential iron–sulfur protein has a $2-/1-$ redox couple at $+0.35$ V NHE.[4] The $3-/2-$ redox potential is known to be influenced by the dielectric properties of the solvent, including pH dependence.[5] In the native ferredoxin a specific solvation as well as NH $\cdots$ S hydrogen bonding are proposed to regulate the $3-/2-$ redox states.[6] The importance of NH $\cdots$ S hydrogen bonding has been established by our study on the redox behavior of a peptide model complex $[Fe_4S_4(Z\text{-cys-Gly-Ala-OMe})_4]^{2-}$ in dichloromethane.[7] Such an interaction is supported by hydrophobic environments. On the other hand, stabilization of the $2-/1-$ redox couple can be realized by various bulky thiolate ligands.[8,9] The hydrophobic environment realized in the complexes with these bulky ligands is found to prevent the $[Fe_4S_4]^{3+}$ core from hydrolysis and to stabilize the couple.

$$[Fe_4S_4(SR)_4]^- \underset{-e^-}{\overset{e^-}{\rightleftarrows}} [Fe_4S_4(SR)_4]^{2-} \underset{-e^-}{\overset{e^-}{\rightleftarrows}} [Fe_4S_4(SR)_4]^{3-} \underset{-e^-}{\overset{e^-}{\rightleftarrows}} [Fe_4S_4(SR)_4]^{4-}$$

Scheme 1

Thus, studies on the model complexes with simple ligands and oligopeptide ligands can help in understanding various weak interactions between metal ions and parts of the peptide chain besides the ligand atoms in the active site.

13-1. Metalloenzymes Having an $[Fe_4S_4]^{2+}$ Core in the Active Site

Aconitase is one of the important enzymes in the citric acid cycle. The enzyme catalyzes isomerization of citrate with dehydration via cis-aconitate as shown in the equation

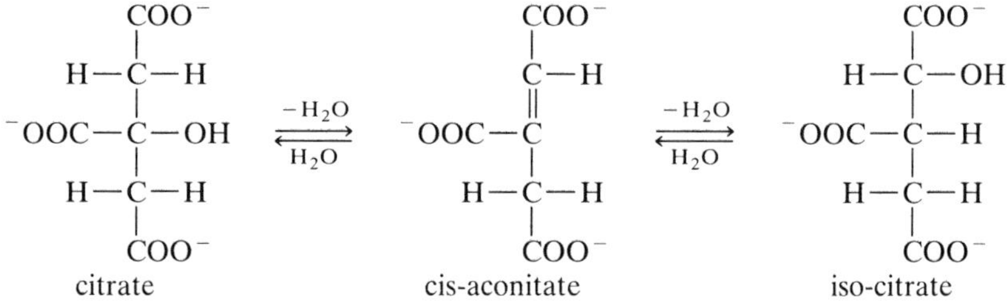

$$(13-1)$$

The requirement of Fe(II) ions for the reconstitution of the active site has led investigators to expect the presence of both mononuclear Fe(II) and the iron–sulfur cluster.[10] Recently, Beinert and co-workers have found the active site to be a $[Fe_4S_4]^+$ core that readily interconverts to $[Fe_3S_4]^+$ via $[Fe_4S_4]^{3+}$ by one-electron oxidation.[11] In this process, one Fe(II) ion of $[Fe_4S_4]^{3+}$ is lost from the core. The [3Fe—4S] core in the oxidized $[Fe_3S_4]^{3+}$ form is known to be inactive as a catalyst (Fig. 13-1).

An isotopic iron atom, ^{56}Fe or ^{57}Fe, can be incorporated into the [3Fe—4S] core in place of the labile Fe_a. Mössbauer analysis of the active site reveals that citrate interacts only with Fe_a but not with the [3Fe—4S] unit.[12] Thus, interconversion of the iron–sulfur cluster of aconitase is summarized in Scheme 2. The observed incorporation of the active site in an iron–sulfur core is extremely interesting because it implies the development of new redox catalysis.

Recently, the structure of activated pig heart aconitase containing a [4Fe—4S] core has been refined at 0.25 nm resolution. The structure was compared with the 0.21 nm resolution structure of the inactive form containing a [3Fe—4S] core.[13] Figure 13-2 shows the stereoviews of the active site residues and Fe—S core in both states of aconitase. The fourth ligand of the Fe_a is water or a hydroxyl group. The results are consistent with those obtained by the electron nuclear double resonance spectroscopic measurements using ^{17}O- and ^{13}C-substituted compounds.[14]

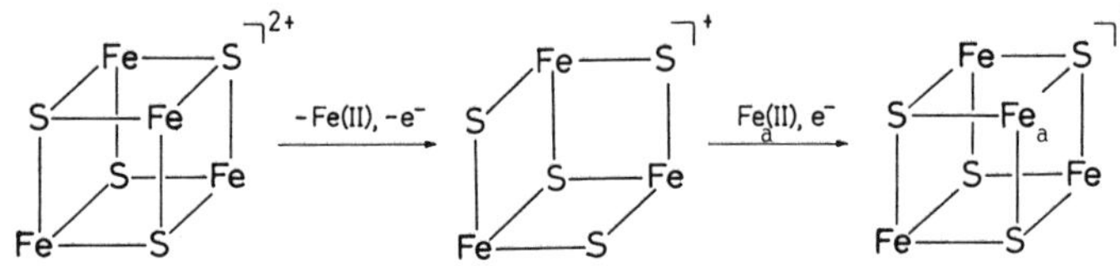

Figure 13-1 ■ Interconversion between [4Fe—4S] and [3Fe—4S] clusters. (Reproduced from ref. 12. Copyright 1983.)

Inactive aconitase

$$[Fe_3S_4]^+ \underset{-e^-}{\overset{e^-}{\rightleftarrows}} [Fe_3S_4]$$

Active aconitase

$$S = 1/2 \qquad\qquad S = 2$$

$$\longrightarrow Fe(II) \longrightarrow$$

$$[Fe_4S_4]^{3+} \underset{-e^-}{\overset{e^-}{\rightleftarrows}} [Fe_4S_4]^{2+} \underset{-e^-}{\overset{e^-}{\rightleftarrows}} [Fe_4S_4]^+$$

$$S = 0 \qquad\qquad S = 1/2$$

Scheme 2

Complexation of nitroisocitrate to the reduced activated aconitase results in simultaneous binding of OH^- group and $H_x^{17}O$ ($x = 1$ or 2).[15] An amino acid residue in a suitable position to act as a ligand, in particular a Cys residue, is lacking. This was also supported by Cys titration and labeling experiments.[16,17] The crystallographic analysis has suggested that one of the three His residues (102, 148, or 168) or Asp residue (166) could serve to abstract a proton from water. They emphasized the

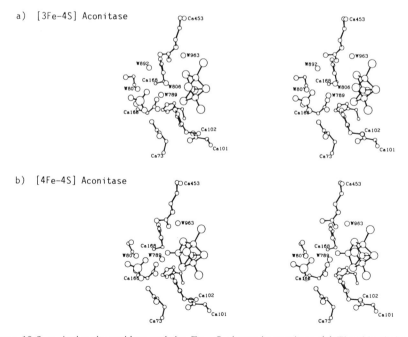

Figure 13-2 ■ Active site residues and the Fe—S cluster in aconitase. (a) The side chains of eight active site residues in the immediate vicinity of the [3Fe—4S](S$_\gamma$)$_3$ cluster, the bound SO_4^{2-} ion, and five water molecules from the 0.21 nm resolution refinement. (b) The amino acid side chains in the [4Fe—4S](S$_\gamma$)$_3$ cluster and four water molecules in the active site in common with the inactive structure (a). (Reproduced from ref. 13. Copyright 1989.)

isomorphism of the active and inactive structures. Interconversion between $[3Fe—4S]$ and $[4Fe—4S]$ can occur without rearrangement of any amino acid residues.

The other important enzymes containing the iron–sulfur cluster in the active site are fumarate reductase and succinate dehydrogenase. Fumarate reductase from *Escherichia coli* is composed of four nonidentical subunits that catalyze the final step in anaerobic respiration as shown in

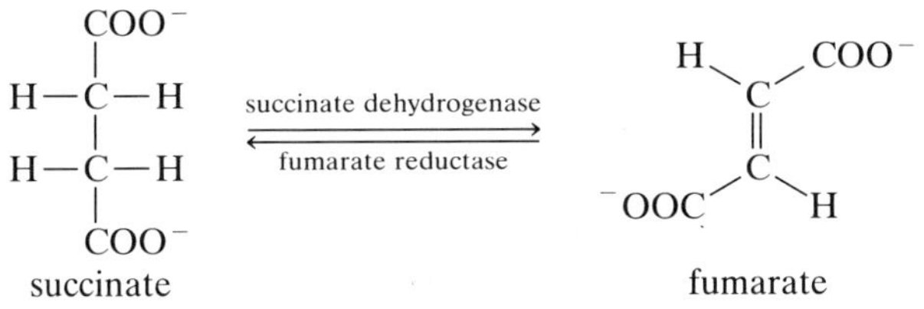

$$
\begin{array}{ccc}
\underset{\text{succinate}}{\begin{array}{c} COO^- \\ | \\ H—C—H \\ | \\ H—C—H \\ | \\ COO^- \end{array}}
&
\underset{\substack{\text{fumarate reductase}}}{\overset{\substack{\text{succinate dehydrogenase}}}{\rightleftharpoons}}
&
\underset{\text{fumarate}}{\begin{array}{c} H \qquad COO^- \\ \diagdown C \diagup \\ \| \\ \diagup C \diagdown \\ {}^-OOC \qquad H \end{array}}
\end{array}
\qquad (13\text{-}2)
$$

The enzyme consists of a flavoprotein subunit (66 kDa) and an iron–sulfur protein subunit (27 kDa).[18, 19] The enzyme is found to contain one FAD and three iron–sulfur clusters having $[3Fe—4S]^{+,0}(E_m = -70$ mV versus NHE$)/[2Fe—2S]^{2+,+}(E_m = -20$ mV versus NHE$)/[4Fe—4S]^{2+,+}(E_m = -320$ mV versus NHE$)$ in a (1:1:1) ratio.[20, 21]

13-2. Hydrogenase

Hydrogenases are found in bacteria, several algae, and a few fungi that catalyze the reactions

$$H_2 \longrightarrow 2H^+ + 2e^- \qquad (13\text{-}3)$$

$$H_2 + {}^2H_2 \longrightarrow 2H\,{}^2H \qquad (13\text{-}4)$$

At least three types of hydrogenases have been reported in hydrogen metabolism. One of them, which consists of only iron–sulfur clusters, is called Fe-hydrogenase and it was isolated from *Clostridium pasteurianum*,[22] *Desulfovibrio vulgaris*,[23] or *Megasphaera elsdenii*.[24] The enzyme has a $[12Fe—12S]$ composition consisting of two ferredoxin-like $[4Fe—4S]$ cores and one novel iron–sulfur cluster (H$_2$-activating center, H center) that exhibits a rhombic EPR signal on oxidation.[25] Still no unified structure has been proposed for the H center. Mössbauer, ENDOR, and EPR studies proposed that the H center has an $S = 1/2$ spin state in the oxidized form and an $S = 0$ state in the reduced form.[26-28] The H center in oxidized Fe-hydrogenase consists of a low spin ($S = 0$) Fe(II) site linked by a bridging ligand X to a novel $Fe—S$ cluster with an $S = 1/2$ ground state as proposed in Fig. 13-3.[29] Dihydrogen binding and activation occur at the low spin Fe(II) center.

The second type of hydrogenase, NiFe-hydrogenase, contains Ni and iron–sulfur clusters, for example, *Desulfovibrio gigas* hydrogenase. The enzyme has a

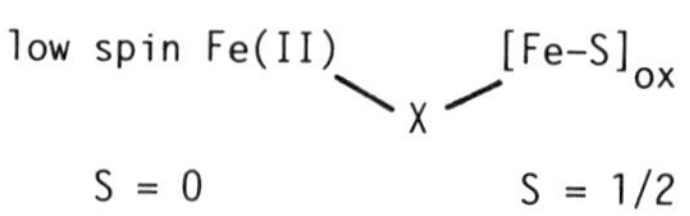

low spin Fe(II)   [Fe-S]$_{ox}$
$$\diagdown_{X}\diagup$$
$$S = 0 \qquad\qquad S = 1/2$$

Figure 13-3 ■ Hydrogen-activating center (H center) of Fe-hydrogenase isolated from *D. vulgaris*. (Reproduced from ref. 29. Copyright 1989, American Society for Biochemistry and Molecular Biology.)

Ni/[3Fe—xS]/[4Fe—4S] (1:1:1) composition.[30] Ni extended X-ray absorption fine structure analysis showed that Ni—S distances for both oxidized and dihydrogen-reduced enzyme are 0.220(2) nm.[31] The Ni site geometry is considered to be pseudooctahedral with a mixture of N, O, S, and Cl ligands. As shown in Fig. 13-4, an EPR signal for Ni(III) is observed in the presence of dioxygen.[32] Removal of dioxygen from the Ni(III) species results in formation of an EPR-inactive state that exhibits the enzymatic activity. The species is speculated to catalyze the ionic fission of a H—H bond via a hydride intermediate.

The third type is NiFeSe-hydrogenase containing an iron–sulfur clusters and equimolar amounts of Ni and Se atoms.[33,34] The enzyme was isolated from cells of the sulfate reducing bacterium *Desulfovibrio baculatus*. Ni and Se X-ray absorption spectroscopic studies showed that the Ni site geometry is pseudooctahedral with a coordinating ligand composition of 3-4 (N, O) at 0.206 nm, 1-2 (S, Cl) at 0.217 nm, and 1 Se at 0.244 nm.[35] The Se atom coordinates to the Ni site presumably as a selenocysteine residue. A recent EPR study using [77]Se-enriched NiFeSe-hydrogenase demonstrates unambiguously that the unpaired electron is shared by the Ni and Se atoms and that Se serves as a ligand to the Ni redox center of the enzyme.[36]

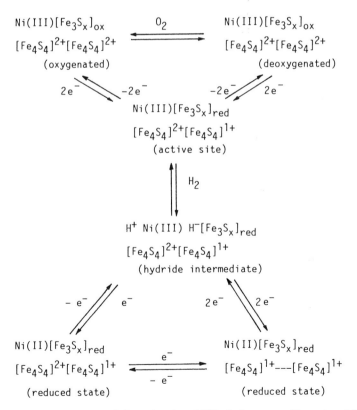

Figure 13-4 ▪ Oxidation states of the active site of NiFe-hydrogenase. (Reproduced from ref. 30. Copyright 1985 American Society for Biochemistry and Molecular Biology.)

Carbon monoxide dehydrogenase from *Clostridium thermoaceticum* is one of this series of enzymes. The enzyme consists of two subunits, one of which contains $Zn/Ni/Fe/S^{2-}$ (1:2:11:14).[37] The redox reaction

$$CO + H_2O \longrightarrow CO_2 + 2e^- + 2H^+ \qquad (13\text{-}5)$$

and water gas shift reaction is catalyzed. Formation of Ni(III)—C bond during the catalytic reaction was detected by EPR spectroscopy.[38]

13-3. Model Complexes of Iron–Sulfur Containing Enzymes and Their Reactions

Many simple thiolate complexes containing a [4Fe—4S] core as a model of ferredoxin have been synthesized by Holm's group.[39-41] Some bulky thiolate complexes serve as a model complex of high potential iron–sulfur protein.[8,9,43] Peptide model complexes containing a partial invariant sequence also have been synthesized.[7,44] All of these complexes possess four thiolate ligands around the [4Fe—4S] core. However, native iron–sulfur enzymes seem to contain a special iron center at a corner of the cluster as the active site of aconitase having a few unusual ligands.[13,16,17] The importance of the synthesis of an unsymmetrical model complex containing a [4Fe—4S] core is now to be emphasized as a structural model of the enzymes.

Catalytic reduction by sodium borohydride in the presence of a ferredoxin model complex $[Fe_4S_4(SPh)_4]^{2-}$ was found in the case of azobenzene to hydrazobenzene

$$\qquad (13\text{-}6)$$

and diphenylacetylene to cis-stilbene.[45] The hydrogenation by $NaBD_4$ indicated that one olefinic hydrogen of stilbene comes from $NaBD_4$ and the other from methanol[46]:

$$\qquad (13\text{-}7)$$

$[Fe_4S_4(SPh)_4]^{2-}$ also served as a catalyst for the reduction of aromatic ketones, aldehydes, cis-stilbene, trans-stilbene, and diphenylacetylene to the corresponding alcohols and hydrocarbon compounds in the presence of excess *n*-butyllithium or phenyllithium and slightly excess thiophenol[47,48]:

$$\qquad (13\text{-}8)$$

$$\qquad (13\text{-}9)$$

Formation of *N*-(*p*-chlorophenyl)ethylthioformimidate from *p*-chlorophenylisocyanide by $[Fe_4S_4(SEt)_4]^{2-}$ was realized in the presence of ethanethiol[49]:

$$Cl-\!\!\!\!\bigcirc\!\!\!\!-NC + [Fe_4S_4(SEt)_4]^{2-} \longrightarrow Cl-\!\!\!\!\bigcirc\!\!\!\!-N=CHSCH_2CH_3$$

(13-10)

$[Fe_4S_4(SPh)_4]^{2-}$ was found to catalyze the reduction of nitrobenzene to aniline, 1,4-benzoquinone to hydroquinone, and amine-N-oxides to amines in the presence of large amounts of thiophenol[50]:

$$\underset{O}{\overset{O}{\bigcirc}} \xrightarrow{PhSH} \underset{OH}{\overset{OH}{\bigcirc}}$$

(13-11)

$$R_3N{-}O \xrightarrow{PhSH} R_3N$$

(13-12)

$$\underset{}{\overset{NO_2}{\bigcirc}} \xrightarrow{PhSH} \underset{}{\overset{NH_2}{\bigcirc}}$$

(13-13)

In the preceding systems, large amounts of thiol react with the cluster as a reductant but also protect the cluster from decomposition.

Oxidative reactivity of the oxidized form of $[Fe_4S_4(SR)_4]^-$ toward ketoalcohols was studied in the presence of 1,4-benzoquinone.[51] The catalytic turnover number depends on the bulkiness of the thiolate ligands (SR). One-electron oxidized states of the [4Fe—4S] complexes having bulky thiolate ligands are thermodynamically more stable. Dehydrogenation of benzoin provides benzil and releases two electrons and two protons as shown in the scheme in Fig. 13-5. In this reaction system, the [4Fe—4S] cluster functions not only as a mediator of electron transfer but also as a carrier of protons. Actually, transport of protons in a heterogeneous water/toluene solvent by $[Fe_4S_4(SPh)_4]^{2-}$ was demonstrated.[52]

Macromolecular $[Fe_4S_4]^{2+}$ clusters connected by m- or p-benzenedithiolate ligands were synthesized as a model of iron–sulfur proteins having many clusters, such

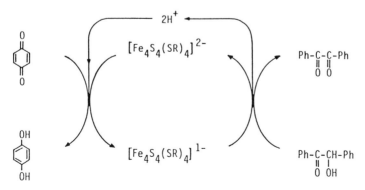

Figure 13-5 ▪ Catalytic oxidation of benzoin by benzoquinone in the presence of the [4Fe—4S] cluster. (Reproduced from ref. 51. Copyright 1986, Royal Society of Chemistry.)

as *Peptococcus aerogenes* ferredoxin.[53] A cooperative effect among $[Fe_4S_4]^{2+}$ cores was found in the catalytic oxidation of benzoin by 1,4-benzoquinone. A systematic study using various bridging ligands revealed that an electronic interaction between two $[Fe_4S_4]^{2+}$ cores in the new cross-linked tridentate $[4Fe-4S]$ complexes depends on the length of the bridging ligand.[54]

By a controlled electrochemical method,[55,56] isonitriles and hydrazine were catalytically reduced by $[Fe_4S_4(SR)_4]^{4-}$:

$$CH_3NC \xrightarrow[\text{in THF/MeOH}]{} CH_3NH_2 + CH_4 + H_2 \qquad (13\text{-}14)$$

$$H_2N-NH_2 \xrightarrow[\text{in MeOH}]{} NH_3 \qquad (13\text{-}15)$$

Electrochemically controlled reduced $[Fe_4S_4(SCH_2Ph)_4]^{2-}$ reacts with carbon dioxide and produces formate and benzoate[57]:

$$CO_2 \xrightarrow{[Fe_4S_4(SCH_2Ph)_4]^{2-}} \left\langle \bigcirc \right\rangle-COOH + HCOOH \qquad (13\text{-}16)$$

Catalytic reaction of nitrite and acetophenone by $[Fe_4S_4(SPh)_4]^{2-}$ in acetonitrile facilitates deprotonation of acetophenone, which forms a carbanion. The anion reacts with carbon dioxide to form ketoacid[58]:

$$8PhCCH_3 + 8CO_2 + 2NO_2^- + 6e^- \longrightarrow 8PhCCH_2COO^- + N_2 + 4H_2O \quad (13\text{-}17)$$
$$\quad\ \ \overset{\|}{O} \qquad\qquad\qquad\qquad\qquad\qquad\quad \overset{\|}{O}$$

The bovine serum albumin/$[Fe_4S_4]^{2+}$ complex catalyzed hydrogen evolution from an aqueous reduced methyl-viologen solution (pH 7) in the presence of dithionite.[59,60] This reaction system simulates that of hydrogenase. $[Fe_4S_4(SPh)_4]^{3-}$ reacts with thiophenol and generates dihydrogen in 30% yield.[61] Recently *Desulfovibrio* Ni(II)-substituted rubredoxin has been reported to exhibit dihydrogen production and deuteron–proton exchange activities.[62] The catalytic activities using these model complexes were still weak because of lack of involvement of other iron–sulfur clusters that probably serve a crucial role in the multielectron-transfer reaction.

13-4. Nitrogenase

Nitrogen fixation is one of the key reactions for plant and animal metabolism. Nitrogenase, which functions in the conversion of dinitrogen to ammonia, is contained in bacteria such as *Azotobacter*. Nitrogenase is composed of dinitrogenase reductase (Fe protein) and dinitrogenase (MoFe protein).[63] Dinitrogenase reductase contains two subunits having a single $[4Fe-4S]$ core in each subunit. Dinitrogenase contains four $[4Fe-4S]$ clusters and a Mo-containing cofactor (FeMoco). FeMoco, a $MoFe_{5-8}S_{8-9}$ component, is the active site of this enzyme and can be readily extracted with *N*-methylformamide.[64,65] An electrochemical study of FeMoco in *N*-methylformamide identified two apparent one-electron redox couples[66,67]:

$$FeMoco_{ox} \underset{}{\overset{-0.3\ V}{\rightleftharpoons}} FeMoco_{sr} \underset{}{\overset{-1.0\ V}{\rightleftharpoons}} FeMoco_{red} \qquad (13\text{-}18)$$
$$\text{EPR silent } (S=0) \quad (S=3/2) \quad \text{EPR silent}$$
$$g = 4.6, 3.4, 2.0$$

The oxidized form ($FeMoco_{ox}$) interconverts to the semireduced form ($FeMoco_{sr}$) and further to the reduced form ($FeMoco_{red}$). The $FeMoco_{sr}$ exhibits an $S = 3/2$ EPR signal.[68] Recent electrochemical and EPR studies demonstrated the existence of multiple forms of $FeMoco_{sr}$.

Nitrogenase catalyzes a variety of reactions involving unsaturated compounds other than dinitrogen reduction. All reactions of nitrogenase substrates are shown in the following equations,[69, 70] where the reactions are fundamentally based on two-electron transfer processes.

1. Two-electron processes:

$$C_2H_2 + 2H^+ + 2e^- \longrightarrow C_2H_4$$
$$2H^+ + 2e^- \longrightarrow H_2$$
$$N_3^- + 3H^+ + 2e^- \longrightarrow NH_3 + N_2$$
$$N_2O + 2H^+ + 2e^- \longrightarrow N_2 + H_2O$$
$$\text{cyclopropene} + 2H^+ + 2e^- \longrightarrow CH_2 = CH - CH_3 \text{ or cyclopropane}$$

2. Four-electron processes:

$$RNC + 4H^+ + 4e^{e^-} \longrightarrow RNHCH_3$$
$$HCN + 4H^+ + 4e^- \longrightarrow CH_3NH_2$$

3. Six-electron processes:

$$N_2 + 6H^+ + 6e^- \longrightarrow 2NH_3$$
$$HCN + 6H^+ + 6e^- \longrightarrow CH_4 + NH_3$$
$$N_3^- + 7H^+ + 6e^- \longrightarrow NH_3 + N_2H_4$$
$$NO_2^- + 7H^+ + 6e^- \longrightarrow NH_3 + 2H_2O$$
$$RCN + 6H^+ + 6e^- \longrightarrow RCH_3 + NH_3$$
$$RNC + 6H^+ + 6e^- \longrightarrow RNH_2 + CH_4$$

One dihydrogen molecule is evolved in each dinitrogen fixation. Dihydrogen evolution is an inherent part of the dinitrogen fixation as shown by the following stoichiometry[71-73]:

$$N_2 + 8H^+ + 8e^- \longrightarrow 2NH_3 + H_2 \tag{13-19}$$

Nitrogenase with multiple iron–sulfur centers and FeMoco serves as an electron sink capable of reducing these substrates. The MoFe proteins have been shown to also exist in at least three redox states just as in the case of FeMoco. The corresponding EPR active protein may be prepared in the presence of dithionite as shown in Scheme 3.[74] $MoFe(S_2O_4^{2-})$, which exists in the isolated state of MoFe protein, undergoes a three-electron oxidation with weak oxidants ($E_{1/2} < -0.125$ V versus NHE), a 6-electron oxidation with mild oxidants ($E_{1/2} > 0$ V versus NHE), a 9-electron oxidation with copper proteins, and 12-electron oxidation with cytochromes ($E_{1/2} = 0.2–0.3$ V versus NHE).[75] It is likely that the reduction of MoFe protein occurs with a

$$MoFe_o \xleftarrow{\text{oxidation}} MoFe(S_2O_4^{2-}) \xrightarrow{\text{reduction}} MoFe_r$$

EPR silent $g = 4.3, 3.6, 2.01$ EPR silent

Scheme 3

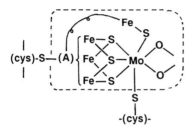

Figure 13-6 ■ Proposed active site around the Mo center in the MoFe protein of nitrogenase. (Reproduced from ref. 76. Copyright 1985, American Chemical Society.)

Table 13-1 ■ **Comparison of Physicochemical Properties of Components of the Three Nitrogenases**[a]

	Metal	Mr (k_D)	Subunit Mr	Specific Activities (nmol product/min/mg protein)		
				H_2	NH_3	C_2H_4
Nitrogenase 1	Mo, Fe	227	60	2138	1521	1924
Nitrogenase 2	V, Fe	210	55	1374	350	516
			50			
Nitrogenase 3	Fe	216	58	253	38	28
			50			

[a]Data are from the refs. 78 and 79.

stoichiometry of one electron transfer per FeMoco. The existence of multi forms of FeMoco is suggested to be associated with the multielectron transfer in MoFe protein. Specific successive electron transfer from reduced Fe protein to oxidized MoFe protein is believed to be required for the reduction of dinitrogen, but these reactions are still unknown.

A cubane-type [MoFe$_3$S$_4$] core is considered to be involved in a part of the FeMoco cluster. Mo K-edge, X-ray absorption edge, and near-edge structure data suggest a local environment of three S and three O (or N) ligands around the Mo and a weak Cys-S-Mo bond in MoFe protein as shown in Fig. 13-6.[76] Substitution of the sulfur ligands in MoFe protein to a citrate ligand in FeMoco provides the difference in [95]Mo ENDOR measurements between the MoFe protein and FeMoco.[77]

Mo has long been considered an essential element for nitrogenase activity of nitrogen fixation. However, this is not true. Nitrogenases having V and Fe in the active site were found in *Azotobacter chroococcum* and in a *nif* HDK deletion strain of *Azotobacter vinelandii*, respectively.[78, 79] Table 13-1 lists the comparison of the properties of three typical nitrogenases. Vanadium K-edge X-ray absorption of thionine-oxidized, super-reduced, and dithionine-reduced VFe protein from the V-containing nitrogenase of *A. chroococcum* indicated the presence of V atoms as part of a VFeS cluster.[80] The V atom was bound to 3O (or N), 3S, and 3Fe atoms at distances of 0.215, 0.231, and 0.275 nm.

13-5. Model Complexes of MoFe Protein and Their Model Reactions

Various MoFeS complexes having a cubane-like or larger clusters have been synthesized as models for the active site of the MoFe protein of nitrogenase. The synthetic developments, which are summarized in a recent review,[81] have stimulated progress of the metal–sulfur cluster chemistry. Recently, a cubane-like VFeS complex was also synthesized.[82]

The model reaction for the reduction of dinitrogen to ammonia has been studied using systems such as molybdate/cysteine/ferredoxin model complex/ATP/sodium borohydride.[83] This reaction is considered to proceed catalytically and requires the presence of $[Fe_4S_4(SR)_4]^{2-}$. The isolated FeMoco has no activity for the reduction of dinitrogen, although it can catalyze the reduction of acetylene to ethylene by $NaBH_4$.[84]

Various reductions for nitrogenase-related substrates were reported. Electrochemically controlled $[Mo_2Fe_6S_8(SR)_9]^{n-}$ reduces hydrazine,[85] azide,[86] isonitrile,[56] and acetylene.[87] The $[MoFe_2S_4(PhSH)_2(OCH_3)_4]^-$ or $[MoFe_6S_8(PhCH_2SH)(OCH_3)_6]^{2-}$ cluster was used for the reconstitution of an apoMoFe protein from nitrogenase and both exhibit a weak activity for acetylene reduction.[88] $[(t\text{-BuS})MoFe_3S_4(SPh)_3]^{2-}$ in DMF/H_2O and $[MoFe_3S_4(S\text{-}p\text{-}n\text{-}C_8H_{17}\text{-}C_6H_4)_3(O_2C_6Cl_4)(solv)]^{2-}$ (solv = DMF or acetone) in aqueous micelle solution have a slight catalytic activity for acetylene and azide, respectively.[89] The related molybdenum-containing complexes, such as $Mo_2O_2S_2L_2$ (L = S_2COPr or Cys-Phe-OMe), also exhibit a high catalytic activity for the reduction of azobenzene to hydrazobenzene and aniline.[90, 91] Polymer supported molybdenum complexes or oligopeptides having an alternating tripeptide sequence of Cys-(γ-Bzl)Glu-(γ-Bzl)Glu catalyze reduction of acetylene or phenylacetylene by $NaBH_4$ in THF/MeOH.[92] The addition of $[Fe_4S_4(SR)_4]^{2-}$ also increases the catalytic activity.

13-6. Overview

The active site of metalloenzymes during the catalytic redox reaction of small unsaturated molecules requires a multielectron flow from iron–sulfur, heme, and flavoproteins in biological systems. Orientation of the active site to the cluster complexes is also crucial for the effective electron flow. Although the actual distance between the active site and the electron-transfer part is still unknown, the construction of novel cluster having both the reaction site and the electron-transfer part provides one of the most promising approaches for the chemical simulation of multielectron-transfer systems. The enhanced redox activity by synthetic metal complexes is not enough for the smooth catalytic reduction of inert small molecules, for example, dinitrogen. Effective macromolecular systems are obviously required. Many studies of metalloenzymes have suggested that a peptide ligand induces an unidentified distortion of metal ion involved in the active site. The energy for the distortion has been ascribed to the whole folding energy of the peptide chain. Recent molecular dynamics study of proteins has taught the importance of the conformationally restricted motion by the local part of the peptide ligand around the active site. Fruitful approaches to clarifying the distortion are now being developed by synthesis of the mutant enzymes that have some alternative amino acid residues or by synthesis of oligopeptide model complexes that have an invariant sequence around the active site.

Synthetic model complexes have been observed to be thermodynamically unstable in an aqueous solution, whereas the native enzymes are quite stable under these conditions. Adequate chelation by the ligands with a preferable conformation of the peptide chain contributes to the large association constant, which is, of course, supported by hydrophobic environments around the active site. However, some redox reactions in metalloenzymes require proton transfer from water, substrate, or amino acid residues with a low activation energy through the hydrophobic layer, where a specific proton transfer pathway is believed to exist in the biological system.

Thus, the active site of metalloenzymes has many unidentified specific structures and functions. A clue to these questions may be obtained by the synthesis of model complexes. In certain cases, suitable synthetic ligands should be designed to have a directional and sequential arrangement of several different metal sites in appropriate oxidation states. Well designed synthetic polymers will be useful for constructing such molecular aggregates, and macromolecule–metal complexes with such functionality are thus highly important for future research.

References

1. E. I. Solomon, *NATO ASI Ser*, *Ser. C*, **140**, 463 (1985).
2. C. W. Carter, Jr., J. Kraut, S. T. Freer, R.l A. Alden, L. C. Sieker, E. Adman, and L. H. Jensen, *Proc. Nat. Acad. Sci. USA*, **68**, 3526 (1972).
3. C. L. Hill, R. H. Renaud, R. H. Holm, and L. E. Mortenson, *J. Am. Chem. Soc.*, **99**, 2549 (1977).
4. W. C. Carter, Jr., in *Iron-Sulfur Proteins*, Vol. III, W. Lovenberg, ed., Academic, New York (1977), p. 158.
5. E. T. Lode, C. L. Murray, and J. C. Rabinowitz, *J. Biol Chem.*, **251**, 1683 (1976).
6. E. Adman, K. D. Watenpaugh, and L. H. Jensen, *Proc. Nat. Acad. Sci. USA*, **72**, 4854 (1975).
7. N. Ueyama, T. Terakawa, M. Nakata, and A. Nakamura, *J. Am. Chem. Soc.*, **105**, 7098 (1983).
8. N. Ueyama, T. Terakawa, T. Sugawara, M. Fuji, and A. Nakamura, *Chem Lett.*, **1984**, 1287 (1984).
9. T. O'Sullivan and M. M. Millar, *J. Am. Chem. Soc.*, **107**, 4096 (1985).
10. S. R. Dickman and A. A. Cloutier, *J. Biol Chem.*, **188**, 379 (1951).
11. T. A. Kent, J.-L. Dreyer, M. C. Kennedy, B. H. Huynh, M. H. Emptage, H. Beinert, and E. Münck, *Proc. Nat. Acad. Sci. USA*, **79**, 1096 (1982).
12. M. H. Emptage, T. A. Kent, M. C. Kennedy, H. Beinert, and E. Münck, *Proc. Nat. Acad. Sci. USA*, **80**, 4674 (1983).
13. A. H. Robbins and C. D. Stout, *Proc. Nat. Acad. Sci. USA*, **86**, 3639 (1989).
14. M. C. Kennedy, M. West, J. Telser, M. H. Emptage, H. Beinert, and B. M. Hoffman, *Proc. Nat. Acad. Sci. USA*, **84**, 8854 (1987).
15. J. Telser, M. H. Emptage, H. Merkle, M. C. Kennedy, H. Beinert, and B. M. Hoffman, *J. Biol. Chem.*, **261**, 4840 (1986).
16. D. W. Plank and J. B. Howard, *J. Biol. Chem.*, **263**, 8184 (1989).
17. M. C. Kennedy and H. Beinert, *J. Biol. Chem.*, **263**, 8194 (1989).
18. W. J. Ingledew and R. K. Poole, *Microbiol. Rev.*, **48**, 222 (1984).
19. S. T. Cole, C. Condon, B. D. Lemire, and J. H. Weiner, *Biochim. Biophys. Acta*, **811**, 381 (1985).
20. M. K. Johnson, J. E. Morningstar, G. Cecchini, B. A. C. Ackrell, and E. B. Keaney, *Biochem. Biophys. Res. Commun.*, **131**, 756 (1985).
21. J. E. Morningstar, M. K. Johnson, G. Cecchini, B. A. C. Ackerell, and E. B. Keaney, *J. Biol. Chem.*, **260**, 13631 (1985).
22. M. W. W. Adams and L. E. Mortenson, *J. Biol. Chem.*, **259**, 7045 (1984).
23. B. H. Huynh, M. H. Czechowski, H.-J. Kruger, D. V. DerVartanian, H. D. Peck, Jr., and J. LeGall, *Proc. Nat. Acad. Sci. USA*, **81**, 3728–3732 (1984).
24. C. van Dijk, S. G. Mayhew, H. J. Grande, and C. Veeger, *Eur. J. Biochem.*, **102**, 317 (1979).

25. M. W. W. Adams, *J. Biol. Chem.*, **262**, 15054 (1987).
26. J. Telser, M. J. Benecky, M. W. W. Adams, L. E. Mortenson, and B. M. Hoffman, *J. Biol. Chem.*, **261**, 13536 (1986).
27. J. Telser, M. J. Benecky, M. W. W. Adams, L. E. Mortenson, and B. M. Hoffman, *J. Biol. Chem.*, **262**, 6589 (1987).
28. G. Wang, M. J. Benecky, B. H. Huynh, J. F. Cline, M. W. W. Adams, L. E. Mortenson, B. M. Hoffmann, and E. Munck, *J. Biol. Chem.*, **259**, 14328 (1984).
29. A. T. Kowal, M. W. W. Adams, and M. K. Johnson, *J. Biol. Chem.*, **264**, 4342 (1989).
30. M. Teixeira, I. Moura, A. V. Xavier, B. H. Huynh, D. V. DerVartian, H. D. Peck, Jr., J. LaGall, and J. J. G. Moura, *J. Biol. Chem.*, **260**, 8942 (1985).
31. R. A. Scott, S. A. Wallin, M. Czechowskii, D. V. DerVartanian, J. LeGall, H. D. Peck, Jr., and I. Moura, *J. Am. Chem. Soc.*, **106**, 6864 (1984).
32. M. Teixeira, G. Fauque, I. Moura, P. A. Lespinat, Y. Berlier, B. Prickril, H. D. Peck, Jr., A. V. Xavier, J. LeGall, and J. J. G. Moura, *Eur. J. Biochem.*, **167**, 47 (1987).
33. L. F. Olsen and H. Degn, *Q. Rev. Biophys.*, **18**, 165 (1985).
34. E. E. Selkov, *Eur. J. Biochem.*, **1**, 79 (1968).
35. M. K. Eidsness, R. A. Scott, B. C. Prickril, D. V. DerVartnian, J. LaGall, I. Moura, J. J. G. Moura, and H. D. Peck, Jr., *Proc. Natl. Acad. Sci. USA*, **86**, 147 (1989).
36. S. H. He, M. Teixeira, J. LeGall, D. S. Patil, I. Moura, J. J. G. Moura, D. V. DerVartanian, B. H. Huynh, and H. D. Peck, Jr., *J. Biol. Chem.*, **264**, 2678 (1989).
37. S. W. Ragsdale, J. E. Clark, L. G. Ljungdahl, L. L. Lundie, and H. L. Drake, *J. Biol. Chem.*, **258**, 2364 (1983).
38. S. W. Ragsdale, H. G. Wood, and W. E. Antholine, *Proc. Nat. Acad. Sci. USA*, **82**, 6811 (1985).
39. T. Herskovitz, B. A. Averill, R. H. Holm, J. A. Ibers, W. D. Phillips, and J. F. Weiner, *Proc. Nat. Acad. Sci. USA*, **69**, 2437 (1972).
40. B. A. Averill, T. Herskovitz, R. H. Holm, and J. A. Ibers, *J. Am. Chem. Soc.*, **95**, 3523 (1973).
41. J. M. Berg and R. H. Holm, "Iron-Sulfur Proteins," in *Metal Ions in Biology*, Vol. 4, T. G. Spiro, ed. Wiley-Interscience, New York (1982), p. 1.
42. T. Okuno, K. Uoto, O. Yonemitsu, and T. Tomoshiro, *J. Chem. Soc. Chem. Commun.*, **1987**, 1018 (1987).
43. M. Nakamoto, K. Tanaka, and T. Tanaka, *J. Chem. Soc. Chem. Commun.*, **1988**, 1422 (1988).
44. N. Ueyama, A. Kajiwara, T. Terakawa, and A. Nakamura, *Inorg. Chem.*, **24**, 4700 (1985).
45. A. Nakamura, M. Kamada, K. Sugihashi, and S. Otsuka, *J. Mol. Catal.*, **8**, 353 (1980).
46. T. Itoh, T. Nagano, and M. Hirobe, *Tetrahedron Lett.*, **21**, 1343 (1980).
47. H. Inoue, M. Suzuki, and N. Fujimoto, *J. Org. Chem.*, **44**, 1722 (1979).
48. H. Inoue and M. Suzuki, *J. Chem. Soc. Chem. Commun.*, **1980**, 817 (1980).
49. A. Schwartz and E. E. van Tamelen, *J. Am. Chem. Soc.*, **99**, 3189 (1977).
50. T. Itoh, T. Nagano, and M. Hirobe, *Chem. Pharm. Bull.*, **34**, 2013 (1986).
51. N. Ueyama, T. Sugawara, A. Kajiwara, and A. Nakamura, *J. Chem. Soc. Chem. Commun.*, **1986**, 434 (1986).
52. H. Tsai, W. V. Sweeney, and C. L. Coyle, *Inorg. Chem.*, **24**, 2796 (1985).
53. N. Ueyama, T. Sugawara, A. Nakamura, and T. Fueno, *Chem. Lett.*, **1988**, 223 (1988).
54. T. D. P. Stack, M. J. Carney, and R. H. Holm, *J. Am. Chem. Soc.*, **111**, 1670 (1989).
55. K. Tanaka, Y. Hozumi, and T. Tanaka, *Chem. Lett.*, **1982**, 1203 (1982).
56. K. Tanaka, Y. Imanaka, M. Tanaka, M. Honjo, and T. Tanaka, *J. Am. Chem. Soc.*, **104**, 4258 (1982).
57. M. Tezuka, T. Yajima, A. Tsuchiya, Y. Matsumoto, Y. Uchida, and M. Hidai, *J. Am. Chem. Soc.*, **104**, 6834 (1982).
58. K. Tanaka, R. Wakita, and T. Tanaka, *J. Am. Chem. Soc.*, **111**, 2428 (1989).
59. I. Okura, S. Nakamura, and K. Nakamura, *J. Mol. Catal.*, **6**, 71 (1979).

60. I. Okura and S. Nakamura, *J. Mol. Catal.*, **9**, 125 (1980).
61. T. Yamamura, G. Christou, and R. H. Holm, *Inorg. Chem.*, **22**, 939 (1983).
62. P. Saint-Martin, P. A. Lespinat, G. Fauque, V. Berlier, J. LaGall, I. Moura, M. Teixeira, A. V. Xavier, and J. J. G. Moura, *Proc. Nat. Acad. Sci. USA*, **85**, 9378 (1988).
63. W. H. Orme-Johnson, *Ann. Rev. Biophys. Biophys. Chem.*, **14**, 419 (1985).
64. S.-S. Yang, W.-H. Pan, G. D. Friesen, B. K. Burgess, J. L. Corbin, E. I. Stiefel, and W. E. Newton, *J. Biol. Chem.*, **257**, 8042 (1982).
65. M. J. Nelson, M. A. Levy, and W. H. Orme-Johnson, *Proc. Nat. Acad. Sci. USA*, **80**, 147 (1983).
66. F. A. Schultz, S. F. Gheller, B. K. Burgess, S. Lough, and W. E. Newton, *J. Am. Chem. Soc.*, **107**, 5364 (1985).
67. F. A. Schultz, S. F. Gheller, and W. E. Newton, *Biochem. Biophys. Res. Commun.*, **152**, 629 (1988).
68. W. E. Newton, S. F. Gheller, B. J. Feldman, W. R. Dunham, and F. A. Schultz, *J. Biol. Chem.*, **264**, 1924 (1989).
69. B. K. Burgess, in *Molybdenum Enzymes*, T. G. Spiro, ed., Wiley, New York (1985), p. 161.
70. S. A. Vaughn and B. K. Burgess, *Biochemistry*, **28**, 419 (1989).
71. F. B. Simpson and R. H. Burgess, *Science*, **224**, 1095 (1984).
72. K. L. Hadfield and W. A. Bulen, *Biochemistry*, **8**, 5103 (1969).
73. J. Liang and R. H. Burris, *Proc. Nat. Acad. Sci. USA*, **85**, 9446 (1988).
74. G. D. Watt, Z.-C. Wang, and R. R. Knotts, *Biochemistry*, **25**, 8156 (1986).
75. G. D. Watt and Z.-C. Wang, *Biochemistry*, **28**, 1844 (1989).
76. S. D. Conradson, B. K. Burgess, W. E. Newton, K. O. Hodgson, J. W. McDonald, J. F. Robinson, S. F. Gheller, L. E. Mortenson, M. W. W. Adams, P. K. Mascharak, W. A. Armstrong, and R. H. Holm, *J. Am. Chem. Soc.*, **107**, 7935 (1985).
77. P. A. McLean, A. E. True, M. J. Nelson, S. Chapman, M. R. Godfrey, B. K. Teo, W. H. Orme-Johnson, and B. M. Hoffman, *J. Am. Chem. Soc.*, **109**, 943 (1987).
78. R. L. Robson, R. R. Eady, T. H. Richardson, R. W. Miller, M. Hawkins, and J. Postage, *Nature*, **322**, 388 (1986).
79. J. R. Chisnell, R. Premakumar, and P. E. Bishop, *J. Bact.*, **170**, 27 (1988).
80. J. M. Arber, B. R. Dobson, R. Robert, S. S. Hasnain, C. D. Garner, T. Matsushita, M. Nomura, and B. E. Smith, *Biochem. J.*, **258**, 733 (1989).
81. R. H. Holm and E. D. Simhon, in *Molybdenum Enzymes*, T. G. Spiro, ed., Wiley, New York (1985), p. 1.
82. S. Ciurli and R. H. Holm, *Inorg. Chem.*, **28**, 1685 (1989).
83. G. N. Schrauzer, *Angew. Chem. Int. Ed. Engl.*, **14**, 514 (1975).
84. C. E. McKenna, J. B. Jones, H. Erans, and C. W. Huang, *Nature*, **280**, 612 (1979).
85. Y. Hozumi, Y. Imasaka, K. Tanaka, and T. Tanaka, *Chem. Lett.*, **1983**, 897 (1983).
86. K. Tanaka, M. Tanaka, and T. Tanaka, *Chem. Lett.*, **1981**, 895 (1981).
87. Y. Imasaka, K. Tanaka, and T. Tanaka, *Chem. Lett.*, **1983**, 1477 (1983).
88. J. Xu, X. Lui, L. Wang, Z. Zhang, S. Niu, S. Li, Y. Fan, F. Wang, P. Lu, and A. Tang, *Fenzi Kexue Yu Huaxue Yanjiu*, **3**, 1 (1983).
89. S. Otsuka and M. Kamata, in *Molybdenum Chemistry of Biological Significance*, W. E. Newton and S. Otsuka, eds., Plenum, New York (1980), p. 229.
90. A. Nakamura, M. Kamata, K. Sugihashi, and S. Otsuka, *J. Mol. Catal.*, **8**, 353 (1980).
91. N. Ueyama, M. Nakata, and A. Nakamura, *J. Mol. Catal.*, **14**, 341 (1982).
92. N. Oguni, S. Shimazu, and A. Nakamura, *Polymer J.*, **12**, 891 (1980).

Part V   Sequential Potential Field Constructed in Macromolecular Complexes

Sequential Potential Field and Electron Transfer in Metal Complex Clusters

Naoki Toshima

Department of Industrial Chemistry, The University of Tokyo, Japan

14-1. Introduction

Electron transfer is undoubtedly one of the most fundamental and important chemical processes. Electron-transfer reactions between transition metal complexes in solution have long been a central focus of mechanistic inorganic chemistry.[1,2] Inorganic redox reactions are also of primary importance in biological system.[3-5] Indeed, many interesting works come from the area of inorganic biochemistry.

Excited state chemistry also provides much interesting data on electron transfer as well as energy transfer.[6] Much research has been devoted to photosynthetic centers. The excellent work on the structure of the reaction center of photosynthetic bacteria[7] has encouraged the investigations on the electron-transfer reaction derived by photoexcitation in biological systems as well as artificial model systems.[8]

Solid-state electron transfer[9,10] is of vital importance in such areas of current interest as inelastic electron tunneling spectroscopy, chemical field-effect transistors, conducting polymers, mixed valency, solid-state electrochemistry, surface-modified electrodes, photoelectrochemistry, and photovoltaics. These have been, of course, of importance to solid-state physicists, but recently have become important to chemists too.

The electron transfer in a solid state or a biological system is quite different from that in a homogeneous solution system, and cannot be understood by the classical theory that has been developed for the electron transfer between transition metal complexes in a homogeneous solution.[11]

The nonclassical theories for the electron transfer in a nonhomogeneous system involve, for example, a superexchange mechanism[12-14] in long distance electron transfer in a biological system and an electron tunneling[15–17] in solid state electron-transfer reactions. However, there is no uniform understanding for the electron transfer in a "nonhomogeneous" system that involves both a biological system and a solid system as semiconductors.

Here, we would like to understand both biological and solid systems as a nonhomogeneous system, which is heterogeneous in a micro scale even though it looks homogeneous in a macro scale. The nonhomogeneous system can be easily understood by considering macromolecules, which often provide a heterogeneous reaction field in a homogeneous solution. In the nonhomogeneous system, electron transfer takes place in a kind of reaction field, which we call a sequential potential field.[18, 19]

In this chapter we will present the concept of a sequential potential field to help explain the electron-transfer reactions in a nonhomogeneous system, and we will sketch some model systems for the approach to the sequential potential field using metal complex clusters.

14-2. Electron Transfer in "Nonhomogeneous" Systems

Electron transfers between transition metal complexes in solutions can be usually understood by pure classical theories, in which one common assumption is that all electron-transfer reactions are adiabatic. In other words, the reaction occurs on a continuous path on a single potential surface, as shown in Fig. 14-1(a) as a simplified, two-dimensional picture. In this figure, the reactant and product potential curves are drawn along a reaction coordinate. In this case, the electron donor (D, reductant) and the electron acceptor (A, oxidant) interact with each other, and the electron-transfer reaction can proceed via an activated complex

$$D + A \rightleftharpoons (DA) \rightleftharpoons (DA)^* \rightleftharpoons (D^+A^-)^* \rightleftharpoons (D^+A^-) \rightleftharpoons D^+ + A^-.$$

$$(14-1)$$

The strong interaction or overlap of the electronic wave functions between D and A causes the "avoided crossing" at the intersection of the two potential curves as shown in Fig. 1(a). Because the splitting of the two surfaces is large enough, the system remains on the lower surface, along which the electron-transfer reaction proceeds.

If the interaction between D and A is weak, or the splitting between the surfaces is small, the system can sometimes jump to the upper surface, passing through the transition states many times without proceeding to products, as shown in Fig. 1(b). Such a system is said to be nonadiabatic. The small interaction can arise from unfavorable orbital symmetries, changes in spin multiplicity, or large separation distances between donor and acceptor.

Because the donor–acceptor separation distances are often large in biological and rigid matrix systems, the long-distance electron transfers in biological systems and the solid-state electron transfers in electro devices undoubtedly proceed via nonadiabatic processes.

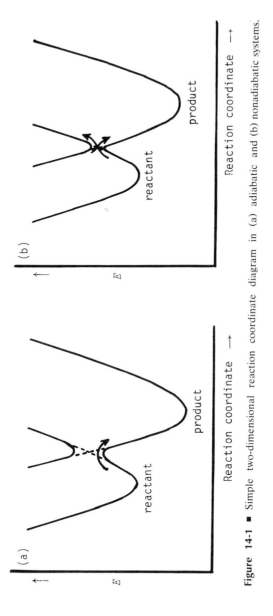

Figure 14-1 ▪ Simple two-dimensional reaction coordinate diagram in (a) adiabatic and (b) nonadiabatic systems.

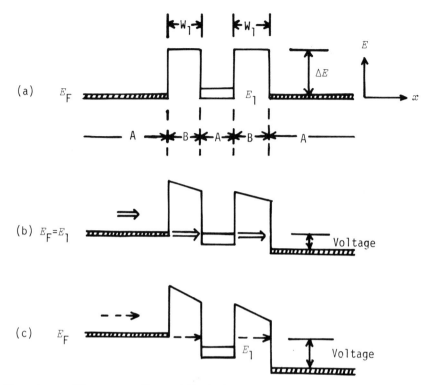

Figure 14-2 ■ Electron tunneling model in the layered system with conductor A and insulator B. (a) The barrier height and width, ΔE and W_1, respectively. Electron tunneling occurs more smoothly when the Fermi level E_F becomes coincident with the energy level of (b) the trapped electron E_1 than (c) at $E_F > E_1$.

Attempts have been made to explain the nonadiabatic process in terms of quantum mechanics such as electron tunneling[15-17] and superexchange.[12-14] Quantum mechanical tunneling is a process where a particle has a certain probability of passing through a barrier. This phenomenon involves both electron and nuclear tunneling. Electron-tunneling can be applied to the electron transfer, where a "space" is an example of a classically impenetrable barrier. For instance, if the reactants are separated in space by a nonconducting medium, the nonadiabatic electron transfer with weak electronic coupling is described in terms of electron-tunneling: The probability of tunneling is related to the barrier height and width.

In the layered system shown in Fig. 14-2(a), the layer A is conducting and the layer B is a barrier, where ΔE and W_1 are the barrier height and width, respectively. When a voltage is applied between terminals, and the Fermi level A of E_F becomes coincident with the energy level E_1 of the trapped electrons that have leaked into the well B, a resonance tunneling occurs and the electron transfers easily through the barriers, as shown in Fig. 14-2(b).

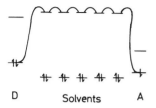

Figure 14-3 ■ Concept of superexchange for electron transfer between electron donor D and acceptor A via LUMO of the solvents.

Another quantum mechanical explanation is based on the principle of superexchange, where the electron transfer can proceed via interactions between similar orbitals of neighboring molecules that mediate donors and acceptors.[14] There are two possible pathways to couple the donor orbital with the acceptor orbital: an electron transfer and a hole transfer, which involve LUMO and HOMO, respectively, of the mediating molecules.

The concept of the superexchange between D and A is shown in Fig. 14-3. The magnitude of the interaction depends on the exchange integral β, the number of exchanges, and the difference in ionization potential between electron donor and mediating molecules.

In sum, the theories make two fundamental predictions. First, electron-transfer rates should decrease rapidly with increasing distance R: $k \propto \exp(-\alpha R)$, where α depends markedly on the medium. Second, at a given distance, the rates should depend significantly on the reaction free energy ΔG. The rates increase rapidly with increasing ΔG, reaching a maximum when the free energy and reorganization energy are equal (and then decreasing at the inverted region).[20]

Over the past years, many examples of electron transfer have been reported, especially "long-distance" electron transfer, in inorganic, organic, biological, electroconducting, and macromolecular chemistry and physics.[5,11] Most of them could be explained by these theories.

However, key questions are still unresolved on the electron transfer in nonhomogeneous systems, for example, selectivity and specification in electron transfer and the relationship between electron transfer and stereostructure. The concept of a sequential potential field could be one of the solutions to these problems.

14-3. Concept of a Sequential Potential Field

Electron transfers in the nonhomogeneous system that was mentioned in the preceding section involve those in a biological system, a rigid matrix system, and a semiconducting inorganic solid system as well as the macromolecular complex system. The macromolecular complex will be a typical model for all these nonhomogeneous systems. The macromolecular complex can form a condensed phase or, in a solution, can provide a heterogeneous phase in micro scale in spite of homogeneity in macro scale.

Electron transfer in a macromolecular complex system is of great importance from the viewpoint of electronic functions of macromolecular complexes, such as catalysis, molecular and ionic recognition, separation, energy conversion, conducting, imaging,

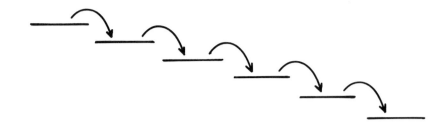

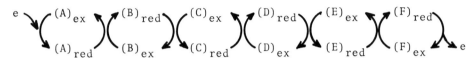

Figure 14-4 ■ Concept of sequential potential field with six redox components A, B, C, D, E, and F.

recording, and so on.[21,22] In macromolecular complexes, the macromolecules surround the central metals, providing a macromolecular reaction field that affects the reactivity and properties of the central metals and their ability to transfer electrons.

In a biological system, selective electron transfers occur often. For example, in a reaction center of photosynthetic bacteria, electrons move from one to another heme complex in a protein system.[16,23] The photoexcited electron moves smoothly along a path, where the redox components are fixed in the protein in the order of the redox potential. In other words, the natural macromolecule, protein, determines both the direction and the path of the electron transfer in a biological system.

Water flows from higher to lower ground along a river. Analogously electrons flow from a place with higher potential to that with lower potential. The sequential potential field can determine the path along which the electrons flow.

The concept of sequential potential field is illustrated in Fig. 14-4. Terms A, B, C, D, E, and F are the symbols of the six redox components like metal complexes with six different redox potentials. If these components stand in line in order of redox potential, the easy electron transfer from A to F via B, C, D, and E will be expected. The macromolecule may play a role in aligning the metal complexes in order and making them work cooperatively.

The redox potential of a metal complex can be determined by (a) the kind and the valency of the central metal, (b) the kind and the number of the ligands that coordinate to the metal, and (c) the environment that surrounds the metal complex. For example, cytochrome c_3, which is an electron carrier in the respiratory chain of sulfate-reducing bacteria, contains four heme moieties in the molecule. The redox potentials of four hemes are slightly different from one another.[24,25] The potentials can be controlled by the degree of the exposure of the hemes or the difference in hydrophilicity of the macromolecule surrounding the hemes. Because the electron flow in cytochrome c_3 is determined by the difference in the redox potentials of the

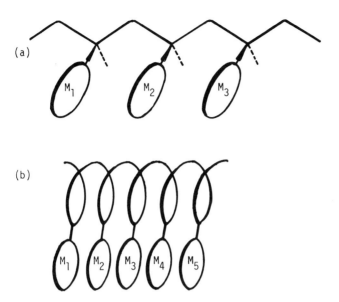

Figure 14-5 ■ Polymer models controlling the position and orientation of the metal complexes $M_1, M_2 \dots$ by configuration and conformation of the polymers.

four hemes, the macromolecules or the proteins, which surrounds the hemes, actually control the direction of electron transfer.

In macromolecular complexes other than proteins, the macromolecule works as the ligands or the materials that control the redox potentials of the metal complexes as well as the framework that arranges the metal complexes in order. In order to control the electron flow among the metal complexes, the orientation of each molecule of the metal complex is of vital importance in addition to the alignment of the metal complexes in the order of the potential. The macromolecule can therefore control the orientation of the complex molecules as well as the ordering.

Examples of the concept are shown in Fig. 14-5, where the configuration and conformation of the polymers determine the position and orientation of the complex molecules.

Orientation of the redox molecules for electron transfer is important because the electron transfer occurs through the overlap between the molecular orbitals. Figure 14-6 illustrates the electron transfer from the excited molecule C to E via D. The overlap between the lowest unoccupied molecular orbitals (LUMO) of molecules C and D can allow the transfer of the photoexcited electron from C to D, which will be followed by the electron transfer from D to E as the result of the LUMO–LUMO overlap of molecules D and E. The back electron transfer from D to C should be avoided, that is, the LUMO–HOMO or the HOMO–HOMO overlap is not desirable. Therefore, a molecular orientation that produces LUMO–LUMO overlap is desirable for reductive electron transfer.

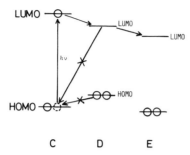

Figure 14-6 ▪ Electron transfer from excited C to E through D.

C   D   E

X-ray structural analysis of the reaction center of the photosynthetic bacteria reveals that the ideal molecular orientation for LUMO–LUMO overlap has not been achieved.[26] If we can improve the position and orientation of the molecules by using the same components as the natural system, much faster and more selective electron transfers than the natural system will be achieved.

14-4. Approaches to Construction of a Sequential Potential Field

The concept of the sequential potential field is very important and effective not only for understanding natural phenomena but also for creating new functional materials that may improve upon nature. However, it is not so easy to construct the complete sequential potential field with synthetic macromolecules and metal complexes. Some approaches to the construction of the sequential potential field which may improve on nature are presented here as examples.

14-4-1. Surface-Separated System

Electrons flow among the components in the order of the potential energy, constructing a geared cyclic system as shown in Fig. 14-4. In the geared system, the back electron transfer should be avoided to allow smooth forward electron flow. For this purpose, a surface-separated system is often used.[28]

Visible light-induced cleavage of water is one of the most popular systems for the conversion of solar photoenergy to chemical energy, and it is understood as a simulation model for artificial photosynthesis.[27] For this reaction, the electron transfer through a geared cyclic system is thought to be necessary. Zinc porphyrin or ruthenium tris (2,2′-bipyridine) is often used as a chromophore for absorbing visible light and viologen as an electron mediator for the simulation model system.

At the initial stage of the investigation solid surfaces were used for the promotion of charge separation. For example, adsorption of methyl viologen on solid cellulose prolongs the lifetime of the viologen radical cation.[29] Visible light-induced hydrogen evolution from water was reported by using EDTA as a sacrificial reductant on the surface of chelate resin on which a ruthenium complex, methyl viologen, and platinum colloid were immobilized.[30] The surface of semiconductors has also been used often for the charge separation in the relay system.[27, 31]

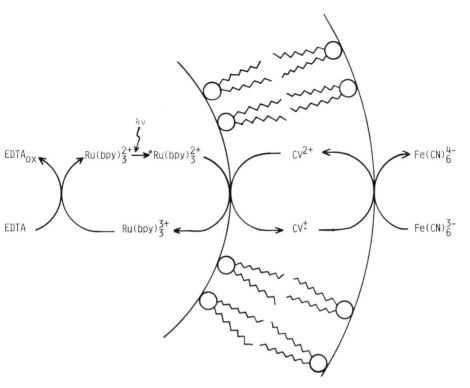

Figure 14-7 ▪ Surface-separated electron transfer in vesicles from excited ruthenium complex $Ru(bpy)_3^{2+}$ to ferricyanide $Fe(CN)_6^{3-}$ through cetyl viologen CV^{2+}.

Instead of solid surfaces, structurally organized molecular assemblies can be used for the surface separation.[32] In these molecular assemblies, the surface separation is expected to be more effective than the solid surface because of micro-scale separation. The molecular assembly systems are also attractive from the viewpoint of their similarity to the natural photosynthetic organisms both in structure and reaction scheme.

An example of the surface-separated systems using vesicles is illustrated in Fig. 14-7, where EDTA and ruthenium complex are located in the inner aqueous cavities and ferricyanide in the outer aqueous phase.[33] Cetyl viologen embedded in the hydrophobic membrane works as an electron mediator to transfer electrons from the inner surface to the outer surface. This vesicle system resulted in the up-hill electron transfer initiated by visible light absorption.

Figure 14-8 illustrates another example, where the two-photon process has been proposed for the visible light-induced up-hill electron transfer in vesicles.[33] In this system the first reaction is the photoelectron transfer from excited zinc porphyrin at the outer surface of the vesicle membrane to methyl viologen in the outer aqueous

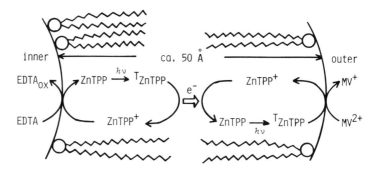

Figure 14-8 ▪ Surface-separated electron transfer from inner surface to outer surface of the vesicles in the two-photon process of the excited triplet state of ZnTPP.

phase, which is followed by electron transfer across the membrane from the other excited zinc porphyrin molecule located near the inner surface of the membrane.

Both examples are concerned with the up-hill electron transfer in vesicles based on visible light excitation. However, both systems do not result in gas evolution as the reaction product. Recently we have successfully prepared micelle-embedded platinum clusters,[34] which can be used as catalysts for visible light-induced hydrogen evolution.

The micelle-embedded platinum clusters were prepared by photoreduction of hexachloroplatinic acid in water in the presence of micelles.[35] The platinum clusters are located in the center of the micelles and are protected by the hydrophobic environment of the micellar phase. This structure is demonstrated by substrate selectivity in the hydrogenation of unsaturated fatty acids catalyzed by the micelle-embedded platinum clusters.[36]

A new system for the visible light-induced hydrogen production was constructed by using micelle-embedded platinum clusters. In this system (Fig. 14-9), ruthenium complex, methyl viologen, and platinum cluster particles are located in an aqueous phase, a hydrophilic intermediate layer, and a hydrophobic micellar phase, respectively. In this way, the three components with different redox potentials are arranged in order of potential, resulting in the construction of a sequential potential field by utilizing a surface-separated micellar system.

The relative rate of hydrogen evolution in this system is shown in Table 14-1. The results indicate that the catalytic system with polymerized micelle is more active than that with linear polymer and that a nonionic system is more active than ionic systems. Therefore, the nonionic polymerized micelle gives the best result among those examined. These results also demonstrate that both the methyl viologen dication and the methyl viologen radical cation are actively located between the aqueous and micellar phases, constituting a sequential potential field. This will be the first example of the controlled arrangement of these components, a photosensitizer, an electron mediator, and a reductive catalyst resulting in the visible light-induced hydrogen evolution from water by construction of a sequential potential field.

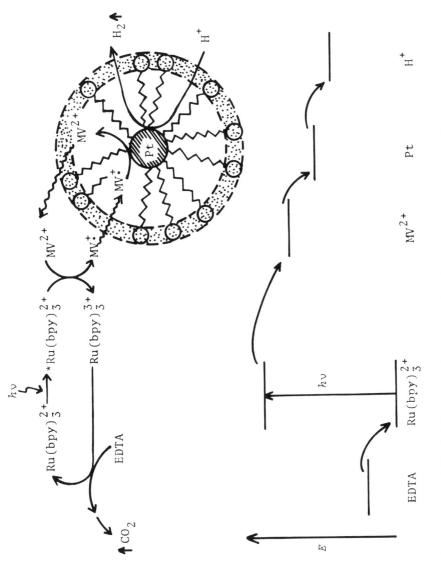

Figure 14-9 ■ Visible light-induced hydrogen evolution in the sequential potential field constructed by micelle-embedded platinum particle, methyl viologen, $Ru(bpy)_3^{2+}$, and EDTA in micellar solutions.

Table 14-1 ■ **Reactive Rate of Visible Light-Induced Hydrogen Evolution from Water in the System EDTA / Ru(bpy)$_3^{2+}$ / MV^{2+} / Pt**

Pt Cluster	Rel. Rate
Nonionic polymerized micelle-embedded Pt	306
Nonionic monomeric micelle-embedded Pt	302
Cationic polymerized micelle-embedded Pt	267
Cationic monomeric micelle-embedded Pt	250
Anionic polymerized micelle-embedded Pt	105
Anionic monomeric micelle-embedded Pt	8
Nonionic linear polymer-surrounded Pt	100
Anionic linear polymer-surrounded Pt	60

14-4-2. Molecular Array and Lamination System

Molecular array is of great importance for the concept of the sequential potential field. In the ideal case the molecular array is expected to be controlled by polymer chains as shown in Fig. 14-5. However, the practical technique has not been established yet.

Lamination of cast films will be a practical method for producing the molecular array for the sequential potential field. The lamination of cast polymer films on an electrode gives a kind of polymer-modified electrode, which is described in a separate chapter in this book. The key features in this system are the bilayer or multilayer structure, the interfilm electron transfer, and the redox splitting by layers or electrode.

Molecular level lamination layers can be achieved by a Lamgmuir–Blodgett (LB) method or a molecular beam deposition method. Both methods are superior to the cast methods from the viewpoint of control of the structure of thin films at the molecular level.

Figure 14-10 illustrates the ideal system of *Y*-type laminated LB films, in which the molecular assemblies A, B, C, and D are arranged in the order of the redox potential, constructing the sequential potential field. In this case, A and C are hydrophilic and shown by round marks and B and D are hydrophobic and shown by square marks. Therefore, the *Y*-type lamination of the A, B, C, and D LB films on the substrate is expected to result in the construction of a sequential potential field.

Practical experiments have been carried out successfully by using amphiphilic linear molecules in which three functional moieties, that is, an electron acceptor, a sensitizer, and an electron donor moiety, are linked covalently by hydrocarbon chains.[37,38] The system is illustrated in Fig. 14-11, where an electron donor D, a photosensitizer S, and an electron acceptor A are ferrocene, pyrene, and viologen moiety, respectively. In these LB film systems, the effective photocurrent was observed for the triad system consisting of a D-S-A monolayer, but not for a diad or a nonarranged system. These facts have demonstrated the successful partial simulation of the photosynthetic reaction by construction of the sequential potential field.

Molecular beam deposition is another method of preparing lamination films at a molecular level under high vacuum. In the molecular beam deposition method, the

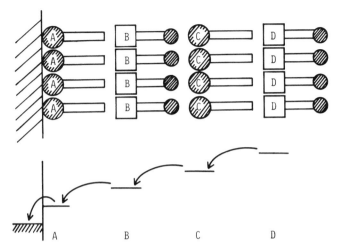

Figure 14-10 ■ Construction of sequential potential field by ideal *Y*-type laminated LB films with components A, B, C, and D.

molecular arrays in the thin film are controlled by the substrate, the temperature, and the deposition rate.[39] The molecular beam deposition has been used often for preparation of thin films of inorganic semiconductors, especially high electronic transducing transistors, quantum well lasers, and superlattice structure. However, utilization of the molecular beam deposition method for organic thin films has not been well developed yet.

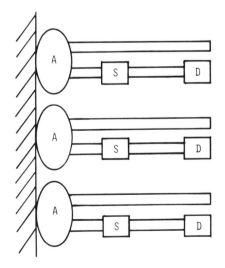

Figure 14-11 ■ Monolayer LB film produced by monomer with donor D, sensitizer S, and acceptor A.

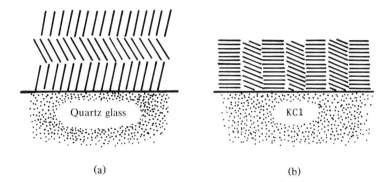

Figure 14-12 ■ Alignment of copper phthalocyanine planar molecules on (a) quartz glass and (b) KCl cleavage substrate by molecular beam deposition.

We have prepared thin films of copper phthalocyanine by molecular beam deposition. Copper phthalocyanine is known as a functional pigment, the crystal of which shows electric conductivity to some extent. When copper phthalocyanine was deposited on the substrate under high vacuum at 10^{-10} Torr by molecular beam deposition, thin films of copper phthalocyanine were produced on the substrates. The molecular planes of copper phthalocyanine in the films were perpendicular to the substrate planes when quartz glass was used as the substrate, whereas the molecular planes were parallel to the planes for the KCl cleavage substrate. These situations are illustrated in Fig. 14-12.[40] These orientations have been demonstrated by X-ray diffraction analyses, FT-IR, and UV-VIS spectra, scanning tunnel micrographs (STM),[41] and scanning electron micrographs (SEM). The SEM photographs are shown in Fig. 14-13, in which planar and linear microcrystals are observed for the films on the KCl and the quartz glass substrate, respectively. These observations are consistent with the spectrochemical results.

The single crystal of copper phthalocyanine is known as an organic semiconductor, and it shows the electric conductivity in the direction perpendicular to the molecular planes. In the film prepared by molecular beam deposition, the electric conductivity in the direction parallel to the substrate plane is higher on the glass substrate than on the KCl cleavage plane. These observations are consistent with the orientation of copper phthalocyanine molecules on the substrates, and again demonstrate the excellent arrangement of the complex molecules in the films.

Lamination films were prepared by cumulative depositions of metal-free phthalocyanine and copper phthalocyanine in this order (Fig. 14-14). The lamination film was active for photoconductivity and the photoconductivity spectrum was consistent with the optical spectrum. The photoconductivity spectrum of the metal-free phthalocyanine film is similar to that of the copper phthalocyanine film. However, copper phthalocyanine has a small but sharp peak at about 1100 nm, which metal-free phthalocyanine does not have. This peak has been assigned to be S-T absorption, which is characteristic only for copper phthalocyanine. The photoconductivity spectrum observed for the lamination film had a sharp peak at 1100 nm, which was

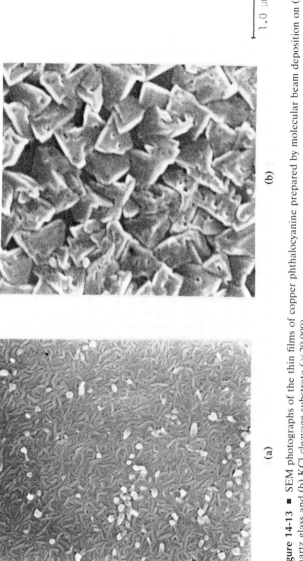

1.0 μm

(a)   (b)

Figure 14-13 ■ SEM photographs of the thin films of copper phthalocyanine prepared by molecular beam deposition on (a) quartz glass and (b) KCl cleavage substrate (×20,000).

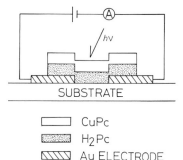

CuPc
H₂Pc
Au ELECTRODE

Figure 14-14 ■ Lamination film of metal-free phthalocyanine H_2Pc and copper phthalocyanine CuPc on a quartz glass substrate.

consistent with the observation for the single film of copper phthalocyanine, but not with that of the single film of metal-free phthalocyanine. This result suggests that electron or hole carriers generated in the copper phthalocyanine film transport to the metal-free phthalocyanine film, demonstrating that the molecular beam deposition can be effective for construction of the sequential potential field by lamination.

A sequential potential field also could be constructed in a molecular array if the redox potential of each molecule in the array changes sequentially in the spatial order. This situation could be satisfied by using mixed valence complexes.

Prussian blue is a mixed valence complex of iron(II, III) with high molecular weight. The Prussian blue films can be prepared by electrodeposition[42-44] or by casting.[45] The Prussian blue film shows electrochromism, where the Fe(II)/Fe(II), Fe(II)/Fe(III), and Fe(III)/Fe(III) states indicate white, blue, and brown color, respectively. When the potential difference is applied across the two terminals of the Prussian blue film electrode, the color gradient or redox state gradient can be achieved on the film.[46, 47] The color changes from yellow at the cathodic terminal to colorless at the anodic terminal through blue at the center. This could be one kind of sequential potential field if the gradient could be maintained after a short-circuit between the terminals. Actually the homogeneous blue color recovers when the applied voltage is cut off.

14-4-3. Atomic Array

A metal cluster is composed of several to several hundreds of metal atoms. It is not so easy to control the structure of the metal cluster producing an atomic array system. Recently, scientists have been interested in planning definite arrangements of the atoms and proposing some potential techniques for preparing them. Several kinds of structures will be possible for the metal clusters composed of the same number of atoms. Therefore, we have to consider a metal cluster with a sense of a "metallic molecule." If the metal atoms composing a cluster have different potentials, a sequential potential field can be constructed in the cluster.

We have investigated metal clusters with small size. They are prepared by alcohol reduction[48, 49] or photoreduction[35, 50] of metal ions in solutions in the presence of a water-soluble polymer like poly(N-vinyl-2-pyrrolidone). The polymers used for the preparation of colloidal dispersions of metal clusters interact both with the metal ions

Figure 14-15 ■ Schematic presentation of the preparation of colloidal dispersions of metal clusters in the presence of polymers. The polymer interacts both with the metal ion before the reduction and with the metal cluster after the reduction.

before the reduction and with the metal clusters produced by the reductions of ions as illustrated in Fig. 14-15.[48,51] The interactions between the polymer and particles of the metal cluster are very important, not only for the stability of polymer-protected colloidal dispersion of metal clusters, but also for the construction of a sequential potential field in the cluster. Such interactions include physical adsorption by van der Waals interaction, hydrophobic interaction, electrostatic interaction, hydrogen bonding, and coordination bonding.[48]

The colloidal dispersions of the noble metal clusters work as very active catalysts. For example, rhodium clusters catalyze hydrogenation of olefins,[52] palladium clusters catalyze partial hydrogenation of dienes to monoenes,[53,54] platinum clusters work as catalysts for hydrogenation of olefins,[55] and hydrogen production from protons,[34,56,57] copper clusters promote hydration of acrylonitrile producing acrylamide selectively,[58] and so on.

One of the most important characteristics of a polymer-protected metal cluster catalyst is its small particle size. Generally speaking, the particle size of a metal catalyst can influence both the catalytic activity and selectivity in chemical reactions. Because the smaller the particle size, the higher the dispersion (the molar ratio of surface atoms in all atoms), catalytic activity increases with decreasing particle size.

In the case of metal clusters with small size prepared by the alcohol reduction or photoreduction, the catalytic activity changes with size. However, the change in catalytic activity is very drastic if only the surface area or the dispersion is considered, because, in the case of a small cluster particle, most of the atoms are located on the surface. For example, the dispersion (the ratio of surfacial atoms) is about 0.8 for a 1.4 nm platinum particle. Therefore, we believe that the drastic change in the activity or selectivity of the metal cluster catalysts is attributable to the effect of the polymer surrounding the cluster particles. Thus, it could be possible that each metal atom in the cluster particle has a different redox potential because of the different environment provided by the surrounding polymer, and that a sequential potential field can be constructed in the cluster. The sequential potential field produced could provide the partial electron transfer among the metal atoms or the electron density gradient, which activates the active site of the cluster.

We will now discuss some examples of the drastic change of catalytic activity or selectivity with particle size for the polymer-surrounded metal cluster.

Poly(vinylpyrrolidone)-surrounded rhodium clusters have been prepared by refluxing solutions of RhCl$_3$ in alcohols,[59] where the particle size varies with the conditions of preparation. Such rhodium clusters work as catalysts for the hydrogenation of olefins, where the observed catalytic activity increases with decreasing particle size of

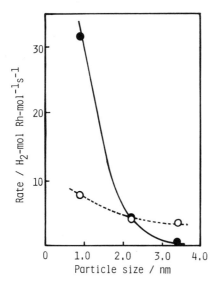

Figure 14-16 ■ Hydrogenation of 4-methyl-3-penten-2-one (●) and 3-buten-2-one (○) over polymer-surrounded rhodium cluster catalysts. The internal olefin was more smoothly hydrogenated by a 0.9 nm size cluster than the terminal olefin.

the cluster catalyst. The rate of increase of activity also depends on the nature of the olefin substrate. Some results are illustrated in Fig. 14-16. The most interesting observation arising from these results is that the rate of hydrogenation of an internal olefin is faster than that of the corresponding terminal olefin when a 0.9 nm size polymer-surrounded rhodium cluster is employed as a catalyst.

Poly(vinylpyrrolidone)-surrounded palladium cluster catalysts have also been prepared by refluxing solutions of $PdCl_2$ in alcohols in the presence of poly(N-vinyl-2-pyrrolidone).[53,54] These catalyzed the selective partial hydrogenation of dienes to monoenes under mild conditions, with cyclooctadiene, for example, giving cyclooctene.[60] The particle size of the palladium clusters, which varies with the

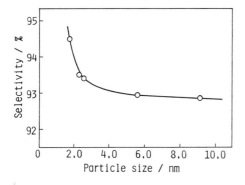

Figure 14-17 ■ Effect of particle size of polymer-surrounded palladium cluster catalysts upon the selectivity of monoene formation in the hydrogenation of 1,3-cyclopentadiene. The selectivity indicates the yield of cyclopentene at 95% hydrogenation.

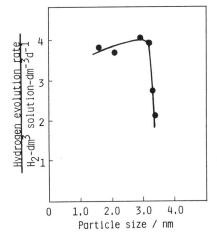

Figure 14-18 ■ Effect of the particle size of polymer-surrounded platinum cluster catalyst upon the visible light-induced hydrogen evolution rate in the system of EDTA/Ru(bpy)$_3^{2+}$/methyl viologen/Pt in water.

preparation conditions, dramatically influenced the selectivity of monoene formation, as shown in Fig. 14-17.

Polymer-surrounded platinum clusters catalyzed the photochemical evolution of hydrogen from Na$_2$EDTA solutions in the presence of Ru(bpy)$_3^{2+}$ and methyl viologen. The rate of hydrogen evolution was dependent on the particle size of the platinum cluster, as shown in Fig. 14-18, a maximum rate being observed with a cluster of ca. 3 nm particle size.[56]

Recently we have succeeded in preparation of bimetallic clusters from mixtures of two different metal ions in the presence of polymers. For example, the colloidal dispersions of Pd-Pt bimetallic clusters were prepared by refluxing mixed solutions of

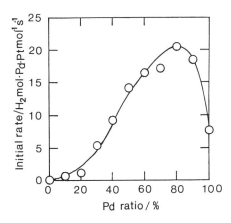

Figure 14-19 ■ Dependence of the catalytic activity of the polymer-surrounded Pd-Pt bimetallic clusters on the metal composition in particle hydrogenation of 1,3-cyclooctadiene.

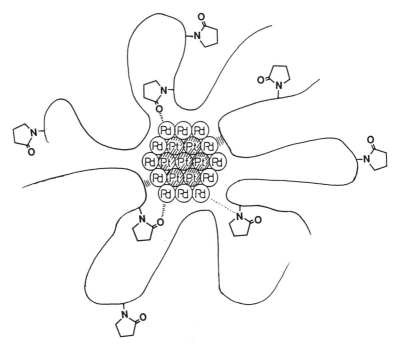

Figure 14-20 ■ A model structure of polyvinylpyrrolidone-surrounded Pd/Pt = 4/1 bimetallic cluster. The EXAFS data suggest the core structure of the cluster.

$PdCl_2$ and H_2PtCl_6 in alcohol/water in the presence of poly(*N*-vinyl-2-pyrrolidone). The atomic ratio of Pd/Pt varies with the relative amounts of the metal ions[61] added. Poly(vinylpyrrolidone)-surrounded Pd-Pt clusters catalyzed the selective partial hydrogenation of dienes to monoenes under mild conditions, with 1,3-cyclooctadiene, for example, giving cyclooctene. Such Pd-Pt clusters had nearly the same particle size of ca. 1.4 nm. The catalytic activity of Pd-Pt clusters depends on the atomic ratio, as illustrated in Fig. 14-19, a maximum activity being observed with a bimetallic cluster with an atomic ratio of Pd/Pt = 4/1.

The structure of poly(vinylpyrrolidone)-surrounded Pd-Pt bimetallic clusters was investigated by EXAFS measurement.[62] Figure 14-20 illustrates a core structure for a polymer-surrounded Pd/Pt = 4/1 bimetallic cluster. Such a cluster demonstrates the magic natural rule that creates the controlled structure from chaos. Coexistence of different atoms in a cluster will assist the construction of a sequential potential field provided by the surrounding polymer.

14-5. Conclusion

The present chapter has described a sequential potential field and electron transfer in metal complex clusters as characteristics of macromolecular complexes. As mentioned in the introduction to the chapter, a sequential potential field is of vital importance for uniform understanding of electron transfer both in a biological and a solid system. A macromolecular complex system provides a simulation model for these nonhomogeneous systems.

For the construction of a sequential potential field, the importance of arrays of phases, molecules, and atoms is emphasized. Thus, the production of such arrayed systems has been reviewed as an approach to the sequential potential field.

Future development of this concept in macromolecular complex systems is expected to be useful and effective both for understanding rapid and selective electron and/or charge transfers in natural and/or artificial systems and for developing new functional materials in advanced technologies. From these viewpoints, macromolecular complexes are likely to receive far more attention in the future and grow into an entirely new field of research both in academia and industry.

References

1. H. Taube, *Electron Transfer Reactions of Complex Ions in Solutions*, Academic, New York (1970).
2. F. Basolo and R. G. Pearson, *Mechanisms of Inorganic Reactions*, 2nd ed., Wiley, New York (1967).
3. T. Guarr and G. McLendon, *Coordination Chem. Rev.*, **68**, 1 (1985).
4. R. A. Marcus and N. Sutin, *Biochim. Biophys. Acta*, **811**, 265 (1985).
5. G. McLendon, *Acc. Chem. Res.*, **21**, 160 (1988).
6. M. Gratzel, ed., *Energy Resources through Photochemistry and Catalysis*, Academic, New York (1983).
7. J. Deisenhofer, O. Epp, K. Miki, R. Huber, and H. Michel, *J. Mol. Biol.*, **180**, 385 (1984); *Nature*, **318**, 618 (1985).
8. T. J. Meyer, *Acc. Chem. Res.*, **22**, 163 (1988).
9. M. D. Newton and N. Sutin, *Ann. Rev. Phys. Chem.*, **35**, 437 (1984).
10. K. V. Mikkelsen and M. A. Ratner, *Chem. Rev.*, **87**, 113 (1987).
11. G. L. Closs and J. R. Miller, *Science*, **240**, 440 (1988).
12. H. M. McCornell, *J. Chem. Phys.*, **35**, 508 (1961).
13. K. I. Zamaraev, R. F. Khairutdinov, and J. R. Miller, *Chem. Phys. Lett.*, **57**, 311 (1978).
14. J. R. Miller and J. V. Beitz, *J. Chem. Phys.*, **74**, 6746 (1981).
15. G. Gamow, *Z. Phys.*, **51**, 204 (1978).
16. B. Chance, D. DeVault, H. Frauenfelder, R. A. Marcus, J. R. Schrieffer, and N. Sutin, eds. *Tunneling in Biological Systems*, Academic, New York (1979).
17. T. C. L. G. Sollner, W. D. Goodhne, P. E. Tannenwald, C. D. Parker, and D. D. Peck, *App. Phys. Lett.*, **43**, 588 (1983).
18. N. Toshima, *Kagaku (Chem. Kyoto)*, **44**, 173 (1989).
19. E. Tsuchida, *Yukigosei Kagaku (J. Synth. Org. Chem. Jpn.)*, **46**, 923 (1988).
20. R. A. Marcus, *Ann. Rev. Phys. Chem.*, **15**, 155 (1964).
21. N. Toshima, M. Kaneko, and Y. Seki, *Macromolecular Complex*, Kyoritsu, Tokyo (1990).
22. E. Tsuchida and M. Kurimura, eds. *Bases of Macromolecular Complex*, Japan Scientific Societies Press, Tokyo (1990).
23. Y. Kurimura, *Adv. Polym. Sci.*, **90**, 105 (1989).
24. K. Niki, Y. Kawasaki, K. Kimura, Y. Higuchi, and N. Yasuoka, *Langumuir*, **3**, 982 (1987).
25. T. Sagara, K. Niwa, A. Sone, C. Hinner, and K. Niki, *Langumuir*, **6**, 254 (1990).
26. M. Plato, K. Mobius, M. E. Michel-Beyerle, M. Bixon, and J. Jortner, *J. Am. Chem. Soc.*, **110**, 7279 (1988).

27. A. Yamada, N. Toshima, and M. Kaneko, eds., *Photo-energy Conversion*, Japan Scientific Societies Press, Tokyo (1983).
28. E. Pelizzetti and N. Serpone, *Homogeneous and Heterogeneous Photocatalysis*, Reidel, Dordrecht (1985).
29. M. Kaneko, J. Motoyoshi, and A. Yamada, *Nature*, **285**, 468 (1980).
30. N. Toshima, Y. Yamada, J. Ishiyama, and H. Hirai, *Nippon Kagaku Kaishi*, **1984**, 368 (1984).
31. M. A. Fox, *Pure Appl. Chem.*, **60**, 1013 (1988).
32. J. H. Fendler, *Chem. Rev.*, **87**, 877 (1987).
33. K. I. Zamaraev, S. V. Lymar, M. Z. Khramov, and V. N. Parmon, *Pure Appl. Chem.*, **60**, 1039 (1988).
34. N. Toshima, T. Takahashi, and H. Hirai, *J. Macromol. Sci.-Chem.*, **A25**, 669 (1988).
35. N. Toshima, T. Takahashi, and H. Hirai, *Chem. Lett.*, **1985**, 1245 (1985).
36. N. Toshima and T. Takahashi, *Chem. Lett.*, **1988**, 573 (1988).
37. M. Fujihira and H. Yamada, *Thin Solid Films*, **160**, 125 (1988).
38. M. Fujihira and H. Sakomura, *Thin Solid Films*, **179**, 471 (1989).
39. N. Toshima, *Kagaku Kogyo* (*Chem. Ind.*), **40**, 959 (1989).
40. S. Tanishima, K. Ohno, Y. Sakakibara, and N. Toshima, *Chem. Express*, **5**, 153 (1990).
41. W. Mizutani, Y. Sakakibara, M. Ono, S. Tanishima, K. Ohno, and N. Toshima, *Jpn. J. Appl. Phys.*, **28**, L1460 (1989).
42. V. D. Neff, *J. Electrochem. Soc.*, **125**, 887 (1978).
43. K. Itaya, H. Akahoshi, and S. Toshima, *J. Electrochem. Soc.*, **129**, 1498 (1982).
44. K. Itaya, I. Uchida, and V. D. Neff, *Acc. Chem. Res.*, **19**, 162 (1988).
45. N. Toshima, R.-J. Lin, and M. Kaneko, *Chem. Lett.*, **1990**, 485 (1990).
46. M. Kaneko, and K. Toyoda, *Makromol. Chem., Rapid. Commun.*, **9**, 407 (1988).
47. N. Toshima, R.-J. Lin, and M. Kaneko, *Chem. Lett.*, **1989**, 1099 (1989).
48. H. Hirai and N. Toshima, *Tailored Metal Catalysts*, Y. Iwasawa, Reidel, Dordrecht (1986), pp. 87–140.
49. H. Hirai, Y. Nakao, and N. Toshima, *J. Macromol. Sci.-Chem.*, **A13**, 727 (1979).
50. M. Ohtaki and N. Toshima, *Chem. Lett.*, **1990**, 489 (1990).
51. M. P. Andrews and G. A. Ozin, *Chem. Materials*, **1**, 174 (1989).
52. H. Hirai, Y. Nakao, and N. Toshima, *J. Macromol. Sci.-Chem.*, **A12**, 1117 (1978); H. Hirai, Y. Nakao, N. Toshima, and K. Adachi, *Chem. Lett.*, **1976**, 905 (1976).
53. H. Hirai, H. Chawanya, and N. Toshima, *Bull. Chem. Soc. Jpn.*, **58**, 682 (1985).
54. H. Hirai, H. Chawanya, and N. Toshima, *Reactive Polym.*, **3**, 127 (1985).
55. N. Toshima and T. Takahashi, *Chem. Lett.*, **1988**, 573 (1988).
56. N. Toshima, M. Kuriyama, Y. Yamada, and H. Hirai, *Chem. Lett.*, **1981**, 793 (1981).
57. N. Toshima, T. Takahashi, and H. Hirai, *Chem. Lett.*, **1987**, 1031 (1987).
58. H. Hirai, H. Wakabayashi, and M. Komiyama, *Bull. Chem. Soc. Jpn.*, **59**, 545 (1986).
59. H. Hirai, Y. Nakao, and N. Toshima, *Chem. Lett.*, **1978**, 545 (1978).
60. Z. Bin and N. Toshima, *Kobunshi Ronbunshu*, **46**, 551 (1989).
61. N. Toshima, K. Kushihashi, T. Yonezawa, and H. Hirai, *Chem. Lett.*, **1989**, 1769 (1989).
62. N. Toshima, T. Yonezawa, M. Harada, K. Asakura, and Y. Iwasawa, *Chem. Lett.*, **1990**, 815 (1990).

Geared Cycle in Photoexcited Multielectron Transfer in Macromolecule–Metal Complexes

M. Kaneko

The Institute of Physical and Chemical Research, Wako, Saitama, 351-01 Japan

15-1. Geared Cycle for Electron Transport

A geared cycle is composed of a series of redox reactions by which an electron is transported from one terminal to another terminal through redox reactions between neighboring compounds (see Fig. 15-1). Geared cycles can work when the redox potentials of the components are arranged in a sequential way according to their magnitudes. The electron is transported from the redox couple having the lowest redox potential to that having the highest potential. This kind of geared cycle plays an essential role when electrons are transported between molecules. This is especially the case for the electron transport in biological systems and in electronic devices based on molecular assemblies. Geared cycles are important in the processes in which electron transfer is involved as follows:

1. Electron transfer between molecules (Fig. 15-1).
2. Catalysis to mediate a reaction between substrate and reactant when the reaction is impossible without the mediating catalyst [Fig. 15-2(a)].
3. Mediation between single- and multielectron transfer reactions. Four-electron oxidation of water is taken as an example and is shown in Fig. 2(b).
4. Mediation of photoelectron processes and conventional electron-transfer reactions [Fig. 15-3(a)].
5. Mediation of one photoelectron process and a multielectron process, such as water oxidation [Fig. 15-3(b)].

Among these items, electron transfer between molecules (1) is described in Chapter 12 and catalysis (2) is discussed in Chapter 14. In this chapter geared cycles

$$x_1^{n+1} \rangle \langle x_2^n \rangle \langle x_3^{n+1} \rangle ----- \langle x_m^n$$
$$x_1^n \quad x_2^{n+1} \quad x_3^n \quad\quad\quad x_m^{n+1}$$

$$\xleftarrow{\quad\quad\quad} e^-$$

Redox potentials; $(E_o)_1 > (E_o)_2 > ----> (E_o)_m$

Figure 15-1 ■ Geared cycles for electron transport.

concerning multielectron transfer (3) and (5) will be described in conjunction with natural and artificial photosynthetic processes.

Multielectron transfer (3) is an important reaction in biological systems, such as water oxidation in photosynthesis, oxidation by dioxygen in respiration, nitrogen fixation by nitrogenase, oxygenation by oxygenase, and other various redox reactions. In order to couple the photoexcitation process with an electron-transfer reaction (4), mediation is important to realize photoinduced electron transport as represented by the photosynthetic electron process. In photosynthesis, a four-electron process of

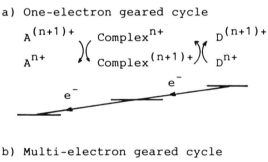

a) One-electron geared cycle

$$A^{(n+1)+} \qquad Complex^{n+} \qquad D^{(n+1)+}$$
$$A^{n+} \qquad Complex^{(n+1)+} \qquad D^{n+}$$

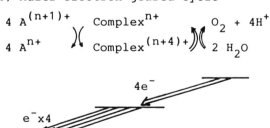

b) Multi-electron geared cycle

$$4\ A^{(n+1)+} \qquad Complex^{n+} \qquad O_2 + 4H^+$$
$$4\ A^{n+} \qquad Complex^{(n+4)+} \qquad 2\ H_2O$$

$$4e^-$$
$$e^- \times 4$$
$$e$$

Figure 15-2 ■ Geared cycles for electron transfer mediated by complex n^+.

a) Photoinduced one-electron geared cycle

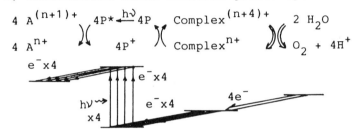

b) Photoinduced multi-electron geared cycle

Figure 15-3 ▪ Photoinduced geared cycles for electron transfer mediated by complex n^+.

water oxidation to give dioxygen is

$$2H_2O \longrightarrow O_2 + 4H^+ + 4e^- \tag{15-1}$$

coupled with a one-electron process of photoexcitation occurring at a chlorophyll compound. Such a process is possible only by utilizing a geared cycle such as Fig. 15-3(b).

In photosynthesis, catalysis for four-electron oxidation of water to evolve dioxygen [Eq. (15-1)] is a very important process to provide electrons to the whole system.[1] Oxygen evolving geared cycles in nature as well as in artificial systems will be treated in Sections 15-2 and 15-3, respectively. Thereafter, geared cycles for synthetic photochemical water cleavage systems as a model for the photosynthesis will be discussed (Section 15-4).

15-2. Photosynthesis and Multielectron Transfer in the Oxygen Evolution

Photosynthesis[2] by the green plants in nature is represented by a photoinduced electron flow from water to carbon dioxide as shown in Fig. 15-4.[1] The electron abstracted from water is pumped up to higher energy by solar irradiation at photosystems II and I (PS II and I) where chlorophyll (Chl) molecules work as photoexcitation centers. After being pumped up, the electron flows to lower energy levels through many geared cycles, and in the final stage it reduces NADP to NADPH, which

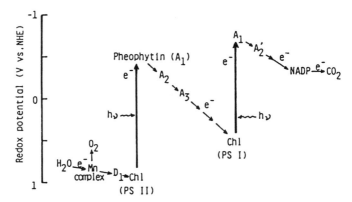

Figure 15-4 ▪ Photoinduced electron flow in the photosynthesis. $A_1, \cdots, A_n$, acceptors; D_1, donor; Chl, chlorophyll; PS I, II, photosystem I, II.

thereafter reduces carbon dioxide to carbohydrate $(C_6H_{12}O_6)$. Photosynthesis is therefore regarded as reduction of CO_2 by water with the help of solar energy (visible light):

$$CO_2 + 2H_2O + 8 \text{ photons} \longrightarrow (1/6)C_6H_{12}O_6 + O_2 + H_2O \qquad (15\text{-}2)$$

Water is the so-called electron source that provides electrons to the photosynthesis. Water oxidation to get electrons [Eq. (15-1)] is a key reaction for the whole photosynthetic system. Dioxygen is evolved as a by-product of this water oxidation. Because almost all life on Earth is supported by photosynthesis, water oxidation is the most important process for life.

In photosynthesis the components for transporting electrons are arranged in a photosynthetic membrane called the thylakoid membrane. The structure at and near

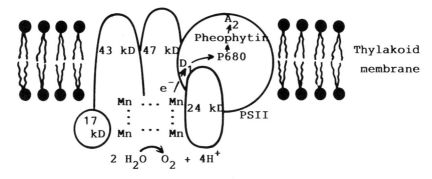

Figure 15-5 ▪ A model of the oxygen evolving center in photosynthesis. The number kD represents the molecular weight (kilodalton) of each protein.

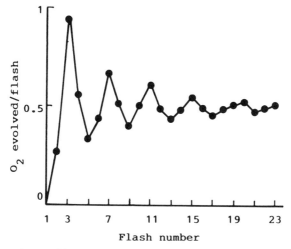

Figure 15-6 ■ Flash experiment for photosynthetic oxygen evolution.

the oxygen evolving center is schematically shown in Fig. 15-5. The central part is called PS II and it is composed of chlorophyll molecules, two molecules of pheophytin (chlorophyll without the central Mg ion), quinones, and a manganese–protein complex.[3] The photoreaction center chlorophyll is called P680. Four-electron oxidation of water is coupled with one-electron excitation through geared cycles composed of the Mn–protein complex and an unknown compound D_1 (represented very often by Z), which is presumably a hydroquinone analogue.

In the Mn–protein complex of photosynthesis,[4–6] at least four Mn ions are contained to achieve four-electron oxidation of two molecules of water evolving one molecule of dioxygen [see Eq. (15-1)]. As for the four-electron oxidation of water in photosynthesis, a famous experiment has been reported.[7] After chloroplasts from spinach were kept in the dark, they were irradiated with pulsed flash lights, and the evolved dioxygen was measured as a function of the number of the flash (Fig. 15-6). The first flash did not induce oxygen evolution, but the third flash brought about maximum oxygen evolution, and, thereafter, every fourth flash (i.e., 7th, 11th, 15th, and so on) gave a peak for oxygen evolution.[7] In order to explain this phenomenon, five redox states (S_0, S_1, S_2, S_3, and S_4) have been assumed for the oxygen evolving center. S_0 is the most reduced state, and every one-electron oxidation gives the states, S_1, S_2, S_3 and S_4, and O_2 is evolved after the S_4 state is reached. The fact that the third flash gives the first maximum O_2 evolution indicates that S_1 is the prevailing state in the dark. Although many proposals have been made as to the structure of these S states, it is still unresolved. They should correspond to different oxidation states of the four Mn atoms involved in the oxygen evolving protein complex.

Among the various Mn compounds proposed as models for the oxygen evolving Mn complex, dinuclear[8, 9] **1** and tetranuclear[10–12] Mn complexes **2** are the typical

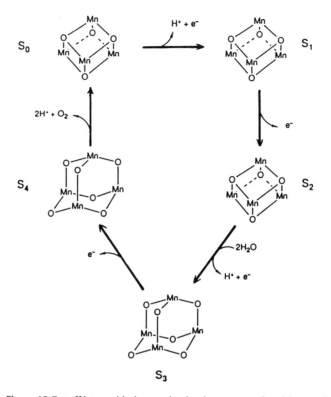

Figure 15-7 ▪ Water oxidation mechanism by a tetranuclear Mn complex.

examples:

$$[(bpy)_2 Mn \underset{O}{\overset{O}{<}} Mn(bpy)_2]^{3+}$$

1

$$[Mn_4O_2(O_2CMe)_7(bpy)_2]^-$$

2

Taking as an example a cubane-type tetranuclear Mn_4O_4, the corresponding S_n states and the water oxidation mechanism have been proposed as shown in Fig. 15-7.[10] In this scheme, O denotes O^{2-} or OH^- ligands. Four-electron oxidation of the complex from S_0 to S_4 is carried out stepwise by the oxidized photoreaction center. One proton is released from the complex in the S_0 to S_1 and in the S_2 to S_3 steps, and two protons are released in the S_4 to S_0 step. The S_4 state releases oxygen and returns to S_0. As for the valence state of the Mn ions, three possibilities have

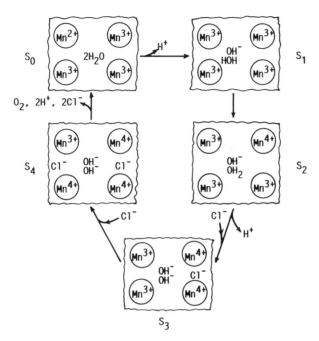

Figure 15-8 ▪ A mechanistic model for the photosynthetic water oxidation by Mn-protein complex (boxed).

been proposed.[10] One possibility is the states

$$S_0: \quad 3[Mn(III)] + [Mn(IV)]$$

$$S_1: \quad 2[Mn(III)] + 2[Mn(IV)]$$

$$S_2: \quad [Mn(III)] + 3[Mn(IV)]$$

$$S_3: \quad 4[Mn(IV)]$$

$$S_4: \quad 3[Mn(IV)] + [Mn(V)]$$

which are in agreement with ESR and UV spectral data.

Anions such as Cl^- and Br^- play an important role in the oxygen evolution of the photosynthesis. High binding affinity of Cl^- to S_0 and S_1 states was not found, but was detected for the binding to S_2 and S_3 states.[13] When these results are included, the mechanism of oxygen evolution in photosynthesis can be schematically represented as in Fig. 15-8. Four-electron oxidation of this oxygen evolving center is achieved stepwise by the oxidized photoreaction center chlorophyll ($P680^+$) as shown in Fig. 15-3(b) where P corresponds to P680. Thus, through the geared cycles composed of Mn–protein complex and an electron relay, four-electron oxidation of water by four photons is realized in the photosynthesis.

15-3. Geared Cycles for Multielectron Transfer in the Synthetic Water Oxidation Systems

15-3-1. Mechanism of Four-Electron Oxidation of Water to Give Oxygen

As described in the last section, water oxidation is a very important reaction in nature because it is the key process for providing electrons to the energy cycles in nature.[1] Three mechanisms are possible for this water oxidation to give oxygen (in the parentheses of each equation, the redox potential is shown as volts versus NHE at pH = 7):

Four-step mechanism (stepwise mechanism)

$$H_2O \longrightarrow HO\cdot + e^- + H^+ \qquad (2.33\ V) \qquad (15\text{-}3)$$

$$HO\cdot + OH \longrightarrow H_2O_2 + e^- \qquad (0.39\ V) \qquad (15\text{-}4)$$

$$H_2O_2 \longrightarrow HO_2\cdot + e^- + H^+ \qquad (1.27\ V) \qquad (15\text{-}5)$$

$$HO_2\cdot \longrightarrow O_2 + e^- + H^+ \qquad (-0.73\ V) \qquad (15\text{-}6)$$

Two-step mechanism

$$2H_2O \longrightarrow H_2O_2 + 2e^- + 2H^+ \qquad (1.36\ V) \qquad (15\text{-}7)$$

$$H_2O_2 \longrightarrow O_2 + 2e^- + 4H^+ \qquad (0.27\ V) \qquad (15\text{-}8)$$

One-step mechanism (synchronous mechanism)

$$2H_2O \longrightarrow O_2 + 4e^- + 4H^+ \qquad (0.82\ V) \qquad (15\text{-}9)$$

In the stepwise four-step mechanism, each intermediate is released after one-electron oxidation. Because the redox potential of the first reaction [Eq. (15-3)] is high (2.33 V), the stepwise mechanism is the most unfavorable. In the two-step mechanism, the oxidized compound is released only after two-electron oxidation as H_2O_2, which is further oxidized by two electrons to release oxygen. The much lower redox potential (1.36 V) of the first step [Eq. (15-7)] allows much easier reaction than the four-electron mechanism. In the synchronous one-step mechanism [Eq. (15-9)], no intermediate is released, and only after four-electron oxidation is the product (O_2) released. This is the most favorable reaction among the three mechanisms because oxygen evolution proceeds at the potential of 0.82 V.

In the photosynthetic water oxidation, only the four-electron mechanism is possible because the redox potential of $P680^+/P680$ lies near 0.85 V. Electrochemical water oxidation proceeds also by this mechanism, although overpotential of about 0.4 V is required in this case. In this synchronous four-electron mechanism, all the intermediates from the one-electron oxidized product to the three-electron oxidized product are trapped in the reaction system and stabilized. Of course the overall free energy change accompanying the reaction is the same in all three mechanisms, but the four-electron mechanism can proceed at the lowest potential.

Catalyst design is important for achieving synthetic water oxidation to evolve oxygen. For this purpose, it is important to design a catalyst system that can realize a synchronous four-electron mechanism. In order to couple this four-electron mechanism with a one-electron oxidation system, the use of a geared cycle such as

Fig. 15-2(b) is essential. In succeeding sections, chemical and electrochemical water oxidation using designed catalysts and geared cycles will be described.

15-3-2. Chemical Oxidation of Water by Geared Cycles

Photochemical energy conversion is attracting attention in conjunction with solar energy conversion, which is becoming more and more important as a new energy resources.[14-16] Recently, the environmental problems arising from increased levels of carbon dioxide, and acid rains in particular, have roused much interest in solar energy conversion and in photosynthesis. In order to produce compounds capable of being used economically as a fuel, water should be used as a reductant in a manner similar to photosynthesis, that is, water should be used as an electron source. This is important also for studying the mechanism of photosynthesis as well as for constructing artificial photosynthetic models. For this purpose, four-electron oxidation of water to produce oxygen is the key process.

Synthetic water oxidation has not been studied much because of the difficulty in realizing four-electron oxidation of two molecules of water. Metal oxide solids, such as RuO_2 and MnO_2 have been known to catalyze water oxidation, but these catalysts cannot be models for the water oxidation process because redox reactions of the catalyst itself are not easy to study. In order to establish a molecule-based catalyst or a photosynthetic model for water oxidation, metal complexes are the best candidate. For four-electron water oxidation, polynuclear metal complexes that might work as multielectron catalysts must be considered promising types of compounds. Only a few metal complexes have been reported as catalysts for water oxidation. One such complex is tetrakis(bipyridine)-μ-oxo-diruthenium(III,III),

$$[(bpy)_2 Ru-O-Ru(bpy)_2]^{4+}$$
$$\underset{H_2O}{|} \qquad \underset{H_2O}{|}$$

3

which yields oxygen in the presence of a strong oxidant such as Ce^{4+} in a homogeneous aqueous solution.[17-18] The redox chemistry of the complex also has been reported.[17]

The μ-oxo-trinuclear rutheniumammine complex (**4**), called Ru-red, catalyzed water oxidation to give oxygen in an aqueous solution[19-21]:

$$[(NH_3)_5 Ru-O-Ru(NH_3)_4-O-Ru(NH_3)_5]^{7+}$$

4

Catalytic water oxidation has been studied only in a homogeneous aqueous solution when a metal complex is used as a catalyst. However, the heterogeneous oxygen evolving center of the photosynthesis as previously described suggests strongly that a heterogeneous environment is important to achieve water oxidation. Moreover, it is important to utilize a heterogeneous phase in order to construct artificial photosynthetic systems,[1,22] because homogeneous catalytic photochemical reactions always results in back reactions; the whole system only returns to the starting state and gives no net output. The present author has been interested in developing

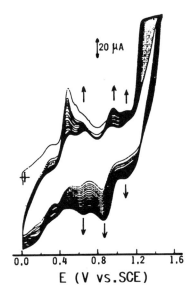

Figure 15-9 ■ Cyclic voltammogram of the dinuclear Ru complex **3** at a Nafion-coated BPG in a 0.1 M H_2SO_4 aqueous solution measured under continuous scanning for 1 hr. Scan rate 50 mV s^{-1}.

heterogeneous metal complex catalyst systems for water oxidation and utilizing them for artificial photoenergy conversion systems.

It was found that the oxo-bridged dinuclear Ru complex **3** gives distinct redox peaks in a cyclic voltammogram (CV) when it is incorporated in a polyanion polymer film[23] that is coated on a basal plane pyrolytic graphite (BPG) electrode. Figure 15-9 shows the CV of **3** incorporated into a Nafion film coated on BPG.[23] The redox waves increase with repeated cyclic scanning, showing that the complex is incorporated into the Nafion film with time. The incorporation is caused by electrostatic binding of the cationic complex to the anionic Nafion. After almost 1 hr the peak height becomes saturated, indicating that the incorporation is complete under these conditions. The three reversible redox waves observed between 0.6 and 1.1 V (versus SCE) have been assigned to the redox reactions of the complex. The narrow peak separation of 20–30 mV of these redox waves suggests strongly that the redox reactions are two-electron processes. The homogeneous catalysis of the complex for oxygen evolution from water indicates that one molecule of this complex is capable of four-electron oxidation of water. Based on these results, the redox reactions and the potentials of this complex are expressed as

$$\text{Ru(V)-Ru(V)} \underset{-2e^-}{\overset{+2e^-}{\rightleftharpoons}} \text{Ru(IV)-Ru(IV)} \underset{-2e^-}{\overset{+2e^-}{\rightleftharpoons}} \text{Ru(III)-Ru(III)} \underset{-2e^-}{\overset{+2e^-}{\rightleftharpoons}} \text{Ru(II)-Ru(II)}$$
$$\quad 1.1\text{ V} \qquad\qquad 0.88\text{ V} \qquad\qquad 0.67\text{ V}$$

$$(15.10)$$

(The potentials are versus SCE.) After the starting complex [Ru(III)-Ru(III)] is oxidized to a Ru(V)-Ru(V) complex by oxidant [here Ce(IV)], it works as a four-electron oxidant of water to yield oxygen. This is a typical geared cycle that connects a one-electron reaction with a multielectron process as shown by Scheme 1.

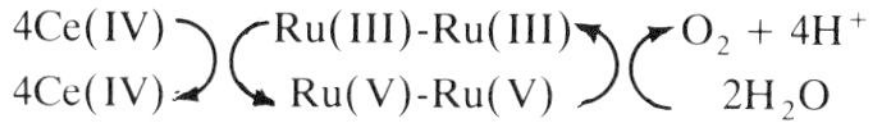

$$4Ce(IV) \searrow \nearrow Ru(III)\text{-}Ru(III) \searrow \nearrow O_2 + 4H^+$$
$$4Ce(IV) \nearrow \searrow Ru(V)\text{-}Ru(V) \nearrow \searrow 2H_2O$$

Scheme 1

It is interesting that the complex **3** is active also as a water oxidation catalyst in a heterogeneous state incorporated into clay[24] or polymeric compounds.

Although this dinuclear Ru complex **3** can catalyze water oxidation, there are only several turnovers for the catalysis, probably because of the instability of the higher oxidation state. Moreover, it is subjected to decomposition under irradiation in its higher oxidation state, such as Ru(V)-Ru(V).[23] On the other hand, the trinuclear Ru complex **4** was much more active than the dinuclear complex **3**, giving a turnover number of more than 50[24] under the experimental conditions studied. The activities of the dinuclear **3** and the trinuclear Ru complexes **4** are compared in Table 15-1. The valence state of the trinuclear Ru complex **4** (Ru-red) is Ru(III)-Ru(IV)-Ru(III). It has an absorption peak at 532 nm with a small peak at 362 nm. When Ru-red is dissolved in an acidic aqueous solution, it undergoes one-electron oxidation with the formation of Ru-brown [Ru(IV)-Ru(III)-Ru(IV)], which shows an absorption peak at 460 nm. From a cyclic voltammogram of the complex, the mechanism for the water oxidation catalysis is represented by Scheme 2.[25]

The starting compound, Ru-red, is oxidized to a Ru(V)-Ru(V)-Ru(V) complex, and then oxidizes water to give oxygen and Ru-brown. During the catalysis the complex cycles between the highest oxidation state and the Ru-brown, which has been observed by visible absorption spectroscopy. The Ru-red undergoes decomposition under irradiation, but the Ru-brown is stable even under irradiation.

The trinuclear Ru complex **4** is active also in a heterogeneous phase of clay[24, 25] or polymer. The activities of **4** in a homogeneous solution, as well as in a heterogeneous state adsorbed into kaolin clay, are shown in Table 15-2. In the latter case, the complex was adsorbed into kaolin powders by suspending the kaolin in an aqueous solution of the complex, filtering off the powders, rinsing with fresh water, and using it as a catalyst by suspension in an aqueous solution of Ce(IV).

It is to be noticed that the activities are almost the same for the homogeneous and the heterogeneous catalysts. The catalysis is composed of two main reactions, that is, oxidation of the catalyst by Ce(IV) and oxidation of water by the highest oxidized state Ru complex, as shown in the Scheme 2. Because the former is an intermolecular

Table 15-1 ■ **Catalytic Activities of Dinuclear 3 and Trinuclear 4 Ru Complexes[a] in a Homogeneous Aqueous Solution for Oxygen Evolution**

Catalyst	Amount (mM)	O₂ Evolved	
		Amount (μl)	Turnover Number
Dinuclear Ru **3**	0.2	88	7.3
Trinuclear Ru **4** (Ru-red)	0.1	1325	54.7

[a]Complex/Ce(IV) = 1/400 (mol. ratio), 25°C, 1 hr.

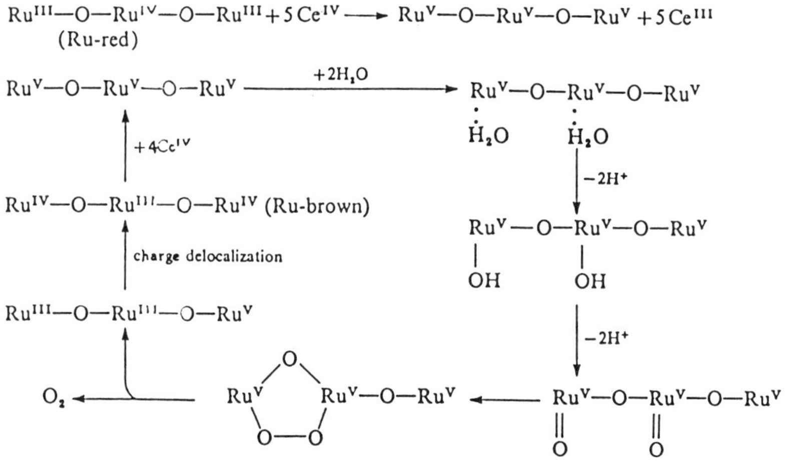
RuIII—O—RuIV—O—RuIII + 5 CeIV → RuV—O—RuV—O—RuV + 5 CeIII
(Ru-red)
RuV—O—RuV—O—RuV
+2H2O
RuV—O—RuV—O—RuV
H2O H2O
+4CeIV
−2H+
RuIV—O—RuIII—O—RuIV (Ru-brown)
RuV—O—RuV—O—RuV
OH OH
charge delocalization
RuIII—O—RuIII—O—RuV
−2H+
O2
RuV RuV—O—RuV
RuV—O—RuV—O—RuV
O O

Scheme 2

reaction, it would be affected much by the diffusion of the reactant. However, because the latter should be an intramolecular reaction when considering oxidation of the coordinated water molecules, it would not be affected much by diffusion factors. The intermolecular oxidation of the Ru complex by Ce(IV) must therefore be prohibited when the catalyst is incorporated in a heterogeneous phase, but the intramolecular oxidation of the coordinated water molecules in the complex might not be hindered much even when the complex is incorporated in a heterogeneous phase. Therefore, the near equality of catalytic activity for the homogeneous and the heterogeneous phases shown in Table 15-2 would indicate that the intramolecular water oxidation step is the rate-determining step for the whole process.

Table 15-2 ▪ Oxygen Evolution by Water Oxidation Using Ru-red 4 under Homogeneous and Heterogeneous Conditions at 25°C[a]

		Ru-Red in Water (Homogeneous)		Ru-Red in Kaolin Clay Heterogeneous)	
Ru-Red:Ce(IV) Molar Ratio	Ce(IV) (10^{-5} mol)	O_2 (10^{-6} dm^3)	Turnover Number	O_2 (10^{-6} dm^3)	Turnover Number
1:100	10.03	459	19.0	—	—
1:200	20.07	858	35.4	870	36.0
1:300	30.10	1073	44.3	1065	44.0
1:400	40.14	1325	54.7	1022	42.2
1:500	50.18	1311	54.1	1044	43.1

[a]Ru-red, 10^{-6} mol; kaolin clay, 300 mg; solution volume, 5 cm^3; reaction time, 1 hr.

Table 15-3 ▪ Catalytic Activity of the Dinuclear Mn Complex 1 in Homogeneous
and Heterogeneous States for Water Oxidation with Excess
Oxidant [Ce(IV) / 10 ml H$_2$O]

State of 1	O$_2$ (μl) Evolved
Homogeneous, saturated	0 (1 day)
Heterogeneous, 181 μmol suspended	148 (4 hr)
Heterogeneous, 1.8 μmol adsorbed on 150 mg kaolin clay	22 (3 hr)

Although a μ-dioxo-dinuclear Mn complex 1 had been proposed as a model compound of the oxygen evolving center in photosynthesis, the complex could not catalyze water oxidation to give oxygen in a homogeneous solution. Interestingly, it has been found that this complex can catalyze water oxidation when used in a heterogeneous state[26, 27]; the results are shown in Table 15-3. For the heterogeneous phase, either excess complex present over its solubility and suspended in water is active or the complex incorporated into a solid, such as clay, can also catalyze water oxidation.

In a cyclic voltammogram, the complex 1 showed two reversible redox peaks at 1.23 and 1.09 V versus SCE, indicating that it can work as a two-electron oxidation catalyst for water oxidation. In order to achieve four-electron water oxidation for oxygen evolution with this complex catalyst, at least two molecules of the complex have to work together on two molecules of water. In a homogeneous solution this would be impossible; only under heterogeneous conditions would such cooperative catalysis by two molecules of the complex be possible. The mechanism for the heterogeneous catalysis by the Mn complex is not known, but the reaction would proceed similarly to a scheme occurring in the photosynthetic oxygen evolution center as shown in Figs. 15-7 and 15-8.

Table 15-4 ▪ Oxygen Evolution by Water Oxidation with Heterogeneous
Dinuclear Mn Complexes[a]

Complex	Medium	O$_2$ Evolved (μl)	
		Without Stirring	With Stirring
[(bpy)$_2$Mn(O)$_n$Mn(bpy)$_2$]$^{3+}$	water	94	141
	0.1 M HNO$_3$	101	171
	0.1 M NaOH	113	144
[(phen)$_2$Mn(O)$_2$Mn(phen)$_2$]$^{3+}$	water	26	32
	0.1 M HNO$_3$	29	39
	0.1 M NaOH	14	32

[a]Mn complex, 200 mg; solution, 10 ml; Ce(IV), 3.6 mM under argon, 2 hr.

Table 15-5 ■ Oxygen Evolution by Heterogeneous Water Oxidation Using Kaolin Clay Adsorbed 1 at 25°C[a]

1 (μmole)	O_2 (μl)	O_2 (μmole)	Decomposition of Catalyst
0.9	19	0.79	No
1.8	22	0.91	No
3.6	35	1.44	Yes
4.5	56	2.32	Yes

[a] Kaolin clay = 150 mg, **1**: Ce(IV) = 1; 100 (molar ratio); reaction time = 3 hr.

The catalytic oxygen evolution by water oxidation is not affected much by the addition of acid and base (Table 15-4), indicating that the oxidized species is the water molecule itself rather than hydroxyl ions.

When the concentration of the dinuclear Mn complex in kaolin clay is too high, it is subjected to decomposition during catalytic water oxidation (Table 15-5).[27] The decomposition product is MnO_4^- as evidenced by the characteristic six absorption peaks observed in the region 600–670 nm of the solution spectrum after the reaction. The decomposition probably occurs because of the disproportionation reaction of the higher oxidation state of the Mn complex

$$2Mn(IV)-Mn(IV) \longrightarrow Mn(VII) + Mn(III) + Mn(III)-Mn(III) \quad (15-11)$$

Because this is a bimolecular reaction, high concentration of the complex in a solid matrix would favor the reaction, thus causing decomposition of the Mn complex.

It has been recognized generally that, for oxygen evolution by water oxidation through catalysis based on a molecule, a catalyst that can work as a four-electron mediator per one molecule is required. For this reason, polynuclear metal complex catalysts have been investigated intensively. However, the result of the dinuclear Mn complex catalysis as previously described suggested strongly that, even if a catalyst is incapable of four-electron oxidation per one molecule, it might work as an oxygen evolving catalyst when more than two molecules of the catalyst can work cooperatively and achieve four-electron oxidation of two molecules of water. Figure 15-10 shows one possible scheme for this cooperative water oxidation by two molecules of catalyst (ML^{n+}).

It was found for the first time that a mononuclear Ru complex (penta-ammineruthenium complex **5**), which is capable of only two-electron oxidation, can

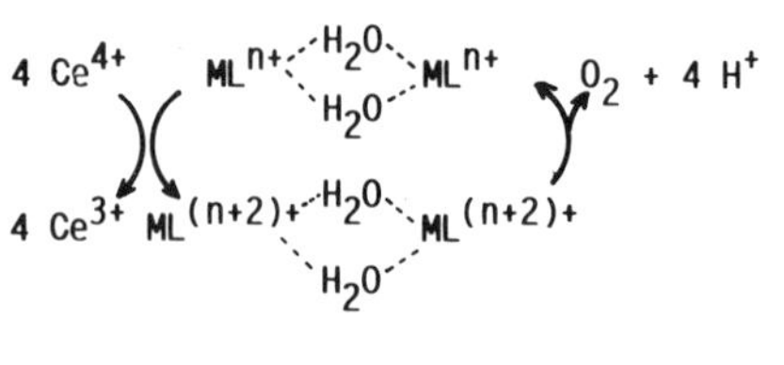

Figure 15-10 ■ Cooperative water oxidation for oxygen evolution by metal complexes through geared cycles.

Table 15-6 ■ **Catalytic Effect of [Ru(NH$_3$)$_5$Cl]$^{2+}$ for the Homogeneous Water Oxidation with Excess Oxidant [Ce(IV)]^a**

[Ru(NH$_3$)$_5$Cl]$^{2+}$ (μM)	O$_2$ Evolved (μl)	Turnover No. of Ru Complex
201	166	6.86
335	131	5.41
660	64	2.63

^{a}Ru/Ce(IV) = 1/200 (mol. ratio): [Ru(NH$_3$)$_5$Cl]$^{2+}$, 2 μmol; 25°C, 90 min.

catalyze water oxidation to give oxygen in a homogeneous reaction.[28, 29]

$$\left[Ru(NH_3)_5X\right]^{m+} \quad \begin{array}{l} \text{(a)} \ X = Cl^-, \ m = 2 \\ \text{(b)} \ X = H_2O, \ m = 3 \end{array}$$

5

Although **5** is a two-electron catalyst (redox potentials 1.28 and 1.09 V versus SCE), it can work as an active catalyst for water oxidation in a homogeneous solution as shown in Table 15-6.[29] Because it is a two-electron oxidation catalyst, some cooperative interaction of the complexes in a homogeneous aqueous solution should be assumed (see Fig. 15-10). An interaction of two molecules of the complex through a peroxo bridge like Fig. 15-11 is suggested. This kind of interaction is supported by the fact that the reaction is retarded remarkably by high ionic concentration (Table 15-7).

This mononuclear Ru complex **5** works as a water oxidation catalyst also in a heterogeneous state, such as when incorporated into clay, and the heterogeneous catalysis is more active than the homogeneous one because the heterogeneous

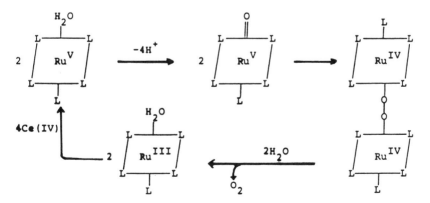

Figure 15-11 ■ Cooperative water oxidation in a homogeneous aqueous solution of mononuclear Ru complex **5**.

Table 15-7 ▪ Effect of Ionic Strength of Homogeneous Water Oxidation with $[Ru(NH_3)_5Cl]^{2+}$ Catalyst[a]

Ionic Strength[b]	O_2 Evolved	
	(μl)	(μmol)
0.48	166	6.86
0.58	135	5.56
0.98	84	3.47
1.48	70	2.89
2.48	50	2.04

[a]$Ru/Ce(IV)$, $1/200$ (mol. ratio); $[Ru(NH_3)_5Cl]^{2+}$, 2 μmol; solution, 10 ml; 25°C, 90 min.
[b]Adjusted with KNO_3.

environment around the complex would favor cooperative catalysis of two molecules of the catalyst.

The Ru and the Mn complexes already described are cationic and can be adsorbed electrostatically into anionic solids such as clay and polyanions. In spite of the electrostatic binding, the complex adsorbed is not desorbed when the complex-incorporating solid is dipped into an aqueous solution containing electrolytes. This can be ascribed to a kind of hydrophobic interaction between the ligand of the complex and the solid matrix.

Thus, heterogeneous four-electron oxidation of water to give oxygen has been realized by utilizing chemical-geared cycles of metal complexes incorporated in a solid phase. These geared cycles are applied to electrochemical water oxidation as described in the following section.

15-3-3. Electrochemical Oxidation of Water by Geared Cycles

Electrochemical water oxidation is usually carried out by using metal electrodes such as platinum. However, if water oxidation has to be coupled with other processes, such as photochemical reactions, a molecule-based water oxidation system should be utilized. Electrochemical water oxidation through a catalyst molecule is possible only by utilizing geared cycles like the Scheme 3.

Such water oxidation for oxygen evolution by utilizing electrochemical geared cycles has not been reported previously. When the water oxidation catalyst described in the last section was coated on a graphite electrode (BPG) by using a polymer membrane, catalytic water oxidation according to Scheme 3 was realized.[30] An anionic polymer membrane such as Nafion was at first coated on a graphite electrode by casting from a solution, and then a complex was adsorbed into the membrane by

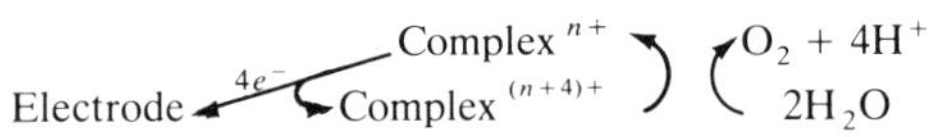

$$Electrode \xleftarrow{4e^-} \begin{matrix} Complex^{n+} \\ Complex^{(n+4)+} \end{matrix} \quad \begin{matrix} O_2 + 4H^+ \\ 2H_2O \end{matrix}$$

Scheme 3

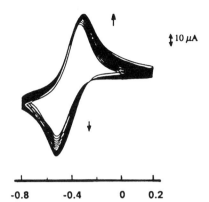

Figure 15-12 ■ Incorporation of Ru complex **4** into a Nafion film coated on BPG represented by repeated scanning cyclic voltammograms. Measured in 10 mM H_2SO_4 under continuous scanning for 1 hr. Scan rate 30 mV s^{-1}.

dipping the polymer-coated electrode into an aqueous solution containing the complex. The incorporation of the complex into the polymer membrane was studied by measuring the cyclic voltammogram. The cyclic voltammogram of Fig. 15-12 shows that Ru-red is incorporated from an aqueous solution into the Nafion membrane during repeated cyclic scanning of the potentials.

The potential-current characteristics of a Ru-red/Nafion-coated BPG dipped in an aqueous electrolyte solution is shown in Fig. 15-13 with those of a bare BPG and a Nafion-coated BPG. The anodic current at over 1 V due to oxygen evolution is much

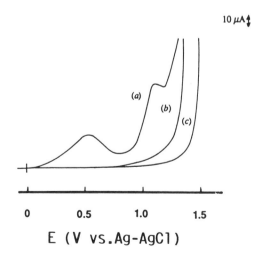

Figure 15-13 ■ Potential-current characteristics of (a) Ru-red **4**/Nafion-coated BPG, (b) bare, and (c) Nafion-coated BPG electrodes dipped in an aqueous solution of 0.1 M KNO_3 and 10 mM HNO_3.

Table 15-8 ■ Electrochemical Oxygen Evolution at Ru-red 4/Nafion-Coated BPG Electrode and at a bare BPG Electrode in an Aqueous Solution Containing 0.1 mol dm^{-3} KNO$_3$ and 0.1 mol dm^{-3} HNO$_3$[a]

Applied Potential	O$_2$ Evolved (mm^3)	
/V vs. Ag, AgCl	4/Nafion-Coated BPG	Bare BPG
1.24	62.7	36.9 (ca. 0)[b]
1.29	80.9	34.1
1.34	90.9	40.9
1.39	128.0	68.8
1.44	155.0	62.0[c]
1.49	119.4	60.5[c]

[a]The reaction was carried out in a cell (volume 55 cm^3) for 1 hr at 25°C.
[b]Nafion-coated BPG without complex.
[c]The BPG electrode was changed.

higher for the Ru-red/Nafion-coated BPG than the other two electrodes. Table 15-8 shows that the amount of oxygen evolved electrochemically is actually much more for the Ru-red/Nafion-coated BPG than for the bare or the Nafion-coated electrode. Even at the potential where the Nafion-coated graphite does not evolve oxygen at all (1.24 V versus SCE), the Ru-red/Nafion-coated electrode can oxidize water to give oxygen. These results indicate clearly that water is oxidized electrochemically through incorporated complex according to the geared cycles of Scheme 3.

The dinuclear Mn complex **1** is also active as a catalyst for electrochemical water oxidation (see Fig. 15-14 and Table 15-9). These electrochemical geared cycles that

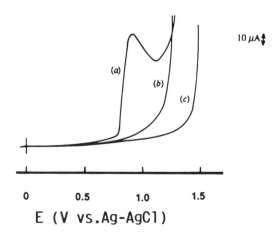

Figure 15-14 ■ Potential-current characteristics of (a) dinuclear Mn complex **1**/Nafion-coated BPG, (b) bare, and (c) Nafion-coated BPG electrodes dipped in aqueous solution of 0.1 M KNO$_3$ and 10 mM HNO$_3$.

Table 15-9 ▪ Electrochemical Oxygen Evolution at Dinuclear Mn Complex
1 /Nafion-Coated BPG Electrode and at a Bare BPG Electrode
in an Aqueous Solution Containing 0.1 mol dm^{-3} KNO$_3$
and 0.1 mol dm^{-3} HNO$_3$ [a]

| Applied Potential | O$_2$ Evolved (mm^3) | |
/V vs. Ag, AgCl	1/Nafion-Coated BPG	Bare BPG
1.09	60.0	37.5
1.19	64.5	42.5 (ca. 0)[b]
1.24	107.5	42.0
1.34	100.5	55.0

[a] The reaction was carried out in a cell (volume 55 cm^3) for 1 hr at 25°C.
[b] Nafion-coated BPG without complex.

realize four-electron oxidation of water have been utilized for water cleavage by visible light as will be described in the next section.

15-4. Geared Cycles for the Artificial Photochemical Energy Conversion Systems

The interest in photochemical solar energy conversion is increasing again because of environmental problems, such as carbon dioxide increase, ozone layer depletion, and so on. All compounds used as fuels have a reduced form. In order to produce reduced compounds, that is, fuels, with solar irradiation, it is important from an economic standpoint to use water as an electron source in a manner similar to photosynthesis. Such a study can lead to construction of a photosynthetic model, which is important as a basic scientific research. Four-electron water oxidation is the key reaction for this purpose, and it has to be coupled with some photoexcitation process to construct geared cycles as shown in Fig. 15-3(b).

It is well known that a liquid-junction n-TiO$_2$ photoanode dipped in an aqueous electrolyte solution can photolyze water under UV light irradiation to give oxygen on the TiO$_2$ and hydrogen on a counter platinum electrode.[31] The n-TiO$_2$ can utilize only UV light below 400 nm because of the large bandgap ($E_g = 3.0$ eV). Because narrow bandgap n semiconductors (bandgap smaller than 3.0 eV) are oxidatively corroded when irradiated in an aqueous electrolyte, they cannot be used for photoelectrochemical purpose.

The present authors have chosen narrow bandgap n semiconductors for conducting photochemical conversion systems and have used them as liquid-junction photoanodes.[1,22] At first, stabilization of the liquid-junction n semiconductors against photocorrosion was dealt with. In order to pick up corrosive holes formed on the semiconductor surface under illumination and to prevent corrosion of the semiconductor, hole carriers were attached on its surface by utilizing a polymer membrane. Tris(2,2'-bipyridine)ruthenium(II) complex [abbreviated to Ru(bpy)$_3^{2+}$] was used as a hole carrier. Polymer-pendant Ru(bpy)$_3^{2+}$ 6 was prepared for use as a stable membrane in water[32,33] (M = styrene):

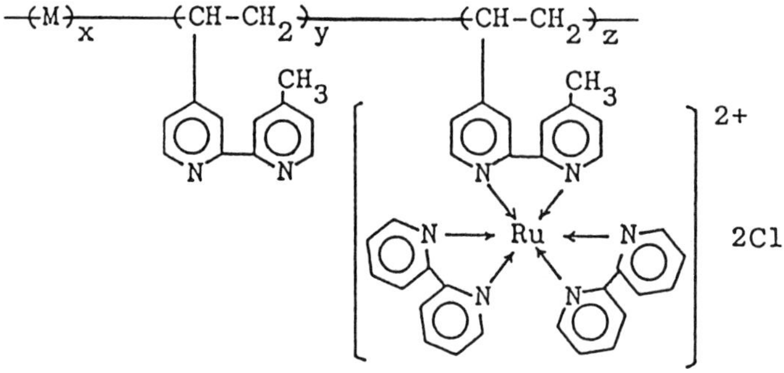

$$-(M)_x \quad -(CH-CH_2)_y \quad -(CH-CH_2)_z-$$

(structure: poly(styrene-vinylpyridine) Ru complex)

$2+$

$2Cl^-$

Hole transport is not sufficient to stabilize a liquid-junction narrow bandgap semiconductor photoanode because the transported hole has to be used for some oxidation reaction (or at least has to be accumulated somehow). In order to achieve multielectron oxidation by the transported hole, RuO_2 was dispersed in the **6** membrane coated on a semiconductor photoanode. At first, oxidation of halide ions was chosen as the oxidation reaction. When an n-CdS ($E_g = 2.4$ eV, excited under 520 nm) electrode is irradiated in an aqueous solution containing chloride ions, it is corroded and, due to corrosion, gives characteristic redox peaks in its cyclic voltammogram measured under irradiation as shown in Fig. 15-15(a).[34] The cathodic peak at around -0.9 V (versus Ag-AgCl) is due to the reduction of Cd^{2+} formed by corrosion of CdS under irradiation. The anodic peak at -0.7 V is due to reoxidation of Cd^0 to Cd^{2+}. However, when an n-CdS photoanode is coated with the **6** membrane and RuO_2 is dispersed in it, it is stable under irradiation and no peak due to corrosion is observed in its cyclic voltammogram under irradiation [see Fig. 15-15(b)]. When either Ru complex group or RuO_2 is missing in the coated membrane, the n-CdS is not stabilized against photocorrosion, indicating that both the Ru complex and RuO_2 are needed for the stabilization.

The mechanism of the stabilization of an n-CdS photoanode by coating with **6** is shown in Fig. 15-16. An n semiconductor, when dipped in an electrolyte solution, forms a kind of Schottky barrier called a liquid-junction and gives band bendings as shown in the figure. On irradiation, the holes formed migrate to the semiconductor surface due to the valence band bending and are transferred to the Ru complex coated on the surface. The holes thus trapped in the membrane are transported to RuO_2 and oxidize X ions to give X_2. The electrons formed under irradiation migrate into the semiconductor bulk due to the conduction band bending, are transferred to the counter electrode, and reduce protons there to give hydrogen. The Cl_2 and H_2 formed under irradiation when HCl was used were confirmed by analysis.[34] The total reaction is therefore photolysis of HCl to give H_2 and Cl_2. This reaction has been realized by utilizing the geared cycles as shown in the Scheme 4. Other hydrogen

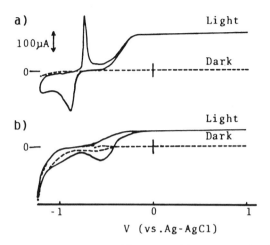

Figure 15-15 ▪ Cyclic voltammograms of (a) bare and (b) polymer-Ru(bpy)$_3^{2+}$ **6**/RuO$_2$-coated
n-CdS photoanodes measured in a 0.5 M KCl aqueous solution. Electrode area 3.03 mm^2; scan
rate 50 mV s^{-1}.

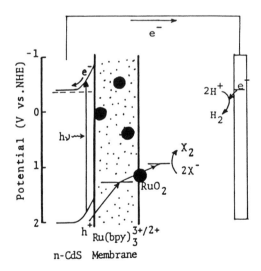

Figure 15-16 ▪ Stabilization of an *n*-CdS photoanode by coating with polymer
Ru(bpy)$_3^{2+}$/RuO$_2$ and photolysis of HX by utilizing the geared cycles.

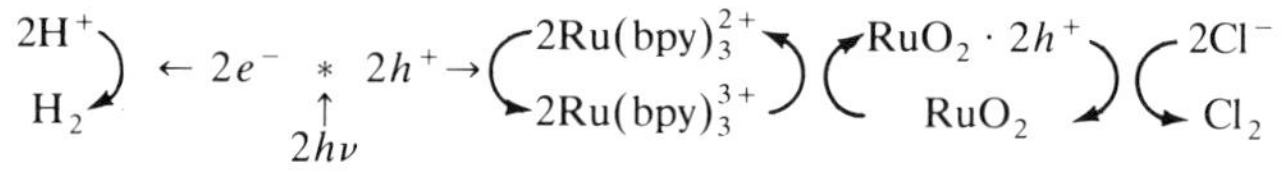

$$2H^+ \Big) \leftarrow 2e^- \underset{\underset{2h\nu}{\uparrow}}{*} 2h^+ \rightarrow \Big(\begin{matrix} 2Ru(bpy)_3^{2+} \\ 2Ru(bpy)_3^{3+} \end{matrix} \Big) \Big(\begin{matrix} RuO_2 \cdot 2h^+ \\ RuO_2 \end{matrix} \Big) \Big(\begin{matrix} 2Cl^- \\ Cl_2 \end{matrix}$$

Scheme 4

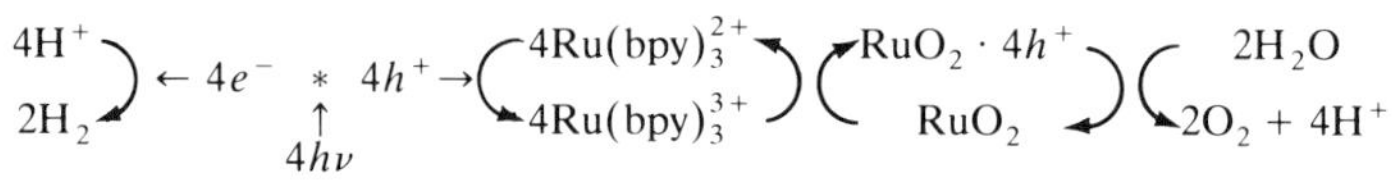

$$4H^+ \Big) \leftarrow 4e^- \underset{\underset{4h\nu}{\uparrow}}{*} 4h^+ \rightarrow \Big(\begin{matrix} 4Ru(bpy)_3^{2+} \\ 4Ru(bpy)_3^{3+} \end{matrix} \Big) \Big(\begin{matrix} RuO_2 \cdot 4h^+ \\ RuO_2 \end{matrix} \Big) \Big(\begin{matrix} 2H_2O \\ 2O_2 + 4H^+ \end{matrix}$$

Scheme 5

halides such as HBr and HI can also be photolyzed with visible light by utilizing these geared cycles. Because oxidation of halide ions to halogen is a two-electron process, geared cycles are required to connect the one-electron process of $Ru(bpy)_3^{2+}$ with the halide ion oxidation.

These photoinduced geared cycles have been applied to water photolysis.[35] In the absence of halide ions, irradiation of an n-CdS coated with **6** membrane in which RuO_2 is dispersed induced water oxidation to give oxygen bubbles on the coated CdS surface, and hydrogen evolved at the counter Pt electrode. The geared cycles are represented by Scheme 5. Both the O_2 and H_2 that evolved have been analyzed by gas chromatography and mass spectroscopy.

In order to establish an efficient water photolysis system, the water oxidation catalyst described in the last section was applied to the modified CdS photoanode. This kind of metal complex can work not only as a water oxidation catalyst but also as a hole carrier. Thus, and n-CdS photoanode was at first coated with a Nafion membrane, and then one of the complexes (e.g., Ru-red) was incorporated into the film by dipping the coated CdS into an aqueous solution of the complex. The CdS thus modified was dipped in an aqueous solution containing HNO_3 and $KNO_{3'}$, and irradiated with 500 nm monochromatic light under an applied potential with a Pt counter and Ag reference electrodes.[36] The modified CdS gave a stable cyclic voltammogram as shown in Fig. 15-17. It shows no stripping peak due to oxidative corrosion of a CdS photoanode as was shown in Fig. 15-15(a). Photocurrent induced

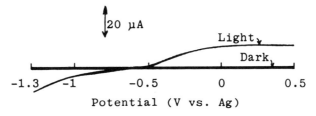

Figure 15-17 ■ Cyclic voltammogram of Ru-red **4**/Nafion-coated CdS (3.22 mm^2) in an aqueous solution containing 0.5 M KNO_3 and 10 mM $HNO_3 \cdot Ag = 0.46$ V versus NHE.

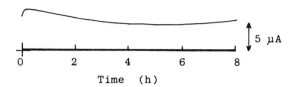

Figure 15-18 ■ Time dependence of the photocurrent generated at Ru-red **4**/Nafion-coated CdS at the applied potential of 0.36 V (versus NHE), irradiated with 496 nm monochromatic light (2.05 mW cm^{-2}).

Table 15-10 ■ **Water Photolysis with *n*-CdS Photoanode Coated with a Nafion Film Containing Adsorbed Ru-Red**[a]

Run	Reaction Time (hr)	Gases Evolved			Efficiency (%)		
		H_2 (μl)	O_2 (μl)	$H_2:O_2$	Photocurrent[b]	A[c]	Photolysis[d]
1	4.0	7.4	3.8	1.9	25	64	16
2	7.9	14.0	7.5	1.9	21	71	15

[a]Electrode area, 4.7 mm^2 (run 1); 3.2 mm^2 (run 2). Applied potential, 0.46 V (run 1); 0.36 V (run 2) (vs. NHE).
[b]Based on the total irradiated 496 nm monochromatic light.
[c]Water cleavage efficiency based on photocurrent.
[d]Total photolysis efficiency based on irradiated light.

under an applied potential was stable (Fig. 15-18), showing almost no decrease during the irradiation, whereas the photocurrent generated at a bare CdS electrode degraded drastically due to corrosion.

The modified CdS evolved O_2 and the counter Pt electrode gave H_2 with a theoretical H_2 to O_2 ratio of almost 2. Table 15-10 shows[36] the photocurrent based on the irradiated 500 nm monochromatic light, the yield of the H_2 and O_2 evolved

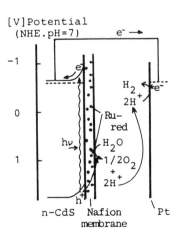

Figure 15-19 ■ Visible light water photolysis by utilizing geared cycles of complex **4**-coated CdS.

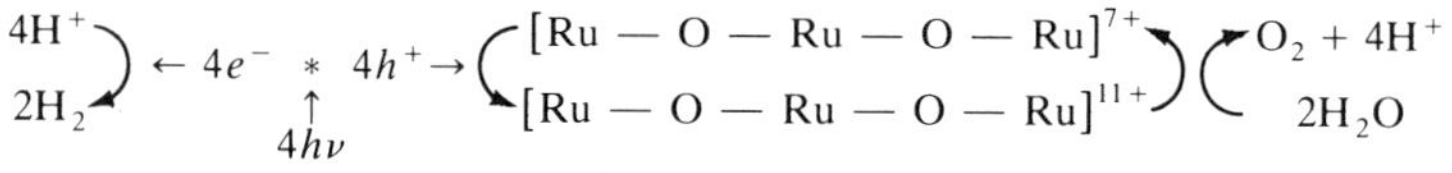

$$\left.\begin{array}{c}4H^+\\2H_2\end{array}\right) \leftarrow 4e^- \quad * \quad 4h^+ \rightarrow \left(\begin{array}{c}[Ru - O - Ru - O - Ru]^{7+}\\[Ru - O - Ru - O - Ru]^{11+}\end{array}\right) \left(\begin{array}{c}O_2 + 4H^+\\2H_2O\end{array}\right)$$
$$\underset{4h\nu}{\uparrow}$$

Scheme 6

based on the photocurrent, and the total water photolysis efficiency based on the irradiated light. This system gives photocurrents with an efficiency of 21–25% based on the total irradiated monochromatic light, reasonably high $H_2:O_2$ yields of 64–71% based on the photocurrent, and high total photolysis efficiency (16%).

The mechanism of the present water photolysis system is illustrated in Fig. 15-19. The metal complex incorporated into the polymer membrane coated on the CdS (Ru-red in this case) is working as both hole carrier and water oxidation catalyst. The photoexcitation is coupled with the four-electron oxidation of water through geared cycles (see Scheme 6), thus realizing efficient water cleavage under visible light irradiation.

In the sense that the present water photolysis system uses water as a reductant to produce a high energy compound (here H_2), it is regarded as a kind of artificial photosynthetic system.[1] It has been realized only by utilizing geared cycles that couple four-electron water oxidation with photoexcitation.

15-5. Concluding Remarks and Future Scopes

Geared cycles in photoexcited multielectron transfer utilizing polymer-metal complexes have been described. This is an important process, especially in conjunction with photosynthesis. Four-electron oxidation of water to give oxygen, which is the most important process in photosynthesis, is made possible only by coupling the multielectron reaction with a one-electron process by using geared cycles. Efficient water oxidation that utilizes geared cycles has been described. Coupling of multielectron transfer, such as four-electron oxidation of water with photoexcitation, is realized also only by utilizing geared cycles. Examples of the coupling of photoexcitation of a semiconductor system with a four-electron water oxidation catalyst through geared cycles to realize visible light water cleavage for simultaneous hydrogen and oxygen evolution have been described.

Carbon dioxide increase in the environment, which might bring about a so-called greenhouse effect on the Earth, will become a serious problem in the next century. This problem is closely related to the use of fossil fuels that produce CO_2 on combustion and, therefore, is related to the need for new energy resources. From both points of view, the study of artificial photosynthetic systems is an important and urgent area of research. Geared cycles in photoexcited multielectron transfer are key processes for accomplishing this objective.

References

1. M. Kaneko and D. Wöhrle, *Adv. Polym. Sci.*, **84**, 141 (1987).
2. R. Govindjee, *Bioenergetics of Photosynthesis*, Academic, New York (1975).
3. A. G. Volkov, *Bioelectrochem. Bioenerg.*, **21**, 3 (1989).

4. V. Ya. Shafirovich, N. K. Khannanov, and A. E. Shilov, *J. Inorg. Biochem.*, **15**, 113 (1981).

5. Govindjee, T. Kambara, and W. Coleman, *Photochem. Photobiol.*, **42**, 187 (1985).

6. G. C. Dismukes, *Photochem. Photobiol.*, **43**, 99 (1986).

7. P. Joliot, G. Barbieri, and R. Chabaud, *Photochem. Photobiol.*, **10**, 309 (1969).

8. G. C. Dismukes and Y. Siderer, *Proc. Nat. Acad. Sci. USA*, **78**, 274 (1981).

9. B. Mabad, J.-P. Tuchagues, Y. T. Hwang, and D. N. Hendrickson, *J. Am. Chem. Soc.*, **107**, 2801 (1985).

10. G. W. Brudvig and R. H. Crabtree, *Proc. Nat. Acad. Sci. USA*, **83**, 4586 (1986).

11. J. B. Vincent, C. Christmas, H.-R. Chang, Q. Li, P. D. W. Boyd, J. C. Huffman, D. N. Hendrickson, and G. Christou, *J. Am. Chem. Soc.*, **111**, 2086 (1989).

12. C. Christmas, J. B. Vincent, J. C. Huffman, G. Christou, H.-R. Chang, and D. N. Hendrickson, *J. Chem. Soc., Chem. Commun.*, **1987**, 1303 (1987).

13. C. Preston and R. J. Pace, *Biochim. Biophys. Acta*, **810**, 388 (!985).

14. J. S. Connolly, ed., *Photochemical Conversion and Storage of Solar Energy*, Academic, New York (1981).

15. M. Graetzel, ed., *Energy Resources through Photochemistry and Catalysis*, Academic, New York (1983).

16. J. R. Norris, Jr. and D. Meisel, ed., *Photochemical Energy Conversion*, Elsevier, New York (1989).

17. J. A. Gilbert, D. S. Eggleston, W. R. Murphy, Jr., D. A. Geselowitz, S. W. Gersten, D. T. Hodgson, and T. J. Meyer, *J. Am. Chem. Soc.*, **107**, 3855 (1985).

18. S. W. Gersten, G. J. Samuels, and T. J. Meyer, *J. Am. Chem. Soc.*, **104**, 4029 (1982).

19. J. E. Earley and T. Fealey, *Inorg. Chem.*, **12**, 323 (1973).

20. J. E. Earley and H. Razavi, *Inorg. Nucl. Chem. Lett.*, **9**, 331 (1973).

21. J. E. Earley, *Inorg. Nucl. Chem. Lett.*, **9**, 487 (1973).

22. M. Kaneko and A. Yamada, *Adv. Polym. Sci.*, **55**, 1 (1984).

23. R. Ramaraj, A. Kira, and M. Kaneko, *J. Chem. Soc., Faraday Trans. 1*, **82**, 3515 (1986).

24. R. Ramaraj, A. Kira, and M. Kaneko, *Angew. Chem. Int. Ed. Engl.*, **25**, 1009 (1986).

25. R. Ramaraj, A. Kira, and M. Kaneko, *J. Chem. Soc., Faraday Trans. 1*, **83**, 1539 (1987).

26. R. Ramaraj, A. Kira, and M. Kaneko, *Angew. Chem. Int. Ed. Engl.*, **25**, 825 (1986).

27. R. Ramaraj, A. Kira, and M. Kaneko, *Chem. Lett.*, **1987**, 261 (1987).

28. R. Ramaraj, A. Kira, and M. Kaneko, *J. Chem. Soc., Chem. Commun.*, **1987**, 227 (1987).

29. M. Kaneko, R. Ramaraj, and A. Kira, *Bull. Chem. Soc. Jpn.*, **61**, 417 (1988).

30. G.-J. Yao, A. Kira, and M. Kaneko, *J. Chem. Soc. Faraday Trans. 1*, **84**, 4451 (1988).

31. A. Fujishima and K. Honda, *Nature*, **238**, 37 (1972).

32. M. Kaneko, A. Yamada, E. Tsuchida, and Y. Kurimura, *J. Polym. Sci., Polym. Lett. Ed.*, **20**, 593 (1982).

33. M. Kaneko, M. Ochiai, K. Kinosita, Jr., and A. Yamada, *J. Polym. Sci., Polym. Chem. Ed.*, **20**, 1011 (1982).

34. K. Rajeshwar, M. Kaneko, A. Yamada, and R. N. Noufi, *J. Phys. Chem.*, **89**, 806 (1985).

35. M. Kaneko, T. Okada, S. Teratani, and K. Taya, *Electrochim. Acta*, **32**, 1405 (1987).

36. M. Kaneko, G.-J. Yao, and A. Kira, *J. Chem. Soc., Chem. Commun.*, **1989**, 1338 (1989).

Sequential Potential Fields in Electrically Conducting Polymers

Takakazu Yamamoto

Research Laboratory of Resources Utilization, Tokyo Institute of Technology, 4259 Nagatsuta, Midori-ku, Yokohama 227, Japan

16-1. Introduction

Electrically conducting polymer materials are a subject of recent interest. Since poly(acetylene) was found to show good electrical conductivity when doped with iodine,[1] the physics and chemistry of electrically conducting π-conjugated polymers has developed rapidly.[2-5] Several industrial applications of the polymers also have been reported.[7-10]

Another class of the electrically conducting polymer materials is a composite prepared by mixing electrically conducting inorganic materials (e.g., metal powder and copper sulfide powder) with common nonconjugated polymers. We have developed a new method for preparing the electrically conducting polymer composite by using organosols of metal sulfides.[11]

When an electric field is applied to the π-conjugated polymer or composite material, a sequential potential field is formed in the polymer material (Fig. 16-1). The electric current in the π-conjugated polymers and composite materials is considered to flow along the sequential potential formed in the polymer materials. In this chapter, an outline of the properties of these electrically conducting materials will be given.

16-2. π-Conjugated Polymers

16-2-1. Electrically Conducting Properties

Figure 16-2 shows examples of the electrically conducting π-conjugated polymers and Table 16-1 summarizes typical electrical conductivity of π-conjugated polymers after doping. As the dopant, oxidizing reagents (e.g., I_2, $FeCl_3$, and AsF_5) or reducing

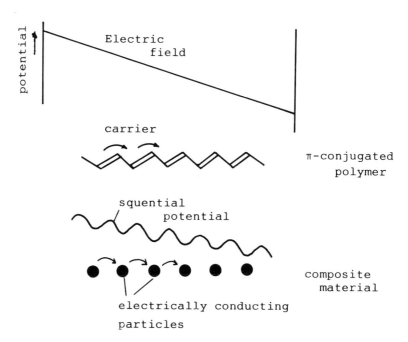

Figure 16-1 ▪ Formation of the sequential potential formed in electrically conducting π-conjugated polymes and composite materials on application of electric field.

reagents (e.g., Na and Li) are employed. Electrical oxidation and reduction also gave doped polymers with anion and cation, respectively (A^-: anion; C^+: cation):

$$\cdots \diagup\!\!\diagdown\!\!\diagup\!\!\diagdown\!\!\diagup\!\!\diagdown\!\!\diagdown \cdots + I_2 \xrightarrow{p\text{ doping}} \cdots \diagup\!\!\diagdown\!\!\diagup\!\!\overset{+}{\diagdown}\!\!\diagup\!\!\diagdown\!\!\diagdown \cdots \qquad (16\text{-}1)$$
$$I_n^-$$

$$\cdots \diagup\!\!\diagdown\!\!\diagup\!\!\diagdown\!\!\diagup\!\!\diagdown\!\!\diagdown \cdots + Na \xrightarrow{n\text{ doping}} \cdots \diagup\!\!\diagdown\!\!\diagup\!\!\overset{-}{\diagdown}\!\!\diagup\!\!\diagdown\!\!\diagdown \cdots \qquad (16\text{-}2)$$
$$Na^+$$

$$\left(\!\!\left(\!\!\begin{array}{c}\\ S \end{array}\!\!\right)\!\!\right)_n + nx A^- \underset{\substack{\text{electrochemical}\\ \text{reduction (undoping)}}}{\overset{\substack{\text{electrochemical}\\ \text{oxidation (p doping)}}}{\rightleftharpoons}} \left(\!\!\left(\!\!\begin{array}{c}+x\\ S \end{array}\!\!\right)\!\!-x A^-\!\!\right)_n + nxe \qquad (16\text{-}3)$$

$$\left(\!\!\left(\!\!\begin{array}{c}\\ S \end{array}\!\!\right)\!\!\right)_n + nx C^- \underset{\substack{\text{electrochemical}\\ \text{oxidation (undoping)}}}{\overset{\substack{\text{electrochemical}\\ \text{reduction (n doping)}}}{\rightleftharpoons}} \left(\!\!\left(\!\!\begin{array}{c}-x\\ S \end{array}\!\!\right)\!\!-x C^+\!\!\right)_n - nxe \qquad (16\text{-}4)$$

Table 16-2 shows examples of the dopant.

The cation and anion centers, which are often assigned to soliton, polaron, or bipolaron, can move along the polymer chain, thus giving rise to electrically conducting properties in the polymer. The *o*- and *n*-doped polymers essentially have charac-

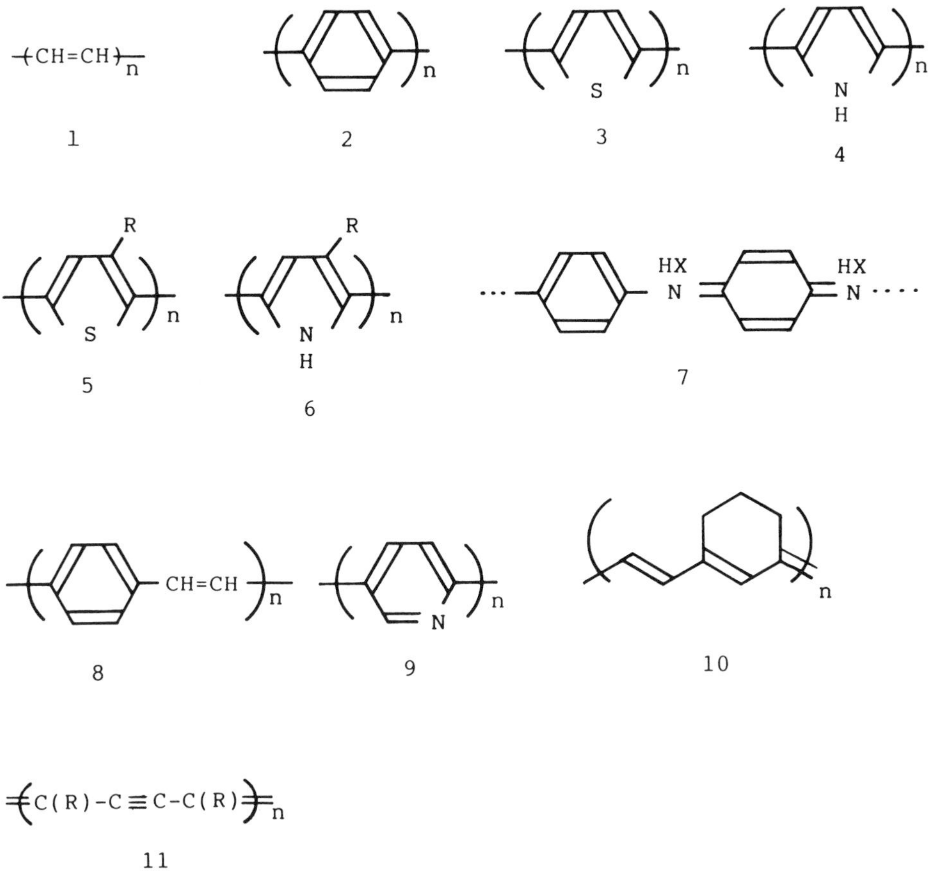

$+(CH=CH)_n$

1

2

3

4

5

6

7

8

9

10

$=(C(R)-C\equiv C-C(R))=_n$

11

Figure 16-2 ■ Examples of electrically conducting π-conjugated polymers. **1**: poly(acetylene): **2**: poly(p-phenylene): **3**: poly(thiophene) or poly(2,5-thienylene); **4**: poly(pyrrole) or poly(2,5-pyrrolylene); **5**: poly(3-substituted thiophene) or poly(3-substituted 2,5-thienylene); **6**: poly (3-substituted pyrrole) or poly(3-substituted 2,5-pyrrolylene); **7**: poly(aniline); **8**: poly(p-phenylenevinylene); **9**: poly(2,5-pyridinediyl); **10**: poly(hexadiyne); **11**: poly(diacetylene)s.

teristics of p- and n-type conductors, respectively, as measured by Hall effect, Seebeck effect, and I-V curves of diodes prepared by using the polymer. Thus, the doped π-conjugated polymers serve as basic materials for making electric devices such as field effect transistors[10] and solar cells.[12]

As for the semiconducting properties of silicon and carbon having the diamond structure, it is known that silicon has much larger mobility of electrons and holes than carbon does. In a chemist's sense, movement of the carrier in silicon and diamond carbon accompanies cleavage of the silicon-silicon and carbon-carbon bonds, respectively (Fig. 16-3), and the greater mobility of carriers in silicon relative to carbon is

Table 16-1 ▪ Electrical Conducting Properties of the π-Conjugated Polymers

Polymer[a]	p or n Type	Electrical Conductivity (S cm^{-1})	Bandgap E_g, (eV)	Ionization Potential I_p (eV)
1	p, n	10^5	1.5	4.7
2	p, n	10^2–10^3	3.2	5.6
3 and 5	p, n	5×10^2	1.6	5.0
4 and 6	p	5×10^2	3.6	3.9
7	p	10	ca. 4	ca. 5
8	p	5×10^3	2.5	5.1
9	n	10^{-1}	3.2	—
10	p	10^{-1}	1.4	4.4
11	p	10^{-2}–10^0	2.1	5.1

[a] Cf. Fig. 16-2.

Table 16-2 ▪ Typical Dopant for π-Conjugated Polymers

Class A: Dopant for p-type conductor (oxidizing reagent) I_2, Br_2, $FeCl_3$, $AlCl_3$, AsF_5, $SnCl_4$, HX, SO_3, conc. H_2SO_4, and counteranion in the electrochemical doping (ClO_4^-, BF_4^-, PF_6^-, $CF_3SO_3^-$, polymer anion, etc.)

Class B: Dopant for n-type conductor (reducing reagent) Li, Na (including Na naphthalide), and countercation in the electrochemical doping (NR_4^+, Li^+, etc.)

related to the lower energy for cleaving the $Si(sp^3)$—$Si(sp^3)$ bond relative to the $C(sp^3)$—$C(sp^3)$ bond. The energy required for rearrangement of the chemical bond by cleavage of the sp^3—sp^3 bond may be related to σ-σ^* excitation energy.

On the other hand, in the case of π-conjugated electrically conducting polymers, their electrical conduction is related to rearrangement of π bonding in the polymer chain. The π-π^* excitation energy is usually much smaller than the σ-σ^* excitation energy. Thus, the mobility of carriers in the π-conjugated polymer chain (μ_a, in Fig. 16-4) is expected to be very large,[12] although in the real polymer material the real mobility is related to movement of the carrier between polymer chains (μ_b in Fig. 16-4) rather than the movement of the carrier along the one-polymer chain.

Poly (1,1'-ferrocenylene),

constituted of recurring sandwich-type ferrocene units, also shows electrically conducting properties when the polymer forms an adduct with electron acceptors such as TCNQ and I_2 (Fig. 16-5).[13] Mössbauer spectroscopic data indicate that a part of Fe(II) in poly(1,1'-ferrocenylene) is oxidized to Fe(III) during adduct formation and a

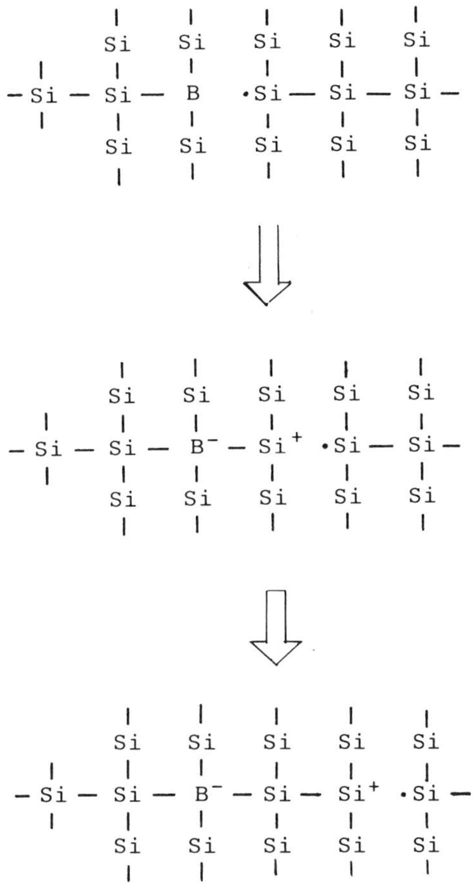

$$
\begin{array}{ccccc}
| & | & | & | & | \\
\mathrm{Si} & \mathrm{Si} & \mathrm{Si} & \mathrm{Si} & \mathrm{Si} \\
| & | & | & | & | \\
-\mathrm{Si}-\mathrm{Si}- & \mathrm{B} & \cdot\mathrm{Si}- & \mathrm{Si}-\mathrm{Si}- \\
| & | & | & | & | \\
\mathrm{Si} & \mathrm{Si} & \mathrm{Si} & \mathrm{Si} & \mathrm{Si} \\
| & | & | & | & |
\end{array}
$$

$$\Downarrow$$

$$
\begin{array}{ccccc}
| & | & | & | & | \\
\mathrm{Si} & \mathrm{Si} & \mathrm{Si} & \mathrm{Si} & \mathrm{Si} \\
| & | & | & | & | \\
-\mathrm{Si}-\mathrm{Si}- & \mathrm{B}^- - \mathrm{Si}^+ & \cdot\mathrm{Si}-\mathrm{Si}- \\
| & | & | & | & | \\
\mathrm{Si} & \mathrm{Si} & \mathrm{Si} & \mathrm{Si} & \mathrm{Si} \\
| & | & | & | & |
\end{array}
$$

$$\Downarrow$$

$$
\begin{array}{ccccc}
| & | & | & | & | \\
\mathrm{Si} & \mathrm{Si} & \mathrm{Si} & \mathrm{Si} & \mathrm{Si} \\
| & | & | & | & | \\
-\mathrm{Si}-\mathrm{Si}- & \mathrm{B}^- - \mathrm{Si} - \mathrm{Si}^+ & \cdot\mathrm{Si}- \\
| & | & | & | & | \\
\mathrm{Si} & \mathrm{Si} & \mathrm{Si} & \mathrm{Si} & \mathrm{Si} \\
| & | & | & | & |
\end{array}
$$

Figure 16-3 ■ Movement of the positive center in p-type silicon with boron as the dopant.

rapid exchange of electron between Fe(II) and Fe(III) on Mössbauer time scale (10^{-7} s) takes place at room temperature.

Mössbauer spectroscopy reveals also that reduction of $FeCl_3$ to $FeCl_2$ takes place when the π-conjugated polymers form adducts with $FeCl_3$:

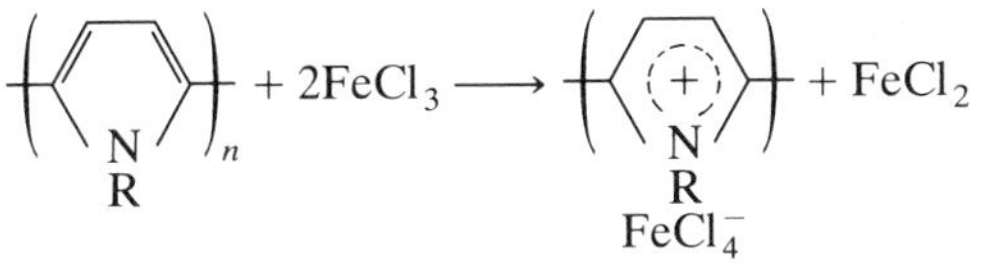

$$
\left(\!\!\left\langle \begin{array}{c} \\ \\ \mathrm{N} \\ \mathrm{R} \end{array} \right\rangle\!\!\right)_n + 2FeCl_3 \longrightarrow \left(\!\!\left\langle \begin{array}{c} (+) \\ \mathrm{N} \\ \mathrm{R} \\ FeCl_4^- \end{array} \right\rangle\!\!\right) + FeCl_2 \qquad (16\text{-}5)
$$

The occurrence of similar redox reactions coinciding with treatment of poly(acetylene) with $FeCl_3$ has been reported. In general, poly(pyrrole) and its derivatives serve as good electron donors.

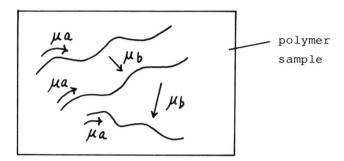

Figure 16-4 ■ Movement of carrier in the polymer chain (mobility $= \mu_a$) and between the polymer chains (mobility $= \mu_b$).

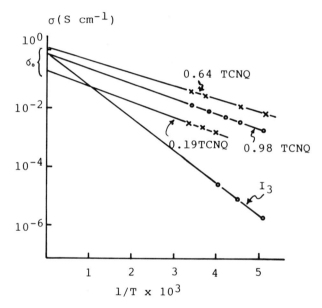

Figure 16-5 ■ Electrical conductivity of the adducts of poly(1,1′-ferrocenylene) with electron acceptors. (TCNQ = tetracyanodimethane.

16-2-2. Preparation and Basic Properties of the π-Conjugated Polymers

Poly(acetylene) is usually obtained by Ziegler–Natta-type polymerization of acetylene: The typical catalyst is a mixture of $Ti(OBu)_4$ and $AlEt_3$. When the polymerization is carried out with a catalytic system with high concentration of catalyst and calm surface, the polymerization takes place on the surface of the catalytic system and gives a film with silver-like color. Aging of the catalyst often affords a polymer film with higher electrical conductivity.

A variety of methods have been developed for preparing the other π-conjugated polymers shown in Fig. 16-2. We now describe typical methods for the preparation of the polymers containing arylene units and basic properties of the polymers.

Poly(p-phenylene) 2. Kovacic and his co-workers developed a method to prepare poly(p-phenylene) by treating benzene with $CuCl_2$ and $AlCl_3$[15]:

$$nC_6H_6 + 2nCuCl_2 \xrightarrow{AlCl_3} (C_6H_4) + 2nHCl + 2nCuCl \qquad (16\text{-}6)$$

$$\mathbf{2}$$

In this polymerization, $CuCl_2$ serves as an oxidizing reagent. Recently preparation of **2** in the presence of a catalytic amount of $CuCl_2$ and $AlCl_3$ was reported; in this case, oxygen served as the oxidizing reagent.

Treatment of p-dihalobenzene with magnesium in the presence of Ni catalyst leads to dehalogenation polycondensation[16]:

$$X\text{—}\langle\bigcirc\rangle\text{—}X + Mg \xrightarrow{NiL_n} \left(\!\left\langle\bigcirc\right\rangle\!\right)_n + MgX_2 \qquad (16\text{-}7)$$

$$\mathbf{2}$$

This method is applicable to m-dihalobenzene, and thus poly(m-phenylene) is obtained. Several other methods using precursor polymers[17] and electrochemical techniques[18] have been reported.

Poly(p-phenylene) has a packing form similar to poly(acetylene) and poly(ethylene). It is converted into electrically conducting material by treatment with an oxidizing (AsF_5) or reducing (Li) reagent and by electrochemical oxidation or reduction.

Poly(2,5-thienylene), poly(2,5-pyrrolylene), and their derivatives (3, 4, 5, and 6). Formal substitution of hydrogens of cis type poly(acetylene) with S[19] and NH afford poly(2,5-thienylene) and poly(2,5-pyrrolylene), respectively,

$$\text{S substitution} \cdots \qquad \qquad (16\text{-}8)$$

$$\mathbf{3}$$

$$\text{NH substitution} \cdots \qquad \qquad (16\text{-}9)$$

$$\mathbf{4}$$

These polymers have π-conjugation systems similar to that of poly(acetylene) and much higher chemical stability than poly(acetylene). The polymers and their derivatives are obtained by various methods, as examplified by[19-22]

$$X-\left\langle{}\right\rangle_S-X + Mg \xrightarrow{\text{Ni complex}} \left(\left\langle{}\right\rangle_S\right)_n + nMgX_2 \qquad (16\text{-}10)$$

$$H-\left\langle{}\right\rangle_{\underset{H}{N}}-H \xrightarrow[\text{[R}_4\text{N]Y}]{\substack{\text{electrochemical} \\ \text{polymerization}}} \left(\left\langle{}\right\rangle_{\underset{H}{\overset{+x}{N}}}\right)_n \cdot (xY^-)_n \qquad (16\text{-}11)$$

$$H-\left\langle{\overset{R}{}}\right\rangle_S-H \xrightarrow{\text{FeCl}_3,\ (\text{NH}_4)_2\text{S}_2\text{O}_8,\ \text{etc.}} \left(\left\langle{\overset{R}{}}\right\rangle_S\right)_n \qquad (16\text{-}12)$$

The electrochemical polymerization [Eq. (16-11)] often affords good doped films. Introduction of substituent R (polymer **5** and **6**) enhances solubility of the polymers without loosing π-conjugation.[22] As the substituents, alkyl, alkoxy $(CH_2)_nSO_3X$ $(X = H, Na)$ have been attached to the polymer. Expansion of the π-conjugation system of **3** by introducing a benzene ring connected at **3** and **4** positions forms poly(isothianaphthene),[22d] which shows interesting optical properties and gives a white and colorless electrically conducting material when doped.

Polymer **3** and its derivatives form complexes with transition metal cations such as Cu^{2+} and Ru^{3+},

$$\left(\left\langle{\overset{R}{}}\right\rangle_S\right)_n$$
$$\text{M}^{n+}$$

and the complex shows a unique electrochemical response.[23]

Poly(aniline) 7. Electrochemical oxidation (EO) of aniline gives poly(aniline)[24]

$$\left\langle{}\right\rangle-NH_2 \xrightarrow{\text{EO}} \cdots -\left\langle{}\right\rangle-NH-\left\langle{}\right\rangle-NH-\cdots \qquad (16\text{-}13)$$
$$\mathbf{7a}$$

$$\xrightarrow{\text{EO}} \cdots -\left\langle{}\right\rangle-\overset{HY}{N}=\left\langle{}\right\rangle=\overset{HY}{N}-\cdots \qquad (16\text{-}14)$$
$$\mathbf{7b}$$

Both polymers **7a** and **7b** undergo acid–base reactions, and the acid–adduct form **7b** shows good electrical conductivity. The positive charge on nitrogen of **7b** is considered to be delocalized along the polymer chain. Poly(aniline) can also be obtained by chemical oxidation of aniline. Preparation of the polymer having solubility in N-methylpyrrolidone is reported.

Poly(phenylenevinylene) 8. Thermal treatment of the precursor

$$+C_6H_4-\underset{\underset{\displaystyle S^+X^-}{\displaystyle |}}{\underset{\displaystyle R_2}{}}CH_2-CH_2+_n \xrightarrow[\text{decomposition}]{\text{thermal}} +C_6H_4-CH=CH+_n \quad (16\text{-}15)$$
$$\textbf{8}$$

which is soluble in water and can be cast as film, gives poly(phenylenevinylene).[25] Treatment of $(C_6H_4-CH(OR)-CH_2)_n$ with acid also affords polymer **8**. Instead of poly(phenylenevinylene), a variety of poly(arylenevinylene)s $(Ar-CH=CH)_n$ can be obtained using their corresponding precursors.

Poly(2,5-pyridinediyl). Dihalogenation polycondensation of 2,5-dibromopyridine using zero valent nickel complexes as the dehalogenation reagent affords poly(2,5-pyridinediyl) **9**[26]:

$$n\,Br\overset{}{-}\!\!\left\langle\!\!\!\begin{array}{c}\\ =N\end{array}\!\!\!\right\rangle\!\!-Br + n\,Ni(O)L_n \longrightarrow \left(\!\!\!\left\langle\!\!\!\begin{array}{c}\\ =N\end{array}\!\!\!\right\rangle\!\!\!\right)_n + NiBr_2L_n \quad (16\text{-}16)$$
$$\textbf{9}$$

The polymer is soluble in formic acid and a film of **9** is obtained by spreading the formic acid solution of **9** on a substrate and removing formic acid by evaporation. Polymer **9** forms adducts with BF_3 and AsF_5, but these adducts show no electrically conducting properties. On the other hand, n doping of **9** with sodium affords an electrically conducting material. BF_3 and AsF_5 seem to be trapped by the lone pair electrons of **9**, whereas the electron deficient nature of the pyridine ring gives rise to the n-doped electrically conducting material.

Polymer **9** has a rigid rod-like structure in formic acid solution, giving a large degree of depolarization, $\rho_v = 0.33$,

$$\rho_v = 0.33$$

$$\alpha_z \gg \alpha_y, \alpha_z$$

Poly(heptadiyne) 10. Polymerization of 1,6-heptadiyne with a Ziegler–Natta-type catalyst affords poly(heptadiyne).[27]

Poly(diacetylene)s 11. Irradiation of γ or UV rays to crystals of diacetylenes or heating of the crystals leads to polymerization of diacetylenes accompanied by deep color change. The polymer shows high mechanical strength and third harmonic nonlinear optic phenomena (THG).

16-2-3. Application of the π-Conjugated Polymers to Devices

The following devices have been made using the π-conjugated polymers: polymer batteries, electric capacitors, electrochromic devices, transistors, diodes, solar cells, nonlinear optic devices, and others including sensors. We will now describe the devices.

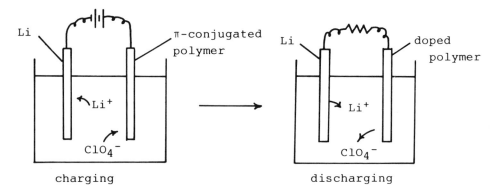

Figure 16-6 ■ Charging and discharging processes for the lithium battery using π-conjugated polymer as the positive electrode.

Polymer Batteries. The oxidized (p-doped) and reduced (n-doped) π-conjugated polymers not only show electrically conducting properties but also serve as oxidizing and reducing reagents, respectively. Due to these properties of the oxidized and reduced π-conjugated polymers, they are useful as active materials for, respectively, positive and negative electrodes of a battery.[7,8] Figure 16-6 shows the process of charging and discharging lithium battery using the π-conjugated polymer as the positive electrode.

In the charging stage, the polymer is electrochemically oxidized to store positive charge in the polymer, and anions Y^- in the electrolytic solution move to the π-conjugated polymer to form an ion pair with the positive charge in the polymer. The positive electrode reaction is

$$\cdots \diagdown\diagup\diagdown\diagup\diagdown \cdots + Y^- \underset{\text{discharging}}{\overset{\text{charging}}{\rightleftharpoons}} \diagdown\diagup\diagdown\underset{\underset{Y^{\cdot-}}{+}}{\diagup}\diagdown\diagup + e^- \quad (16\text{-}17)$$

In the discharging state, a reverse reaction takes place. In the case of poly(aniline), occurrence of the following positive electrode reaction is assumed:

$$\cdots \text{—}\bigcirc\text{—NH—}\bigcirc\text{—NH—}\bigcirc\text{—NH—} \cdots + nY^- \left(\text{ClO}_4^- \; etc.\right)$$

$$\underset{\text{discharging}}{\overset{\text{charging}}{\rightleftharpoons}} \bigcirc\overset{\text{HY}}{\underset{\text{—N}=}{}}\bigcirc\overset{\text{HY}}{\underset{=\text{N—}}{}}\bigcirc\overset{\text{HY}}{\underset{\text{—N}=}{}} \cdots + ne^- \quad (16\text{-}18)$$

In general, the π-conjugated polymer can store about 0.12–0.17 units of positive charge per C=C double bond. For example, poly(acetylene), poly(2,5-thienylene), and poly(aniline) can store about +0.12, +0.3, and +0.5 units of charge, respectively, per monomer unit. Further oxidation usually leads to degradation of the polymer.

Table 16-3 shows examples of the polymer battery using the π-conjugated polymer. A polymer battery using poly(aniline) as the positive electrode and Li as the negative

Table 16-3 ▪ **Examples of Polymer Battery**

Positive Electrode[b]	Negative Electrode[b]	Electrolyte	Voltage[a] (v)
1	Li	$LiClO_4$	3.1
1	1	$[NBu_4]Y$	2.5
1	1	Polymer electrolyte	2.9
2	Li	$LiAsF_6$	3.8
2	Li or 1	LiY	2.4–4.4
3	Li	$LiClO_4$	3.4
3	3	$[NBu_4][BF_4]$	2.7
4	Li	LiY	ca. 3
7	Li	$LiClO_4$	3.3
7	Zn	$ZnSO_4$	ca. 1

[a]Open circuit or closed circuit voltage.
[b]For **1–7**, see Fig. 16-1.

electrode is now commercially available. An "all plastic battery" using the π-conjugated polymers for both the positive and negative electrode materials is also basically possible. However, the all plastic battery so far does not show good charge–discharge cyclability.

Electrolytic Capacitors. Electrochemical etching of aluminum sheet forms a porous tough thin layer of Al_2O_3 with large dielectric constant on the surface. This aluminum sheet with porous Al_2O_3 layer is now used to make an electric capacitor, which is called an electrolytic capacitor. Recently use of poly(2,5-pyrrolylene) **4** as the electrode for the porous Al_2O_3 layer side has been developed. The electrolytic capacitor using the polymer electrode shows excellent electric response.

Electrochromic Devices. On oxidation or reduction, most of the π-conjugated polymers change color. Examples are:

Polymer	Neutral	Oxidized	Reduced
3	red	blue	dark green
7	light yellow	deep blue	
9	light yellow		reddish purple

When oxidation and reduction are carried out electrochemically, the color changing phenomenon is called electrochromism. Electrochromism is expected to be useful for making displaying and memory devices.

Transistors. Poly(2,5-thienylene) serves as a material for making polymer-based field effect transistor.[10] Figure 16-7 shows the structure of a field effect transistor (FET). When no electric field is applied to poly(2,5-thienylene), electric current between source and drain is negligible. However, when a negative electric field is applied through the gate in Fig. 16-7, carrier concentration near the SiO_2 layer

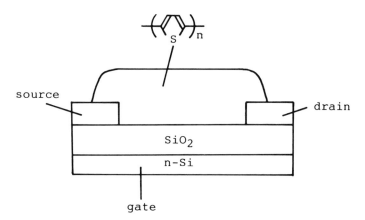

Figure 16-7 ■ Structure of the field effect transistor (FET) obtained by using poly(2,5-thienylene).

markedly increases and flow of electric current is observed. Therefore, by using this transistor, on–off control of electric current becomes possible. This FET transistor seems to be suitable for controlling a large surface display.

Diodes. Contacting p-doped π-conjugated polymer with n-doped π-conjugated polymer affords a p-n diode[28] and contacting doped π-conjugated polymer with metals forms a Schottky-type diode.[10,29] Both types of diodes show good rectification of electric current. π-conjugated polymers and inorganic semiconductors also form diodes.[30]

Solar Cells. A p-n junction prepared using p-doped poly(acetylene) and n-doped Si serves as a solar cell with power efficiency of 4.3%, $V_{oc} = 0.53$, $I_{sc} = 18$ mA, and Fill factor of 0.32.[31] A p-doped poly(2,5-thienylene) and n-GaAs system also forms a solar cell and a GaAs polymer Au system shows a corrected power efficiency of 17.5%.[32]

Nonlinear Optical Materials. When strong light (laser) with frequency ν_0 is used for irradiation, some inorganic and organic materials emit double harmonic and triple harmonic light with frequencies of $2\nu_0$ and $3\nu_0$, respectively:

$$\text{emitted light} = a\nu_0 + b2\nu_0 + c3\nu_0 + \tag{16-19}$$

Emissions of $2\nu_0$ and $3\nu_0$ light are called SHG and THG, respectively.
Poly(diacetylene), poly(acetylene), poly(2,5-thienylene), and poly(arylenevinylene) show the THG phenomena.[3] A potential application of nonlinear optic phenomena (especially THG) is a light computer using the polymer devices.

Others. Sensors for gases[34] and urea[35] have been made using the π-conjugated polymers and various types of polymer alloys of the π-conjugated polymers with common polymers, such as poly(vinyl alcohol), poly(vinyl chloride), and poly(urethane), have been reported.[36]

16-3. Electrically Conducting Metal Sulfide–Polymer Composites

A variety of composites of electrically conducting materials with nonconjugated common polymers have been developed. The author recently found that electrically conducting polymer composites were obtained using organosols of metal sulfide like CuS[37,38] in DMF (N,N-dimethylformamide) and DMSO (dimethyl sulfoxide).

Some semiconducting metal sulfides are now used as photoconductors, phosphors, electrodes of photogalvanic cells, catalysts, and so forth. Exploration of the chemistry and physics of metal sulfides is becoming more important in view of the recent rapid development in the practical uses of compound semiconductors.[39,40] Utilization of the metal sulfides, however, has been somewhat restricted because of difficulties in molding them and their low solubilities. By using the organosol method, several disadvantage points in utilizing the metal sulfide can be overcome, and we obtain semiconducting polymer films by the method shown in Scheme 1. The organosols of CuS and CdS contain CuS and CdS particles with diameters of 140 and 10 nm, respectively (Table 16-4). In the case of HgS organosol, the microcrystal structure of HgS particles (α-HgS or β-HgS) strongly depends on the kind of solvent.[41]

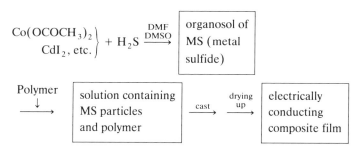

Scheme 1 ■ Organosol method.

Figure 16-8 shows the dependence of electrical conductivity (σ) on the wt% of CuS in the composite film. The electrical current seems to flow through chains formed by contact of the electrically conducting particles, CuS particles in this case (Fig. 16-9). Even if the electrically conducting particles do not have direct contact with each other, tunnel current flows between the particles smoothly when the distance between the two particles is less than about 10 Å.[42]

Table 16-4 ■ Particle Size of Metal Sulfide in the Organosol

Metal Sulfide	Solvent	Particle Size (nm)
CuS	DMSO	134
CuS	DMF	140
CdS	DMF	10
ZnS	DMF	30

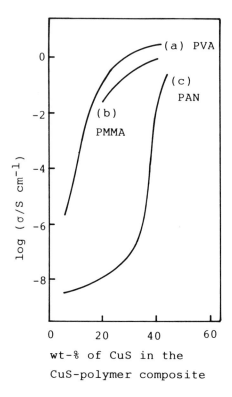

Figure 16-8 ▪ Dependence of electrical conductivity (σ) on the wt% of CuS in the composite film prepared according to Scheme 1.

   As shown in Fig. 16-8, the electrical conductivity of the CuS–polymer composite strongly depends on the kind of polymer. It is generally recognized that if the electrically conducting particles form condensed structure in the polymer composite, the composite shows high electrical conductivity due to ease of formation of the conductive chain of the conducting particles.[43] The ease of formation of the condensed structure is related to surface energy between the electrically conducting

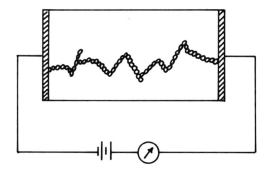

Figure 16-9 ▪ Electrical conduction through the chain formed by electrically conductive particles.

Table 16-5 ▪ Electrical Conducting Properties of Metal Sulfide–Polymer Composites

Metal Sulfide (MS)	wt% of MS	Conductivity σ (S cm^{-1})	$E_a{}^a$ (eV)	Mobility of Carrier (cm^2 V^{-1} s^{-1})	σ Obtained on Use of Other Polymers
CuS	44	2×10^{-1}	-0.04	50 (hole)	$10^{-10}\text{--}10^{1.5}$
CdS	59	1×10	0.47	10 (electron)	$10^{-9}\text{--}10^{-2.5}$
NiS	29	9×10	0.11		$10^{-6}\text{--}10^{-3}$
HgS (black, β type)	65	3×10	0.01	50 (electron)	
HgS (red, β type)	63	1×10^{-7}	0.36		
ZnS	54	1×10^{-8}	0.36		
PdS	47	5×10^{-8}	0.26		
MnS	42	1×10^{-9}			

^{a}Activation energy for the electrical conduction.

particles and the polymer. In cases of CuS–PVA, –PMMA, –PAN composite films, the electrical conductivity increases with decreasing surface energy (γ) of the polymer:

$$\sigma: \quad \text{PVA} > \text{PMMA} > \text{PAN}$$
$$\gamma: \quad \text{PVA}(37 \text{ dyn cm}^{-1}) < \text{PMMA}(39) < \text{PAN}(50)$$

The metal sulfide polymer–composite films not only show good electrical conductivity

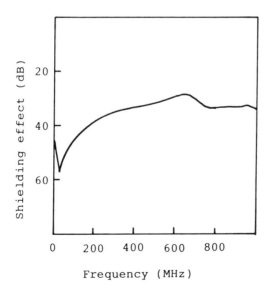

Figure 16-10 ▪ Shielding of electromagnetic waves achieved by CuS-poly(parabanic acid) composite film. d.b. $= 20 \log(I_0/I)$.

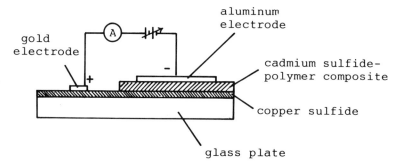

Figure 16-11 ▪ Structure of the heterojunction. Area: 1.2×1.2 cm for gold electrode, 1.2×2.0 cm for aluminum electrode, and 2.0×2.5 cm for cadmium sulfide–polymer composite film. Surface resistivity of the copper sulfide substrate film = 100 Ω.

but also are controlled to *p*- or *n*-type conductors. The type of electroconduction of the composite film essentially coincides with the type of electroconduction of the metal sulfide (CuS = *p* conductor; CdS = *n* conductor). Table 16-5 summarizes electrical conducting properties of the various metal sulfide–polymer composite films.

Because the CuS–polymer composite shows good electrical conductivity, it is useful as the shielding material of electromagnetic waves. The shielding effect is shown in Fig. 16-10. When a thin layer of *n*-type CdS–polymer composite film

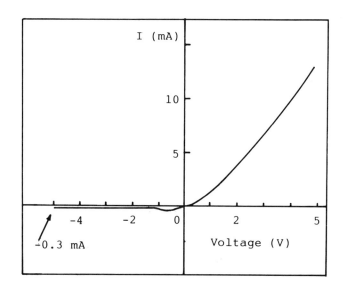

Figure 16-12 ▪ *I-V* characteristic of the heterojunction shown in Fig. 16-11. Polymer = 80:20 copolymer of vinylidene chloride and acrylonitrile (see text). At room temperature (ca. 25°C).

is formed on the surface of a p-type copper sulfide layer (Fig. 16-11), the junction shows rectification of electric current (Fig. 16-12). The rectification ratio is about 10^2.

References

1. H. Shirakawa, E. J. Louise, A. G. MacDiarmid, C. K. Chiang, and A. J. Heeger, *J. Chem. Soc., Chem. Commun.*, **1977**, 578 (1977).
2. T. A. Skotheim, ed., *Handbook of Conducting Polymers*, Vol. 1 and 2, Marcel Dekker, New York (1986).
3. R. H. Baughman, J. L. Bredas, R. R. Chance, R. L. Elsenbaumer, and L. W. Shacklette, *Chem. Rev.*, **82**, 209 (1982).
4. P. M. Lahti, J. Obrzut, and F. E. Karasz, *Macromolecules*, **20**, 2023 (1987).
5. J. C. Scott, P. Pfluger, M. T. Krounbi, and G. B. Street, *Phys. Rev. B*, **28**, 2140 (1983).
6. C & EN, July 21, 28 (1986).
7. T. Yamamoto, *J. Chem. Soc., Chem. Commun.*, **1981**, 187 (1981).
8. D. MacInnes, Jr., M. A. Druy, P. N. Nigrey, D. P. Nairns, A. G. MacDiarmid, and A. J. Heeger, *J. Chem. Soc., Chem. Commun.*, **1981**, 317 (1981).
9. A. G. MacDiarmid, L. S. Yang, W. S. Huang, and B. D. Humphrey, *Synthetic Metals*, **18**, 393 (1987).
10. A. Tsumura, H. Koezuka, and T. Ando, *Appl. Phys. Lett.*, **49**, 1210 (1986).
11. T. Yamamoto, A. Taniguchi, K. Kubota, and Y. Tominaga, *Inorg. Chim. Acta*, **104**, L1 (1985); *J. Mater. Sci. Lett.*, **5**, 132 (1986); *Chem. Ind. (London)*, 187 (1986).
12. S. Kivelson and A. J. Heeger, *Synthetic Metals*, **22**, 371 (1988).
13. K. Sanechika, T. Yamamoto, and A. Yamamoto, *Polym. J.*, **13**, 255 (1981); T. Yamamoto, K. Sanechika, A. Yamamoto, M. Katada, I. Motoyama, and H. Sano, *Inorg. Chim. Acta*, **73f**, 75 (1983).
14. H. Sakai, Y. Maeda, T. Kobayashi, and H. Shirakawa, *Bull. Chem. Soc. Jpn.*, **56**, 1616 (1983); T. Yamamoto and H. Sakai, *J. Mater. Sci. Lett.*, **4**, 916 (1985).
15. P. Kovacic and J. Oziomek, *Macromol. Synth.*, **2**, 23 (1966).
16. T. Yamamoto and A. Yamamoto, *Chem. Lett.*, 353 (1977); T. Yamamoto, Y. Hayashi, and A. Yamamoto, *Bull. Chem. Soc. Jpn.*, **51**, 2091 (1978).
17. D. H. G. Ballard, A. Courtis, L. M. Shirley, and S. C. Taylor, *J. Chem. Soc., Chem. Commun.*, **1983**, 954 (1983); D. R. McKean, and J. K. Stille, *Macromolecules*, **20**, 1787 (1987).
18. E. Tsuchida, K. Yamamoto, T. Asada, and H. Nishide, *Chem. Lett.*, 1541 (1987); M. Pelamar, R. C. Lacaze, J. Y. Dumoussau, and J. E. Dubois, *Electrochim. Acta*, **27**, 61 (1982).
19. K. Sanechika, T. Yamamoto, and A. Yamamoto, *Polym. Prepr. Jpn.*, **28**, 966 (1979); T. Yamamoto, K. Sanechika, and A. Yamamoto, *J. Polym. Lett. Ed.*, **18**, 9 (1980); *Bull. Chem. Soc. Jpn.*, **56**, 1497, 1503 (1983).
20. K. K. Kanazawa, A. F. Diaz, R. H. Geiss, W. D. Gill, J. F. Kwak, J. A. Logan, J. F. Rabolt, and G. B. Street, *J. Chem. Soc., Chem. Commun.*, **1979**, 854 (1979).
21. S. Machida, S. Miyata, and J. Han, *Polym. Prepr. Jpn.*, **37**, 737 (1988); G. P. Gardini, *Adv. Heterocycle Chem.*, **15**, 67 (1973).
22. (a) T. Yamamoto and K. Sanechika, *Chem. Ind. (London)*, 301 (1982), U.S. Pat. 4, 521, 589 (1985); (b) K.-Y. Jen, G. G. Miller, and R. L. Elsenbaumer, *J. Chem. Soc., Chem. Commun.*, **1986**, 1346 (1986); (c) M. Sato, S. Tanaka, and K. Kaeriyama, *Synthetic Metals*, **14**, 279 (1986); (d) M. Kobayashi, N. Colaneri, M. Boysel, F. Wudl, and A. Heeger, *J. Chem. Phys.*, **82**, 5717 (1985).

23. A. Czerwinski, D. D. Cunningham, A. Amer. J. R. Schrader, C. V. Pham, H. Zimmer, and H. B. Mark, Jr., *J. Electrochem. Soc.*, **134**, 1158 (1987).
24. A. F. Diaz and J. A. Logan, *J. Electroanal. Chem.*, **111**, 111 (1981); T. Nakajima, M. Harada, R. Osawa, T. Kawagoe, Y. Furukawa, and I. Harada, *Macromolecules*, **22**, 2644 (1989), and references therein.
25. H. H. Horhold, H. Rathe, M. Helbig, and J. Opfermann, *Makromol. Chem.*, **188**, 2083 (1987); T. Momii, S. Takito, T. Tsutsumi, and S. Saito, *Chem. Lett.*, 1201 (1988).
26. T. Yamamoto, T. Ito, and K. Kobota, *Chem. Lett.*, 153 (1988); T. Yamamoto, T. Ito, K. Sanechika, and M. Hishinuma, *Synthetic Metals*, **25**, 103 (1988).
27. H. W. Gilson, F. C. Balley, A. J. Epstein, H. Rommelmann, and J. P. Pochan, *J. Chem. Soc., Chem. Commun.*, **1980**, 426 (1980).
28. K. Kaneto, S. Takeda, and K. Yoshino, *Jpn. J. Appl. Phys.*, **24**, L553 (1985); M. Aizawa et al., *J. Chem. Soc., Chem. Commun.*, **1986**, 1315 (1986).
29. S. Chao and M. S. Wrighton, *J. Am. Chem. Soc.*, **109**, 2197 (1987).
30. M. Ozaki, D. L. Peebles, B. R. Weinberger, C. K. Chiang, S. C. Gau, A. J. Heeger, and A. G. MacDiarmid, *Appl. Phys. Lett.*, **35**, 83 (1979).
31. H. Shirakawa and S. Ikeda, *Kobūnshi*, **28**, 369 (1979).
32. F. Garnier and G. Horowitz, *Synthetic Metals*, **18**, 693 (1987).
33. H. Nakanishi, *Kobūnshi*, **38**, 350 (1989).
34. J. P. Banc, G. Blasquez, J. P. Germain, A. Larbi, C. Maleysson, and H. Robert, *Sens. Actuators*, **14**, 143 (1988).
35. P. C. Pandey and A. P. Mishra, *Analyst*, **113**, 329 (1988).
36. A. Pron, M. Zagorska, W. Fabianowski, and S. Lefrant, *Polym. Commun.*, **28**, 193 (1987); O. Niwa and T. Tamura, *Synthetic Metals*, **20**, 235 (1987); O. Niwa, M. Kikuchi, and T. Tamura, *Macromolecules*, **20**, 749 (1987); X. Bi and Q. Pei, *Synthetic Metals*, **22**, 145 (1987).
37. T. Yamamoto, A. Taniguchi, K. Kubota, and Y. Tominaga, *Inorg. Chim. Acta*, **104**, L1 (1985).
38. T. Yamamoto, E. Kubota, A. Taniguchi, K. Kubota, and Y. Tominaga, *J. Mater. Sci. Lett.*, **5**, 132 (1986).
39. N. B. Hannay, *Semiconductors*, Reinhold, New York (1959); H. K. Henish, *Electrochemiluminescence*, Pergamon, Oxford (1962); G. Beghi, ed., *Performance of Solar Energy Converters: Thermal Collectors and Photovoltaic Cells*, Reidel, Dordrecht (1983).
40. R. K. Willardson and A. C. Beer, ed., *Semiconductors and Semimetals*, Vols. 2, 3, 4, 11, and 18, Academic, New York (1966, 1967, 1968, 1975, and 1981).
41. S. Dev, A. Taniguchi, T. Yamamoto, K. Kubota, and Y. Tominaga, *Colloid Polym. Sci.*, **265**, 922 (1987).
42. J. G. Simmons, *J. Appl. Phys.*, **34**, 1793 (1963).
43. K. Miyasaka, K. Watanabe, E. Jojima, H. Aida, M. Sumita, and K. Ishikawa, *J. Mater. Sci.*, **17**, 1610 (1982).

Index

Acid and base hydrolyses 103
Acid dissociation constant 98, 101
Aconitase 316
Alizarine red S 297
Amino acid complex 180
Anaerobic bacteria 205
Angular dispersive method 177
Apparent diffusion coefficient 279
Arthropodan 200
Artificial photosynthetic system 376
Atomic array 346
Azo compounds 39

Battery 388
Bimetallic cluster 350
Biocatalyst electrode 229
Bioelectrocatalysis 229
Biofuel cells 246
Bioreactors 246
Bohr effect 35

Cadmium-sulfide photoanode 372, 375
Carboxypeptidase A 181
Carrier-photogeneration efficiency 72
Catalytic activity 315, 321, 324, 347
Chelate formation constant 101
Chiral polymer 104
Chiral ruthenium complex 106
Chronoamperometry 277
Chronocoulometry 277
Cleavage of water 338
Clostridium pasteurianum 205
Cobalt porphyrin 121, 140
Colloidal metal 347
Complex moiety 4
Configuration and conformation of
    polymer 337

Cooperative coordination 16, 21
Cooperativity 21
Coordinated water molecules 166
Coordination dynamics 29
Coordination equilibrium constant 124, 126,
    131
Copper proteins 217
18-Crown-6 182
c-type cytochromes 213 ff.
Cyclam (1,4,8,11-tetraazacyclotetradecane) 181
Cyclopentadienylmanganese 129
Cytochrome c 190, 251, 274, 336
Cytochrome c_3
    assignment of NMR signals 271, 272
    differential pulse polarogram 255
    interacting potentials 264, 273
    intermolecular electron exchange rate
        259
    intramolecular electron exchange rate
        259
    macroscopic redox potentials 253, 262
    microscopic redox potentials 257, 273
    normal pulse polarogram 254
    nuclear magnetic resonance (NMR)
        256–273
    optically transparent thin-layer electrode
        259, 260
Cytochrome c' 194, 198, 217
Cytochrome c oxidase 190, 206
Cytochrome c peroxidase 218, 221
Cytochrome P-450 190, 194

Dahms–Ruff equation 285
Debye–Waller factor 55
Deoxyhemoglobin 194
Deoxymyoglobin 194
Devices 387
Diaphorase 242, 245
Diffuse reflectance laser photolysis 66
Diffusion coefficient in macromolecular
        complexes 128, 135, 140
Dilatometric study 170
Dimer cation 85

Diode 390, 394
Diporphinatometal 20
Distance 218, 225
DNA 106, 108
Dopant 382
Doping 379
Driving force 216
Dual-mode transport theory 134
Dynamic interaction 4, 5, 9
Dynamics 222, 225, 226

Effect of polymer matrix 126
Electrochemical oxidation of water 368
Electrochemical pulse methods 277
Electrochromic devices 389
Electrolytic capacitor 389
Electronic function 8
Electronic process 7
Electron–lattice interaction 52
Electron transfer 107, 117
Electron transfer, in enzyme electrodes
    229, 242
Electron transfer in nonhomogeneous
    systems 332
Electron-transfer mediator
    in enzyme electrodes 230, 235
  in glucose-oxidase electrode 235, 240
Electron transport 353
Electron tunneling 334
Electrooptical measurement 25
Electropolymerized polymers 304
Energy dispersive method 177
Energy transfer 155
Enhancement of luminescence intensity 105
Enzyme electrode
    as biosensors 242, 246
    current-potential transfer in 240, 242
    electrode process at 229, 235
    limiting current in 233
    Thiele modulus in 232
    with electron-transfer mediator 230, 235
Enzyme model complex 320, 324
EXAFS 175, 182, 185, 202
    in enzyme electrodes 230, 235
    in glucose-oxidase electrode 235, 240
Excimer 76, 80
Exciplex 70, 77
Excitation and emission spectra of
        polymer–lanthanide ion complex 148
Exterplex 76

Facilitated transport 132, 136
Fick's diffusion law 278
Flash photolysis 125, 131

Flavodoxin semiquinone 217, 223
Flourescence intensity 104
Formal redox potential 286
Foster equation 156
Four-consecutive one-electron transfer
    254, 256, 259
Four-electron oxidation of water 353, 357, 360
Free-volume distribution 39
Fructose 185

Geared cycle 353
Geminate recombination 75
Gluconate 185
Gluconate dehydrogenase 240, 242
D-glucosamine 187
Glucose-oxidase electrode
    effect of mediators 236, 240
    film-coated mediator-mixed carbon
        paste 235

Hemocyanine 200
Hemoglobin 15, 119, 190, 194
Higher-order structure 3
HIPIP 217
H_2O isotope effect 167
Hole formation efficiency 47
Hole migration 83
Homogeneous linewidth 52
Hydrogenase 318, 324
Hydrogenation 348
Hydrogen generation 340, 349

In situ polymerization 29
In situ quartz crystal microbalance 309
Intervening media 224, 226
Intramolecular electron transfer 117, 213 ff.
Ion binding properties 160
Ion pair 68, 75
Iron porphyrin 190, 194
Iron–sulfur cluster 316, 318, 320

Kaolin clay 366
Kinetic parameters 306
Kinetic profile of coordination 31

Lanthanide ion 145
Lanthanide ion luminescence 146

Lanthanide ion–monomeric carboxylate complexes 170
Lanthanide ion–nitrogen bond 145
Lanthanide ion–oxygen bond 145
Lanthanide ion–polymeric carboxylate complexes 170
Lanthanide metal probe 147
LB film 342
Lipid–heme 23
Liquid junction 371
Low-energy excitation modes 55

Macromolecular assembly 23
Macromolecular cluster 322
Macromolecular complex 3
Macromolecular metal complex 3
Manganese complex 357–358
Mannitol 185
Marcus theory 215, 225, 345
Metal cluster 315, 325
Metal complex 7, 8
Metallic molecule 346
Metalloenzyme 316–317
Metalloprotein 116
Metal sulfide–polymer composite 391
Microscopic redox potentials 261, 271
Mobility 383, 393
Model reaction 321, 324
Molecular beam deposition 342
Molecular complex 7, 8
Molecular function 3
Molecular orientation 337
Molecular array 342
Molluscan 200
Monaspartatozinc (II) complex 181
Morphology of Nation 155
Multielectron transfer 8, 11, 103, 353
Multiple interaction 4, 9, 103, 119, 138
Multistep complexation 93
Myoglobin 190, 194, 220, 225

NiFe-Hydrogenase 319
Nitrogenase 322–324
Nitrogen coordination 129
Nitrogen transport 137
Nonadiabatic process 332
Normal pulse voltammetry 277

Oligosaccharide dehydrogenase 242
Ordered structure 28

Organized macromolecules 308
Organosol 391
Orientation 220, 225
Orienting angle 26
Oxygen-coordinating affinity 16, 30, 35
Oxygen coordination 120, 123
Oxygen evolution 355, 363
Oxygen-permeation coefficient 133, 136, 140
Oxygen transport 132, 140
Oxygen transporter 16
Oxyhemoglobin 195
Oxymyoglobin 195

Peptococcus aerogenes rubredoxin 205
Phase transition 24, 32, 34
Phonon side band 52
Phospholipid bilayer 24
Photochemical energy conversion 371
Photochemical hole burning (PHB) 29, 44
Photoconductivity 342, 344
Photoinduced charge reparation 82
Photoinduced electron transfer 70, 84
Photoinduced geared cycles 374
Photoisomerization 39
Photosensitized reaction 114
Photosynthesis 355
Phthalocyanine 344
π-conjugated polymers 379, 381, 384
Platinum cluster 340
Poly(aniline) 386
Poly(N,N-dialkylaniline) 301
Poly(1,1'-ferrocenylene) 382
Poly(2,5-thienylene) 381, 385
Poly(N-vinylcarbazole) 70, 72, 75, 79, 83
Polyacrylic acid 94, 95
Polyelectrolyte 160
Polymer-coated electrodes 275
Polymer matrix 4
Polymerized liposome 27
Polymer-pendant Ru(bpy)$_3^{2+}$ 371
Polyvinylpyridine 95, 96, 103
Polyviologen 281
Porphinatoiron complex 15
Porphyrins 44
Potential energy profile 31
Prussian blue 346

Quantum admixed state 194

Reactions at low temperatures 41, 49
Reaction centers 219, 221, 224
Reactions in polymer solids 39

D-ribose 187
Rubredoxin 205
Ruthenium complex 361, 363

Sauerbrey equation 310
Secondary binding force 4, 5
Sequential potential 379
Sequential potential field 11
Shielding of electromagnetic wave 393
Soft interaction 4, 9, 104, 106
Solution X-ray diffraction 176
Specific interheme interaction 273
Spin crossover complex 194
Spiny lobster *Panulirus interrupts* 202
Stabilization of unstable complex 107
Structure of ionomers 147
Successive formation constant 95
Sulfate-reducing bacteria 251
Superexchange 335
Superoxide dismutase (SOD) 202
Surface-separated system 338
$S_n \leftarrow S_1$ transition 79

Tetraheme protein 251
Thermodynamic function 102

Thermodynamic parameter 288
Thermolysin 181
Time-of-flight measurement 68
Time-resolved fluorescence spectrum 77
Transistor 389
Transmittance laser photolysis 63
2,2′-bipyridine 293
Tris(2,2′-bipyridine)ruthenium(II) 371
Two-electron transfer process 289
^2H_2O isotope effect 167
Two-photon photolysis 66

Ultrahigh-density optical data storage 39

Water oxidation 355, 361
Water photolysis 374
Weak coordination 17
Weak hydrogen bonding 18

XANES 175, 185
Xanthine oxidase 205